中　外　物　理　学　精　品　书　系

本书出版得到“国家出版基金”资助

中 外 物 理 学 精 品 书 系

引 进 系 列 · 3

Transport in Multilayered Nanostructures:

The Dynamical Mean-Field Theory Approach

多层纳米结构中的输运

——动力学平均场方法

(影印版)

〔美〕弗雷里克斯（J. K. Freericks）著

著作权合同登记号　图字:01-2012-2826
图书在版编目(CIP)数据

Transport in Multilayered Nanostructures: The Dynamical Mean-Field Theory Approach = 多层纳米结构中的输运: 动力学平均场方法:英文/(美)弗雷里克斯(Freericks,J. K.)著. —影印本. —北京:北京大学出版社,2012. 12
(中外物理学精品书系·引进系列)
ISBN 978-7-301-21662-0

Ⅰ. ①多…　Ⅱ. ①弗…　Ⅲ. ①纳米材料-薄膜-输运过程-研究-英文　Ⅳ. ①TB383 ②O484. 3

中国版本图书馆 CIP 数据核字(2012)第 282023 号

书　　　名: **Transport in Multilayered Nanostructures: The Dynamical Mean-Field Theory Approach(多层纳米结构中的输运——动力学平均场方法)**(影印版)
著作责任者:〔美〕弗雷里克斯(J. K. Freericks)　著
责 任 编 辑: 刘　啸
标 准 书 号: ISBN 978-7-301-21662-0/O·0901
出 版 发 行: 北京大学出版社
地　　　址: 北京市海淀区成府路 205 号　100871
网　　　址: http://www.pup.cn
新 浪 微 博: @北京大学出版社
电 子 信 箱: zpup@pup.cn
电　　　话: 邮购部 62752015　发行部 62750672　编辑部 62752038
出版部 62754962
印　刷　者: 北京中科印刷有限公司
经　销　者: 新华书店
730 毫米×980 毫米　16 开本　21.75 印张　402 千字
2012 年 12 月第 1 版　2012 年 12 月第 1 次印刷
定　　　价: 87.00 元

《中外物理学精品书系》
编 委 会

序　言

物理学是研究物质、能量以及它们之间相互作用的科学。她不仅是化学、生命、材料、信息、能源和环境等相关学科的基础，同时还是许多新兴学科和交叉学科的前沿。在科技发展日新月异和国际竞争日趋激烈的今天，物理学不仅囿于基础科学和技术应用研究的范畴，而且在社会发展与人类进步的历史进程中发挥着越来越关键的作用。

我们欣喜地看到，改革开放三十多年来，随着中国政治、经济、教育、文化等领域各项事业的持续稳定发展，我国物理学取得了跨越式的进步，做出了很多为世界瞩目的研究成果。今日的中国物理正在经历一个历史上少有的黄金时代。

在我国物理学科快速发展的背景下，近年来物理学相关书籍也呈现百花齐放的良好态势，在知识传承、学术交流、人才培养等方面发挥着无可替代的作用。从另一方面看，尽管国内各出版社相继推出了一些质量很高的物理教材和图书，但系统总结物理学各门类知识和发展，深入浅出地介绍其与现代科学技术之间的渊源，并针对不同层次的读者提供有价值的教材和研究参考，仍是我国科学传播与出版界面临的一个极富挑战性的课题。

为有力推动我国物理学研究、加快相关学科的建设与发展，特别是展现近年来中国物理学者的研究水平和成果，北京大学出版社在国家出版基金的支持下推出了《中外物理学精品书系》，试图对以上难题进行大胆的尝试和探索。该书系编委会集结了数十位来自内地和香港顶尖高校及科研院所的知名专家学者。他们都是目前该领域十分活跃的专家，确保了整套丛书的权威性和前瞻性。

这套书系内容丰富，涵盖面广，可读性强，其中既有对我国传统物理学发展的梳理和总结，也有对正在蓬勃发展的物理学前沿的全面展示；既引进和介绍了世界物理学研究的发展动态，也面向国际主流领域传播中国物理的优秀专著。可以说，《中外物理学精品书系》力图完整呈现近现代世界和中国物理科学发展的全貌，是一部目前国内为数不多的兼具学术价值和阅读乐趣的经典物理丛书。

《中外物理学精品书系》另一个突出特点是，在把西方物理的精华要义“请进来”的同时，也将我国近现代物理的优秀成果“送出去”。物理学科在世界范围内的重要性不言而喻，引进和翻译世界物理的经典著作和前沿动态，可以满足当前国内物理教学和科研工作的迫切需求。另一方面，改革开放几十年来，我国的物理学研究取得了长足发展，一大批具有较高学术价值的著作相继问世。这套丛书首次将一些中国物理学者的优秀论著以英文版的形式直接推向国际相关研究的主流领域，使世界对中国物理学的过去和现状有更多的深入了解，不仅充分展示出中国物理学研究和积累的“硬实力”，也向世界主动传播我国科技文化领域不断创新的“软实力”，对全面提升中国科学、教育和文化领域的国际形象起到重要的促进作用。

值得一提的是，《中外物理学精品书系》还对中国近现代物理学科的经典著作进行了全面收录。20世纪以来，中国物理界诞生了很多经典作品，但当时大都分散出版，如今很多代表性的作品已经淹没在浩瀚的图书海洋中，读者们对这些论著也都是“只闻其声，未见其真”。该书系的编者们在这方面下了很大工夫，对中国物理学科不同时期、不同分支的经典著作进行了系统的整理和收录。这项工作具有非常重要的学术意义和社会价值，不仅可以很好地保护和传承我国物理学的经典文献，充分发挥其应有的传世育人的作用，更能使广大物理学人和青年学子切身体会我国物理学研究的发展脉络和优良传统，真正领悟到老一辈科学家严谨求实、追求卓越、博大精深的治学之美。

温家宝总理在2006年中国科学技术大会上指出，“加强基础研究是提升国家创新能力、积累智力资本的重要途径，是我国跻身世界科技强国的必要条件”。中国的发展在于创新，而基础研究正是一切创新的根本和源泉。我相信，这套《中外物理学精品书系》的出版，不仅可以使所有热爱和研究物理学的人们从中获取思维的启迪、智力的挑战和阅读的乐趣，也将进一步推动其他相关基础科学更好更快地发展，为我国今后的科技创新和社会进步做出应有的贡献。

《中外物理学精品书系》编委会 主任

中国科学院院士，北京大学教授

王恩哥

2010年5月于燕园

transport in multilayered nanostructures

the dynamical mean-field theory approach

James K Freericks

Georgetown University, USA

ICP

Imperial College Press

For Susan, Carl, and Samuel

Preface

Multilayered nanostructures and thin films form the building blocks of most of the devices employed in electronics, ranging from semiconductor transistors and laser heterostructures, to Josephson junctions and magnetic tunnel junctions. Recently, there has been an interest in examining new classes of these devices that employ strongly correlated electron materials, where the electron-electron interaction cannot be treated in an average way. This text is designed to train graduate students, postdoctoral fellows, or researchers (who have mastered first-year graduate-level quantum mechanics and undergraduate-level solid state physics) in how to solve inhomogeneous many-body-physics problems with the dynamical mean-field approximation. The formalism is developed from an equation-of-motion technique, and much attention is paid to discussing computational algorithms that solve the resulting nonlinear equations. The dynamical mean-field approximation assumes that the self-energy is local (although it can vary from site to site due to the inhomogeneity), which becomes exact in the limit of large spatial dimensions and is an accurate approximation for three-dimensional systems. Dynamical mean-field theory was introduced in 1989 and has revolutionized the many-body-physics community, solving a number of the classical problems of strong electron correlations, and being employed in real materials calculations that do not yield to the density functional theory in the local density approximation or the generalized gradient expansion.

This book starts with an introduction to devices, strongly correlated electrons and multilayered nanostructures. Next the dynamical mean-field theory is developed for bulk systems, including discussions of how to calculate the electronic Green's functions and the linear-response transport. This is generalized to multilayered nanostructures with inhomogeneous dynamical mean-field theory in Chapter 3. Transport is analyzed in the context of a generalized Thouless energy, which can be thought of as an energy that is extracted from the resistance of a device, in Chapter 4. The theory is applied to Josephson junctions in Chapter 5 and thermoelectric devices in Chapter 6. Chapter 7 provides concluding remarks that briefly discuss extensions to different types of devices (spintronics) and to the nonlinear and nonequilibrium response. A set of thirty-seven problems is included in the Appendix. Readers who can master the material in the Appendix will have developed a set of tools that will enable them to contribute to current research in the field. Indeed, it is the hope that this book will help train people in the dynamical mean-field theory approach to multilayered nanostructures.

The material in this text is suitable for a one-semester advanced graduate course. A subset of the material (most of Chapter 2 and 3) was taught at Georgetown University in a one-half semester short course in the Fall of 2002. The class was composed of two graduate students, one postdoctoral fellow, and one senior researcher. Within six months of completing the course all participants published refereed journal articles based on extensions of material learned in the course. A full semester course should be able to achieve similar results.

Finally, a comment on what is not in this book. Because many-body physics is treated using exact methods that are evaluated numerically, we do not include any perturbation theory or Feynman diagrams. Also there is no proof of Wick's theorem, no derivation of the linked-cluster expansion, and so on. Similarly, there is no treatment of path integrals, as all of our formalism is developed from equations of motion. This choice has been made to find a "path of least resistance" for preparing the reader to contribute to research in dynamical mean-field theory.

J. K. Freericks
Washington, D.C.
May 2006

Acknowledgments

I have benefitted from collaborations with many talented individuals since I started working in dynamical mean-field theory in 1992. I am indebted to all of these remarkable scientists, as well as many colleagues who helped shape the field with influential work. I cannot list everyone who played a role here, but I would like to thank some individuals directly. First, I would like to express gratitude to Leo Falicov who trained me in solid-state theory research and introduced me to the Falicov-Kimball model in 1989. His scientific legacy continues to have an impact with many researchers. Second, I would like to thank my first postdoctoral adviser Doug Scalapino, and my long-time collaborator Mark Jarrell, who prepared me for advanced numerical work in dynamical mean-field theory, as we contributed to the development of the field. Third, I want to thank Walter Metzner and Dieter Vollhardt for inventing dynamical mean-field theory, Uwe Brandt and his collaborators for solving the Falicov-Kimball model, and Michael Potthoff and Wolfgang Nolting for developing the algorithm to solve inhomogeneous dynamical mean-field theory. Fourth, I would like to thank my other collaborators and colleagues in dynamical mean-field theory and multilayered nanostructures, including I. Aviani, R. Buhrman, R. Bulla, A. Chattopadhyay, L. Chen, W. Chung, G. Czycholl, D. Demchenko, T. Devereaux, J. Eckstein, A. Georges, M. Hettler, A. Hewson, J. Hirsch, V. Janis, M. Jarrell, J. Jedrezejewski, B. Jones, A. Joura, T. Klapwijk, G. Kotliar, R. Lemański, E. Lieb, A. Liu, G. Mahan, J. Mannhart, P. Miller, A. Millis, E. Müller-Hartmann, N. Newman, B. Nikolić, M. Očko, Th. Pruschke, J. Rowell, D. Scalapino, J. Serene, S. Shafraniuk, L. Sham, A. Shvaika, A. N. Tahvildar-Zadeh, V. Turkowski, D. Ueltschi, G. Uhrig, P. van Dongen, T. Van Duzer, M. Varela and V. Zlatić. I thank those researchers who shared figures with me and granted me permission to publish or republish them here. They include Sean Boocock, Nigel Browning, Bob Buhrman, Ralf Bulla, Jim Eckstein, Antoine Georges, Claas Grenzebach, Alexander Joura, Gabriel Kotliar, Jochen Mannhart, Andrew Millis, Nate Newman, Branislav Nikolić, Stephen Pennycook, Ilan Schnell, Serhii Shafraniuk, David Smith, Niki Tahvildar-Zadeh, Ted Van Duzer, Maria Varela, Dieter Vollhardt, Joe Wong and Xia-Xing Xi. I also thank the funding agencies and program officers who have supported my research over the years; this work received support from the National Science Foundation under grant number DMR-0210717, the Office of Naval Research under grants numbered N00014-99-1-0328 and N00014-05-1-0078, and supercomputer time was provided by the High Performance Computer Modernization Program at the Arctic Region Supercomputer Center and the Mississippi Engineering Research and Development Center. Finally, I thank my wife and children who supported me through this project.

Contents

Chapter 1

Introduction to Multilayered Nanostructures

On December 29, 1959, Richard Feynman addressed physicists at the banquet of the annual meeting of the American Physical Society. The title of his talk was "There's plenty of room at the bottom" [Feynman (1961)]. There was much conjecture amongst the audience as to what a talk with that title would be about, but Feynman kept it secret. When he delivered his speech, Feynman described the new field of nanotechnology, although he did not coin that term. He described how one could write all of the information published in all the books in the world on the head of a pin using manipulation of atoms in three dimensions. At the time, the talk seemed to be more science fiction than fact (see Chapter 4 of [Regis (1995)] for a historical account), even though the scientific press published many articles about the presentation; the field of nanoscience has only blossomed since the early 1990s and now there are many devices that work with or manipulate the properties of individual atoms, molecules, or small groups of atoms or molecules.

The semiconductor industry has been reducing the size of structures in its microprocessors at a rapid rate; they now create line features and transistors that are smaller than 100 nm. Current research on quantum dots treat quantum-mechanical boxes that contain a few hundred to a few thousand electrons in a small spatial region. Fabrication techniques have become so sophisticated that novel devices can be made that involve the transport of current through single molecules trapped between metallic electrodes. The discovery of conducting carbon nanotubes has provided the nano world with a possible electrical wiring system. It is clear that the future will hold many surprises and technological advancements coming from nanotechnology.

As device features are made smaller and smaller, in particular, as they become on the order of a few atoms (or nanometers) in size, quantum-mechanical effects begin to take over, and ultimately determine the device performance. It is the job of theorists to understand how to explain, model, and design devices when quantum-mechanical effects cannot be ignored. In this book we discuss one particular kind of nanotechnology—the field of multilayered nanostructures, which are composed of stacked atomic planes of different materials, with the thickness of some of the layers in the nanometer regime. Usually these devices are operated by attaching them to a voltage (or current) source, which transports electrical or heat current perpendicular to the stacked planes.

The approach and focus of this book are different from those of others. Most work on nanostructures focuses on devices that are small in all (or all but one) dimensions, so it is appropriate to start from an atomic or molecular picture and build up to the nanoscale devices (like quantum dots or wires). This class of nanoscale devices usually have strong surface effects, because the surface-to-volume ratio is usually large. Here we take an alternative "top-down" approach as opposed to the more traditional "bottom-up" approach, and consider systems in the thermodynamic limit that have only one dimension on the nanoscale (more precisely only one dimension has nanoscale inhomogeneity). This allows us to employ dynamical mean-field theory to solve the many-body problem because this technique is accurate when the number of nearest neighbors for each lattice site is large. In a multilayered nanostructure, there are no surfaces, so every lattice site maintains approximately the same number of neighbors as in the bulk. Furthermore, multilayered nanostructures are already being employed in technology, and are easier to manufacture and to use in devices than systems that are nanoscopic in all dimensions. Hence, it is likely that most applications that are commercially viable will involve multilayered nanostructures (at least for the not-too-distant future). Indeed, this is the motivation for producing this work.

1.1 Thin Film Growth and Multilayered Nanostructures

Multilayered nanostructures are the most common electronics devices that have at least one length scale in the nano realm. They have been in use for over five decades! The original devices are based mainly on semiconductors and the so-called *pn* junction. But research has been performed on superconducting variants for over four decades, and there are commercial devices in use for niche markets.

Electronics devices often rely on nonlinearities to function. Either it is the nonlinear current-voltage relation that determines the functionality of the device (like in a *pn* junction where current flows in essentially one direction), or it is the avalanche breakdown, or other nonlinear behavior, that ultimately determines when the device ceases to work. The classic multilayered nanostructure is a tunnel junction, consisting of a sandwich of two metallic electrodes separated by a thin layer of insulator. They can be easy to manufacture if the insulator is formed by exposing the metal surface to air (or other oxygen containing gas mixtures like oxygen and argon) where a native oxide layer will form. Since the two metallic regions are connected by a "weak link" due to the proximity or tunneling effect (described in Section I.3), the connection is inherently due to quantum-mechanical effects and the uncertainty principle: electrons in the metal cannot remain localized within the metal, but can leak through the barrier into the other metal. If the electrodes are superconducting and the barrier is thin enough, then the device is a Josephson junction.

A quantum-mechanical wavefunction is highly nonlinear. In classically allowed regions, it will oscillate and have nodes, while in classically forbidden regions, it will exponentially decay. Both behaviors are nonlinear, and ultimately lead to the nonlinear behavior of multilayered nanostructures. We will not discuss nonlinearities much in this work, but we mention this fact to remind the reader that whenever quantum-mechanical behavior governs the transport through a device, it is likely to have some underlying nonlinear features. Tuning and controlling these nonlinear features is often necessary to make the device useful. Examples of nonlinear current-voltage characteristics in Josephson junctions are shown in Fig. 1.1.

Another useful feature in devices is controllability. Many semiconductor devices have a voltage gate which can be varied to change the behavior of the device. Strongly correlated materials (described in Section I.2) often have properties that can be sharply tuned by external fields, pressure or chemical doping, and provide an interesting alternative of materials to use in devices from the conventional metals, semiconductors, and insulators currently in use. They are of particular interest when one considers controlling the transport of the spin of the electron (so-called spintronics devices), since magnetism is inherently quantum mechanical in nature, and many strongly correlated systems also display interesting magnetic properties. But, due to their quantum-mechanical behavior, involving correlated motion of electrons, they are less well understood than semiconductors, and fewer devices have been made from them. At the moment they hold

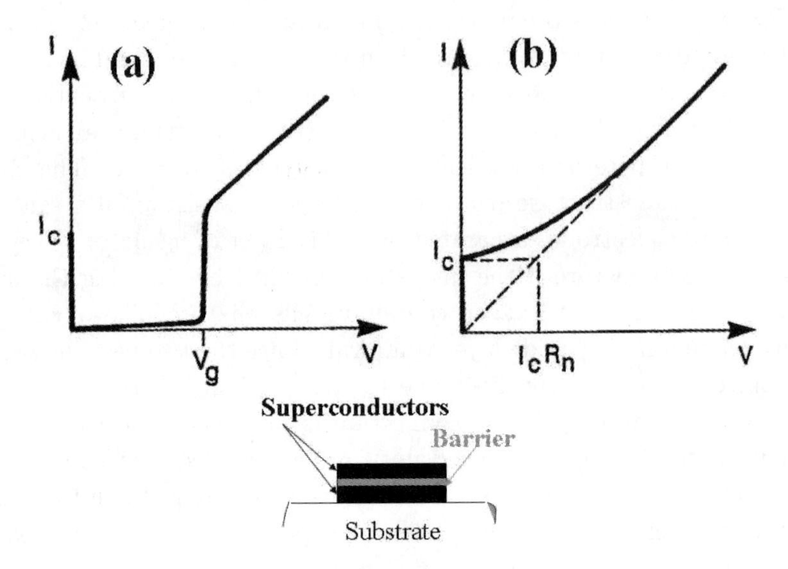

Fig. 1.1 Current-voltage curve for (a) a hysteretic Josephson junction and (b) a nonhysteretic Josephson junction. The bottom figure is a schematic of a Josephson junction which corresponds to a superconductor-barrier-superconductor sandwich; the superconductor "leaks" through the barrier from one superconductor to the other carrying current with a nonlinear current-voltage relation. The Josephson junction can carry current at zero voltage up to the critical current I_c, and then it moves into a resistive state. If the current-voltage curve is multivalued (left panel), then it is a hysteretic junction, while a single-valued curve (right panel) corresponds to a nonhysteretic junction. Both curves ultimately join up to the linear curve of Ohm's law ($I = V/R_n$) at high voltage (R_n is the normal-state resistance). The characteristic voltage where the current-voltage curve starts to become linear is $V_c \approx I_c R_n$ which is typically no larger than a few meV.

great promise and interest. This work hopes to aid with the design of novel devices that use strongly correlated materials by enabling one to calculate properties based on the underlying features of the materials that comprise the device.

Modern science has made great strides in its ability to artificially grow multilayered nanostructures. There are a number of different growth techniques that are used, and they each have their set of advantages and disadvantages. All growth processes start with a substrate material that is chosen either for the lattice match with the candidate material to be grown (to serve as a template and to relieve strain), for the chemical inertness with respect to the growth material (to reduce interdiffusion and creation of unwanted chemical species at the interface), or for practicality in sub-

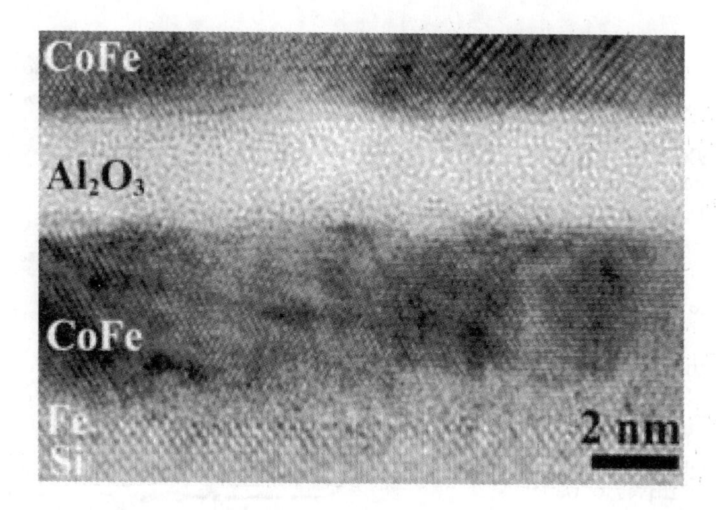

Fig. 1.2 Transmission electron micrograph of a sputtered device for use in spintronics. The TEM image allows us to see individual atomic planes, and is able to discern the chemical composition of each layer. *Figure reprinted with permission from* [Wang, *et al.* (2005)] (©2005 American Institute of Physics).

sequent device processing. The ultimate goal of material growth is to lay down atomically flat planes of each desired material, one plane at a time, and modify the constitution of the growth planes as desired to make the device of interest. In reality, this is never fully achieved with any technique, but in current state-of-the-art device growth, it is possible to achieve almost atomic flatness of the epitaxial growth planes, and in some cases the interface regions can be nearly atomically flat with limited interdiffusion or chemical reactions.

The simplest way to grow materials is via sputtering, which involves bombarding a target with inert ions, forcing the target atoms to be expelled and shower onto the substrate where the thin film will be grown (the word sputtering comes from the Greek verb *sputare* which means to spit). Sputtering is a simple growth process because one need not worry about the relative vapor pressures of the constituents, since the material grows in a nonequilibrium fashion. It also grows with the same stoichiometry as that of the target (essentially because the atoms that are emitted all come from the surface of the target). Sputtering is generally not believed to be able to grow atomically sharp interfaces, and it can be difficult to guarantee uniform coverage during the growth process; its main advantages are

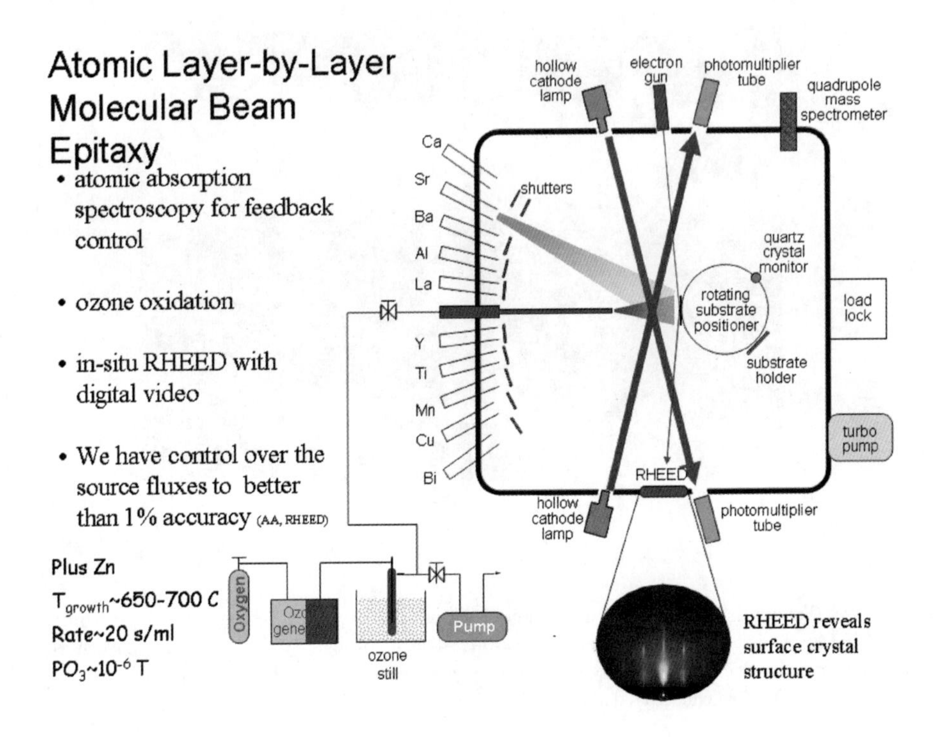

Fig. 1.3 Schematic of a molecular beam epitaxy growth chamber. The MBE growth takes place in ultra high vacuum. Different sources are introduced by opening shutters that allow the heated material to evaporate into the chamber. Many different means to characterize the sample during growth are possible. For example, RHEED oscillations show when a monolayer of growth is completed. *Figure adapted with permission from* [Eckstein and Bozović (1995)].

that it grows stoichiometrically and it is fast, so impurities may not have a chance to enter the device in high concentrations. It can achieve high quality growth, as illustrated in a spintronics device grown via sputtering that has nearly atomically flat interfaces for a variety of magnetic and nonmagnetic multilayers [Wang, *et al.* (2005)].

Molecular beam epitaxy (MBE) is arguably the most precise of the growing techniques. An MBE machine has growth conditions controlled to high precision. The growth chamber is inside an ultra high vacuum (UHV) chamber that has a sample holder and a series of growth materials inside separate furnaces; shutters in front of the furnaces open to allow the evaporated vapors of the different materials into the chamber, which will hit the sample and stick. A schematic of such a device is given in Fig. 1.3.

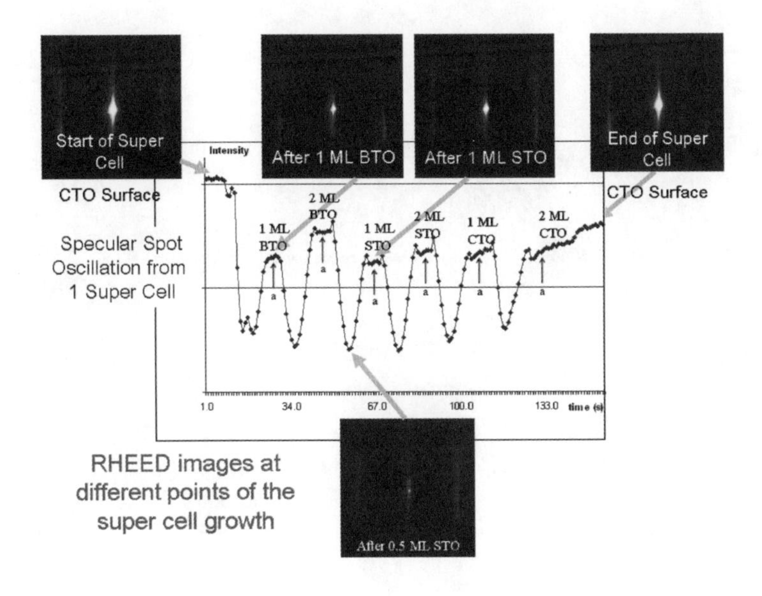

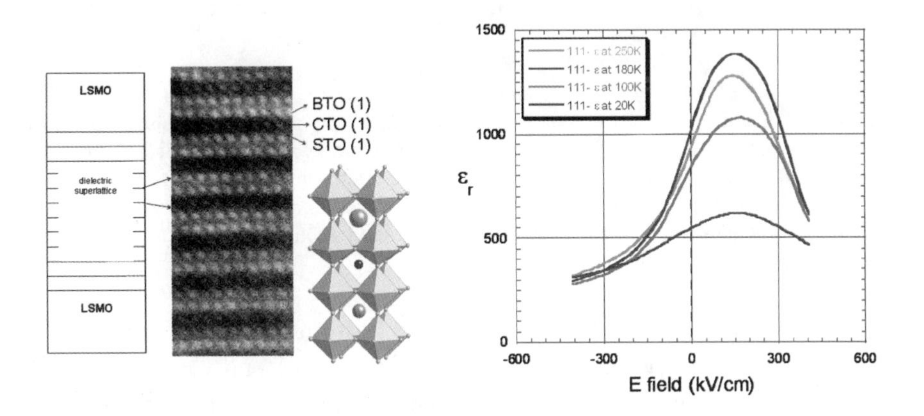

Fig. 1.4 (Top) RHEED oscillations during growth showing the completion of each monolayer. (Bottom) Left panel: schematic of the complex dielectric oxide formed from $CaTiO_3$, $BaTiO_3$, and $SrTiO_3$ along with a TEM image; right panel: dielectric response of different devices. *Figure adapted with permission from* [Warusawithana, *et al.* (2003)] (original figure © 2003 the American Physical Society) and [Warusawithana, Chen, O'Keefe, Zuo, Weissman and Eckstein (unpublished)].

The growth process can be monitored by RHEED oscillations which repeat as each atomic monolayer is placed down. The growth is usually slow, with perhaps a few seconds for each atomic layer. An example of the growth of an artificially engineered dielectric is given in Fig. 1.4. The top panel

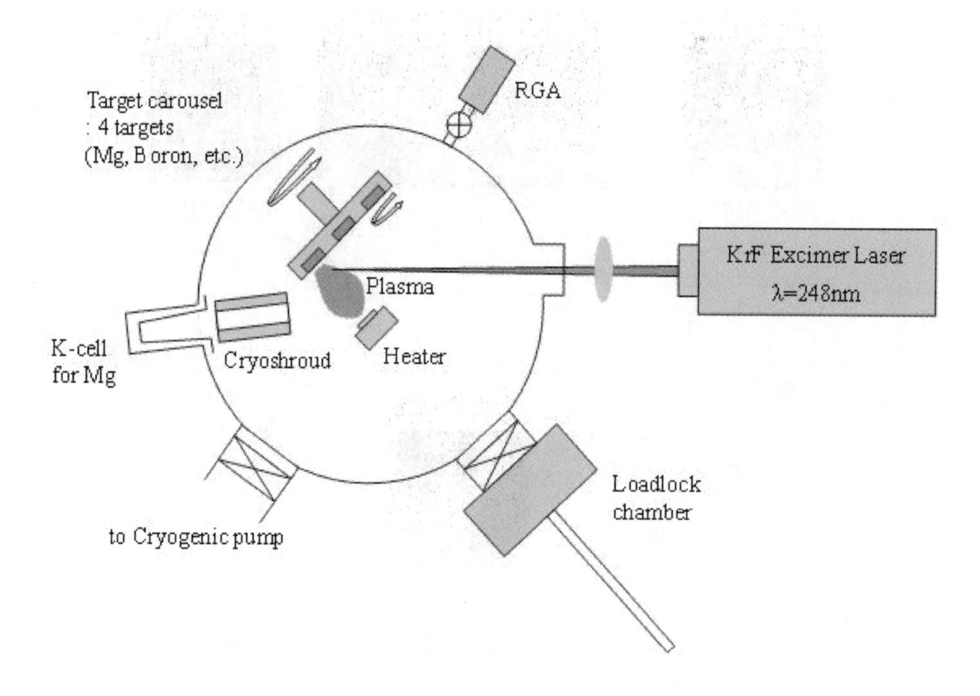

Fig. 1.5 Schematic of a PLD system for growing MgB_2. The magnesium and boron targets (heated up by the UV laser pulse) are supplemented by a so-called Knudsen (or effusion) cell which is an evaporator of a beam of magnesium to maintain high enough Mg pressure for stoichiometric growth. A residual gas analyzer monitors the gases in the chamber, where the growth takes place in vacuum. *Figure reprinted with permission from* [Kim and Newman (unpublished)].

shows the RHEED oscillations, while the bottom left panel is a TEM of the different layers (with a schematic of the device) and the bottom right is an example of the dielectric response as a function of the applied field.

Pulsed laser deposition (PLD) is another high precision growth technique. It involves ablating materials targets with a high power UV laser pulse, which creates a plume that is directed at the sample. The growth proceeds in spurts, in this fashion, and can achieve nearly atomic flatness, but it is not as common to monitor the layer-by-layer growth as in MBE. It is, however, typically much faster than growth in an MBE system, and has emerged as a popular choice for thin-film device growth in research laboratories because of its speed combined with its innate ability to preserve the target's stoichiometry. An example of a PLD system is shown in the schematic picture of Fig. 1.5. A trilayered TiNbN-Ta_xN-TiNbN sample grown with PLD is imaged with a TEM in Fig. 1.6.

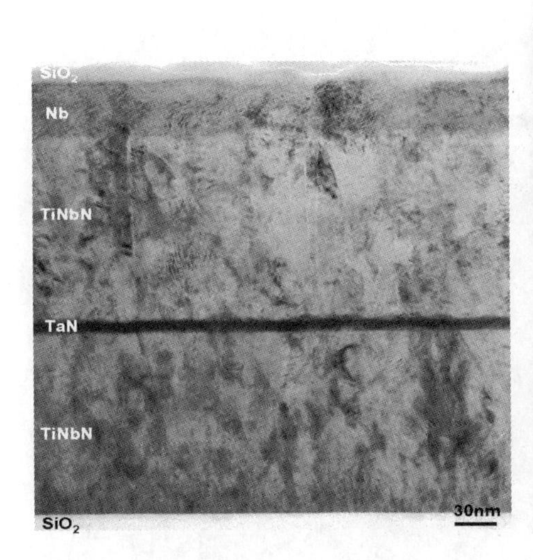

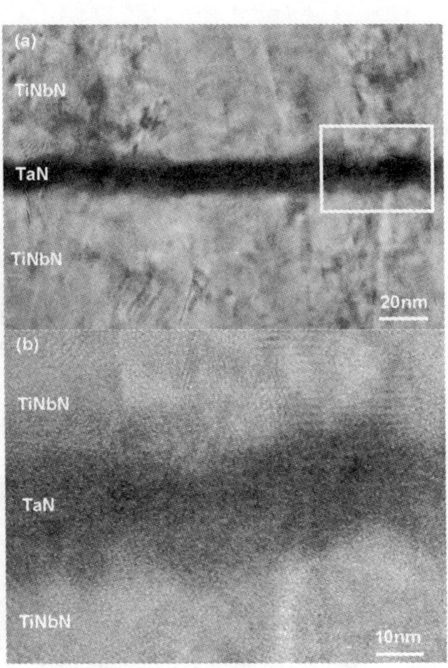

Fig. 1.6 TEM images of a trilayered TiNbN-Ta_xN-TiNbN sample suitable for processing into a Josephson junction. The sample was made with the PLD process. The left panel has the widest field of view, which is blown up in the upper right and then lower right images. Note that although the interfaces meander across the sample, the barrier width is quite uniform throughout the growth process. *Figure reprinted with permission from* [Yu, *et al.* (2006)].

Chemical vapor deposition (CVD) is a technique often used in industrial manufacturing. A series of different gaseous phases of materials are directed toward the sample, where a chemical reaction takes place at the surface, facilitating the growth. CVD is complicated by the need to find the right precursor chemical gases for a given growth process. It can be combined with other techniques, such as in the growth of MgB_2 a recently discovered 40 K conventional electron-phonon superconductor, which uses a gaseous phase for the boron, but thermal evaporation of solid metal for the magnesium.

A schematic of this hybrid physical chemical vapor deposition (HPCVD) procedure is illustrated in the left panel of Fig. 1.7 and is the process used in making high quality MgB_2 films [Zeng, *et al.* (2002)]. It shows the sample substrate region in black, atop the red sample holder. The boron gaseous precursor flows continuously past the sample, and Mg vapor is generated around the sample by the heating of solid Mg. The quality of the films can

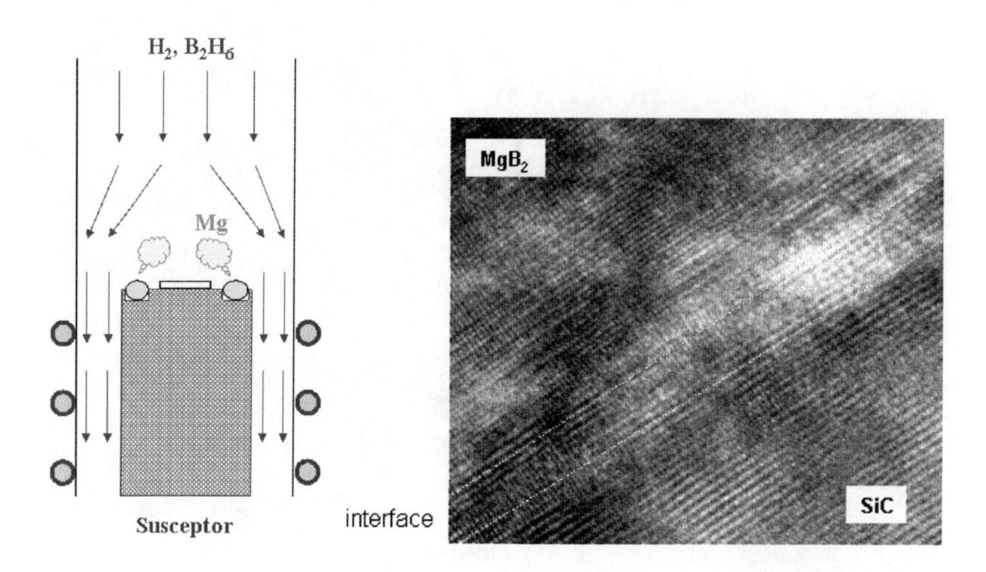

Fig. 1.7 Left panel: schematic diagram of the hybrid physical chemical vapor deposition process used to make ultra high quality MgB_2 films. Right panel: cross-sectional TEM image of the films showing a narrow interface region, where the sample quality is degraded (diagonal region about five atomic planes thick near center of figure). *Right panel reprinted with permission from* [Xi (unpublished)]. *Left panel reprinted with permission from* [Progrebnyakov *et al.* (2004)] (© 2004 the American Physical Society).

be seen in the cross-sectional TEM image in the right panel, which shows the substrate (SiC), the high quality atomically flat layers of MgB_2, and a thin interface region (about five atomic planes thick) where substrate steps and dislocation defects are located and degrade the sample quality. These films are such high quality because the degraded region is so thin.

There are many ways to characterize the quality of the final device that has been grown. We have already shown a number of TEM images, which can determine where the atoms sit, and thereby provides information on the flatness of the interfaces, and of interdiffusion or chemical reactions at the interfaces. But a TEM image is a destructive process, because one needs to slice, polish, and thin the sample until it can be imaged. Furthermore, we are often interested in understanding properties of the transport in a device, and such information cannot be revealed by TEM measurements. Another technique that is quite useful is called ballistic electron emission microscopy or BEEM for short. This measurement is shown schematically in Fig. 1.8. A

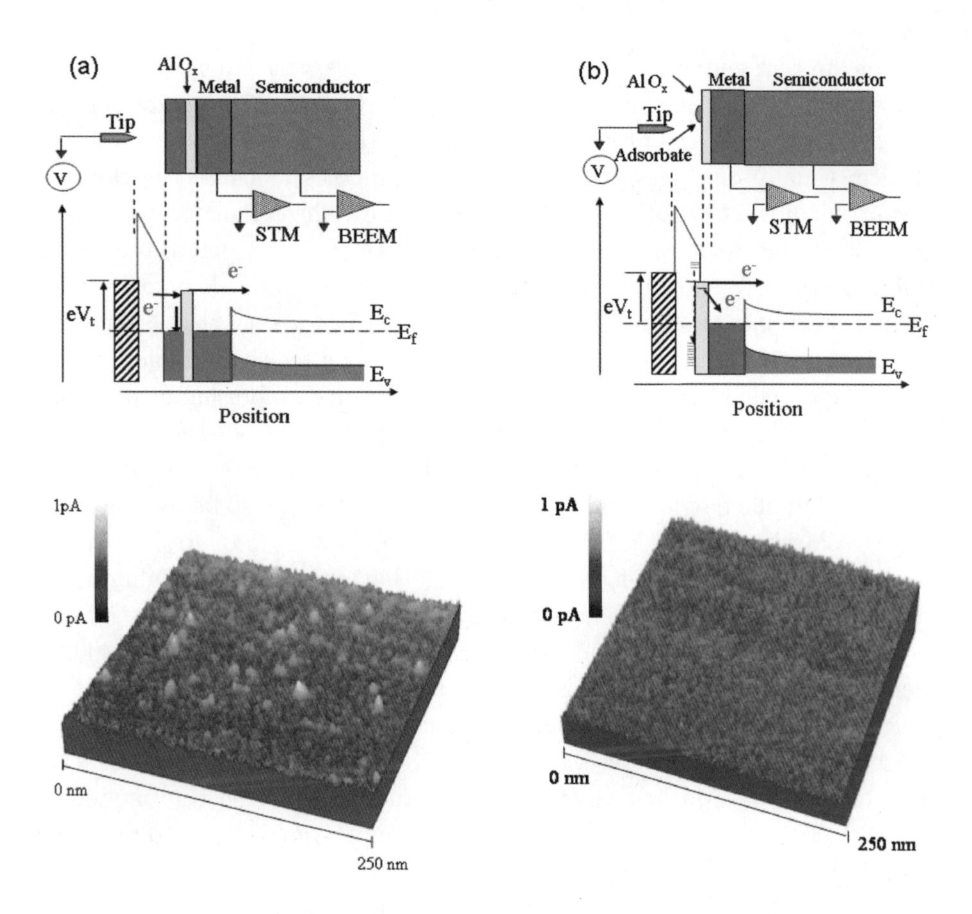

Fig. 1.8 Top panel: BEEM schematic for two different types of samples. The STM is always scanned over the surface with a bias voltage applied to it, and there always is a Schottky barrier formed by the electronic charge reconstruction at a metal-semiconductor interface to provide a barrier to electrons moving through the device. Energetic electrons will pass over the barrier and be collected. In this fashion, one can determine the local contributions to current flow through the device. Such a map is presented in the bottom two panels for a thin (left) and thick (right) disordered AlO_x barrier. The thin barrier has pinholes, while the thick barrier is pinhole free, and has nearly uniform current flow. *Reprinted with permission from* [Buhrman (unpublished)].

scanning tunneling microscope (STM) tip is scanned over the surface of the sample with a voltage difference applied so that it can eject electrons into the sample. Since the sample sits on top of a metal-semiconductor interface, the electron needs to have enough energy to get over the Schottky barrier that forms due to an electronic charge reconstruction at the interface, in order to be collected. By monitoring this collection current versus the position of the STM tip, one can directly measure the uniformity of the sample for perpendicular transport. In other words, one can actually image the so-called pinholes, which are "hot spots" in the device that allow current to flow more easily and provide an inhomogeneous current flow through the device; usually one does not want to have pinholes, because the random nature for how they form can significantly effect the uniformity of device parameters across a chip. Two BEEM images of a disordered aluminum oxide barrier are shown in the bottom panels of Fig. 1.8 [Rippard, *et al.* (2002); Perrella, *et al.* (2002)]. The left panel has a very thin layer, and the right panel has a thicker layer. One can clearly see the pinholes on the left (bright yellow regions), which then become much more uniform on the right. In both cases, however, the barrier is still quite disordered, because the aluminum oxide is not stoichiometric. This can be inferred, in part, from the fact that the barrier height to tunneling, which can also be measured in the BEEM experiment, is far below half of the band gap of Al_2O_3. What is interesting from a device standpoint is that the disordered aluminum oxide barrier creates a uniform tunnel barrier for transport, even if it is nonstoichiometric, as long as it is thick enough [Rippard, *et al.* (2002); Perrella, *et al.* (2002)]. This is one reason why it is so useful in so many different types of multilayered nanostructures.

There is a simple model that explains why the oxygen defects form in aluminum oxide [Mather, *et al.* (2005)], and we describe this model in Fig. 1.9. The common way to form an aluminum oxide layer is to first put down a layer of aluminum, and then to introduce oxygen gas for a certain period of time at a certain pressure to allow the aluminum to oxidize. In some devices, like Josephson junctions, there is no device degradation if some unoxidized aluminum remains, because it will be made superconducting by the proximity effect, while in other cases, like in magnetic tunnel junctions for spintronics, one wants all of the aluminum to oxidize, because metallic aluminum will degrade the tunnel magnetoresistance. The model for the oxidation process is that the oxygen first sits on the surface of the aluminum before it is driven into the sample. After some oxygen has moved in, the oxygen vacancies reach a steady state with the chemisorbed oxygen surface layer, and no more oxygen will flow through to oxidize the aluminum further. When the device is then processed to add additional layers, the oxygen surface layer will either be driven in (due to the processing conditions) or will react with the new layers being added on top, which can potentially degrade the top interface. Heating the sample prior to additional growth of multilayers can drive the chemisorbed oxygen into the aluminum and reduce the number of defects. Indeed, if the device is

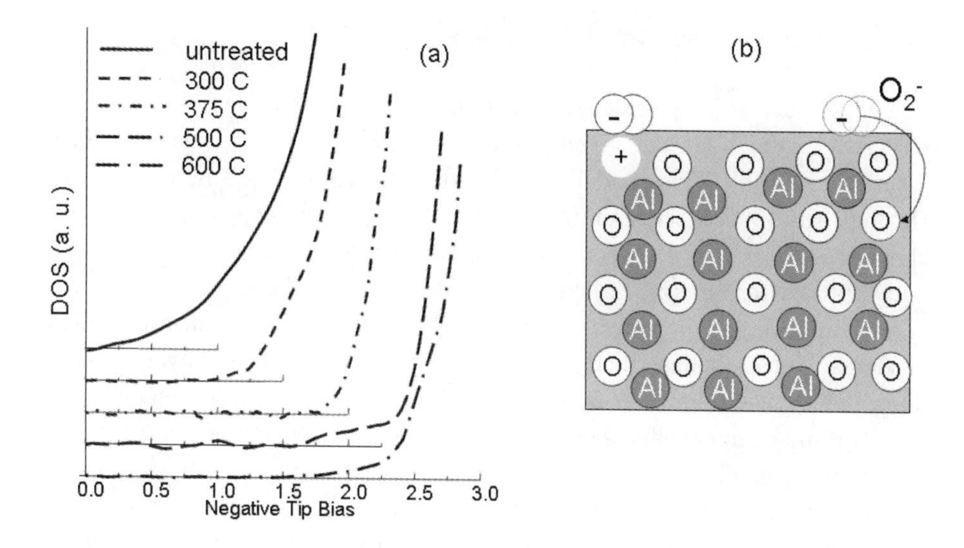

Fig. 1.9 Model for aluminum-oxide growth by exposing a thin film of aluminum to oxygen. On the right, one can see how a chemisorbed oxygen layer can form on the surface, by binding electronically to the defect sites; this chemisorbed layer does not allow further oxygen to flow into the barrier. By heating the sample, one can thermally activate the oxygen to move over the barrier and be driven into the aluminum layer. This is confirmed in the left panel, which shows how the tunnel current turns on at a higher and higher voltage as the sample is annealed at higher temperatures, and eventually a barrier height equal to half the Al_2O_3 band gap develops. *Left panel reprinted with permission from* [Mather, *et al.* (2005)] (©2005 American Institute of Physics) and *right panel reprinted with permission from* [Buhrman (unpublished)].

annealed at higher and higher temperatures, one sees the expected barrier height for Al_2O_3 begin to develop (see the left panel of Fig. 1.9).

In this section, we have described a number of different growth processes and characterization tools for multilayered nanostructures. The growth process is often quite complex, and significant care must be taken to achieve high quality results, but the state-of-the-art does allow quite good devices to be grown in research laboratories. Characterization tools used both during growth and after growth allow the device properties to be determined and understood, helping to find new ways to grow even better devices in the future. We will be concentrating on describing the theoretical and numerical formalisms for how to determine the transport through such devices throughout this book.

1.2 Strongly Correlated Materials

The first successful semiclassical attempt to describe the conduction of electrons in metals was given by Paul Drude in 1900 [Drude (1900a); Drude (1900b)]. This model assumes that electrons move independently through the crystal without feeling the effects of the other electrons but they do scatter off of defects, impurities, lattice vibrations, *etc.*, with a constant scattering time called the relaxation time. From this simple assumption, one can produce a constant electrical current from an applied electrical field (as described in virtually every solid state physics text). This theory was modified by Arnold Sommerfeld in 1927 to include the quantum-mechanical effects of the Fermi-Dirac distribution of electrons and the Pauli principle [Sommerfeld (1927)]. In spite of its incredible simplicity, the Drude-Sommerfeld model works remarkably well in describing the behavior of a wide variety of metals. The theoretical basis for understanding why such a simple model works so well was established by Lev Landau with the introduction of Fermi-liquid theory [Landau (1956)]. Fermi-liquid theory maps the elementary excitations of the interacting electronic system onto the excitations of a noninteracting system, and describes the residual weak interactions with a small set of phenomenological parameters. Nearly all metals can be described by Fermi-liquid theory (or "dirty" Fermi liquid theory, which corresponds to Fermi liquids with some additional static disorder that creates a finite relaxation time at the Fermi energy when $T = 0$). The basic result of Landau's Fermi-liquid theory is that some fraction of the electrons, corresponding to the electrons with the lowest available energies, behave like noninteracting electrons with an infinite relaxation time at the Fermi energy when $T = 0$. Hence they can be described well by semiclassical approaches at finite temperature even though the electrons do feel an electron-electron repulsion from the other electrons in the material.

Strongly correlated electrons are, in general, different from these "garden-variety" electrons found in most metals. In strongly correlated electron materials, the electrons feel strong effects of the other electrons, and hence their motion is constrained by the positions of the neighboring electrons, which can lead to interesting phenomena, most notably a metal-insulator transition, as was first described by Nevill Mott [Mott (1949)].

The Mott metal-insulator transition is easiest to describe with an artificial material of atomic hydrogen placed on a crystal lattice with a continuously varying lattice parameter. If we assume that the electrons do not congregate between the hydrogen nuclei, and hence rule out the formation of molecular hydrogen, then the system can be described by electrons that hop on a lattice constructed by the periodic arrangement of the hydrogen nuclei. If the lattice parameter is very large, then each electron is tightly bound to a nucleus, and we have a collection of isolated hydrogen atoms, which will not conduct electricity because the electrons are localized, and cannot be unbound by applying a small electric field. This state is an insulator. If we now shrink the lattice spacing, bringing the atoms closer together, then the wavefunctions of the electrons will begin to overlap. When this occurs, the electrons can hop from one hydrogen atom site to a neighboring hydrogen atom site if the electrons have opposite spins. Once such a process is allowed, the electrons become delocalized, and then they can screen out the bare Coulomb attraction with the nuclei, which will tend to make them even more delocalized, and eventually they will become metallic, easily conducting electricity when a small electric field is applied. The change in character from a metal to an insulator as the lattice spacing increases is the classic example of the Mott metal-insulator transition.

Strongly correlated electrons are a little bit different from the hydrogen example above, because it is the repulsion of the electrons with each other that determines their behavior, rather than the attraction with the ion cores (which in most crystals determines the band structure). Hubbard devised the simplest model for this behavior [Hubbard (1963)]. In his model, which is described in detail in Chapter 2, we have electrons that move in a single band on a lattice. They can hop to their nearest neighbors with a hopping integral t. When two electrons sit on the same lattice site, there is a screened Coulomb repulsion U. All other long-range Coulomb interactions are neglected. If we have on average one electron per site, then if $U \ll t$, the electrons are delocalized in a band and their motion is only slightly modified by the electron-electron interaction. If, on the other hand, we have $U \gg t$, then the Coulomb repulsion is so strong we cannot have two electrons (of opposite spin) occupy the same lattice site. Hence we have exactly one electron per site, and this configuration is frozen with respect to charge excitations, so the system is an insulator. This implies that there is a Mott-Hubbard metal-insulator transition as a function of U. The transition occurs at $U \to 0^+$ in one dimension [Lieb and Wu (1968)], but at finite U values for higher dimensions.

Predicting when a real material will display Mott-Hubbard insulating behavior is quite difficult. One simple rule is that if a density functional theory calculation predicts the system is metallic, but experiment shows it to be insulating, then it is a strongly correlated insulator. But such a

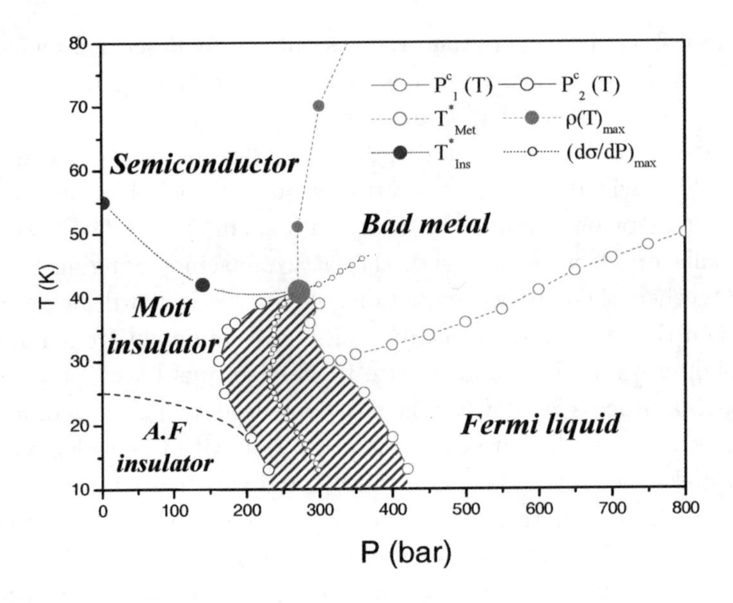

Fig. 1.10 Pressure-temperature phase diagram of the κ-Cl material. Transport measurements on this system identified four regions: (1) a Mott insulator; (2) a semiconductor; (3) a bad or anomalous metal; and (4) a Fermi-liquid metal. These four regions, along with the antiferromagnetic phase are shown in the phase diagram (the superconducting phase, which is also present, has not been depicted). The general character of this phase diagram, in particular, the first-order phase transition between the metal and insulator at intermediate temperatures, can be explained by numerical solutions of the Hubbard model using dynamical mean-field theory. *Figure reprinted with permission from* [Limelette *et al.* (2003)] (© 2003 the American Physical Society).

definition is neither rigorous, nor does it allow for much predictive power in finding new Mott insulators. Gebhard goes to great lengths to carefully describe conditions under which one has a Mott insulator [Gebhard (1997)], and the interested reader is referred there. More recently, a combination of density functional theory plus dynamical mean field theory shows promise in being able to provide a numerical framework for predicting Mott insulators and determining their properties, but the current techniques require huge investments in computer time, so it is not yet a practical tool for numerically exploring new materials (see Sec. 1.7).

We end this section by giving a recent explicit example of experimental work and calculations that illustrate the Mott insulating behavior of a strongly correlated material. This new material is of high interest, because the transition to different regions of the phase diagram can be reached by relatively small changes in either pressure or temperature. Experiments on the organic material κ-(BEDT-TTF)$_2$Cu[N(CN)$_2$]Cl (called κ-Cl) show that it can be tuned through the Mott transition by varying the pressure over a range of about 1 kbar and temperatures up to 80 K [Limelette, *et al.* (2003)]. Results for the phase diagram are shown in Fig. 1.10. As the pressure increases, the ratio U/t decreases, so we see a Fermi-liquid metal on the lower right and a Mott insulator (plus an antiferromagnetic ordered phase) on the lower left. When the system is heated up, the insulating phase becomes more semiconducting, and the Fermi-liquid behavior disappears above the renormalized Fermi temperature; as the system goes into this incoherent phase it is metallic, but with anomalous properties, and typically poor conductivity. If the temperature is raised even further, the first-order transition between the metallic and insulating phases disappears at a classical critical point, above which, the system can undergo a smooth crossover from a metal to an insulator as a function of pressure. The general behavior of this metal-insulator transition is similar to that of the liquid-gas phase diagram of many liquids.

1.3 The Proximity Effect

In quantum mechanics, a "box" determined by a finite potential barrier is a leaky box, because the wavefunction of the electron always extends out of the box boundaries with an exponentially decaying wavefunction. This is shown schematically for a one-dimensional box in Fig. 1.11, where the wavefunctions of the two lowest bound states are plotted (centered on their respective eigenenergies). Note how there is always a finite probability to find the electron lying outside the box due to the uncertainty principle.

In many-body physics, a similar phenomenon occurs whenever two different materials are joined together at an interface; the wavefunctions of the right material leak into the left and *vice versa.* This mild sounding observation leads to some amazing quantum-mechanical effects; indeed the rest of this book focuses on investigating such effects. This "leakage of electrons" across a barrier is called tunneling. It is in many respects a mature subject. Esaki [Esaki (1958)] described a tunnel diode made out of semiconductors in the late 1950s, which was shortly followed by the superconducting version studied by Giaever [Giaever and Megerle (1960)]. Josephson [Josephson (1962)] showed that one gets surprising effects in a superconducting tunnel junction when the barrier is made thin enough. All three shared the Nobel prize in 1973 for their work on tunneling.

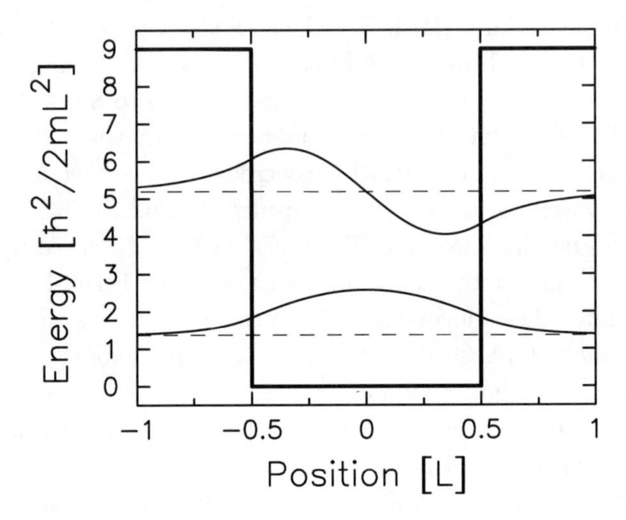

Fig. 1.11 Lowest two wavefunctions for a particle in a one-dimensional box depicted by the thick solid lines (these are the only bound states for a box of this depth). The dashed lines are the values of the respective energy levels. Note how the wavefunction for each case leaks out of the "boundary" of the box.

The best known proximity effect occurs in a Josephson junction [Josephson (1962)] [Anderson and Rowell (1963)], which is a sandwich structure composed of a superconductor-barrier-superconductor. A superconductor is a metal that has a net electron-electron attraction mediated by a phonon (in conventional low-temperature superconductors), which causes electrons with opposite momentum and spin to pair together (due to the so-called superconducting correlations). The physical picture is similar to two marbles on a rubber sheet—each feels the depression of the other marble, and they roll toward each other. In real superconductors, the electrons also repel each other because they have the same electronic charge; the superconductivity occurs because there is a time delay for the interaction with the phonons, which allows them to pair electrons together that are not located at the same position at the same time. The pairing leads to an energy gap, so the superconductor has no low-energy excitations below the energy of the superconducting energy gap (typically on the order of 1 meV). In the Josephson junction, the pairing correlations of the superconductor on the left leak into the nonsuperconducting barrier region in the middle (be it a metal or an insulator), and join up with the superconducting correlations in the superconductor on the right. This weak link between the two superconductors can carry current across it if the macroscopic quantum-mechanical phase changes across the barrier region. This leads to the Josephson supercurrent—a finite current carried by superconducting pairs with zero voltage across the barrier. There is also a corresponding inverse proximity effect, where the pairing correlations in the superconductor are weakened by the closeness to the interface with the barrier.

The physical picture for the proximity-effect coupling of Josephson junctions is different for insulating and metallic barriers. In metallic barriers, the barrier has low-energy states, but the superconductor has none. As a superconducting pair approaches the interface with the barrier, it meets a hole in the metal, which annihilates one of the electrons, while the other electron moves through the barrier to the next interface. There, the electron is retro-reflected as a hole (the hole has the opposite momentum and energy of the electron), leaving behind a superconducting pair to travel through the superconducting lead on the right. This process is called Andreev reflection [Andreev (1964)] (see Fig. 1.12); it takes place over a time scale on the order of $\hbar/\Delta$ independent of the barrier thickness L. In insulating barriers, the barrier has no low-energy states, so the electron pairs must tunnel through the barrier, which occurs due to the quantum-mechanical "leakage" through the barrier. Obviously the supercurrent decreases faster with the thickness of the barrier when it is an insulator than when it is a metal (although both decay exponentially with the thickness).

In a normal-metal–barrier–normal-metal nanostructure, there is also a proximity effect, and it is similar to the problem of a quantum-mechanical particle in a box (Fig. 1.11) when the barrier is an insulator, because the metallic wavefunctions see a potential barrier at the interface, since there are no low-energy states in the insulator. Hence the wavefunctions decay exponentially until they reach the center of the barrier, and then they grow until they reach the metallic interface on the other side. Since the wavefunction connects the two metallic leads, the electrons can directly tunnel from the right to the left (or *vice versa*). In the metallic case, the proximity effect is more subtle, dominated by generating oscillations (with the Fermi wavelength) in the metallic leads due to the mismatch of the wavefunctions between the two metals. Similar effects can also occur in the barrier.

The study of multilayered nanostructures relies heavily on understanding proximity effects between dissimilar materials brought close together in a heterostructure. This forms a significant part of the final five chapters.

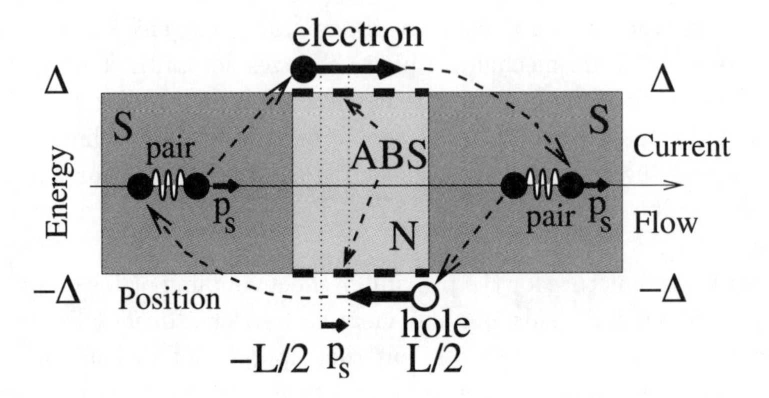

Fig. 1.12 Schematic plot of the Andreev reflection process. The low-energy electrons in the metal are confined to the barrier due to the energy gap Δ in the superconductors (the energy of the superconducting ground state is chosen as the zero in this diagram), so they form an electron-hole bound state, which allows a superconducting pair to travel from the left to the right through the Josephson junction. A similar process allows for current to travel from right to left. The symbol $\mathbf{p}_s$ denotes the momentum of the superconducting pair. *Figure adapted with permission from* [Shafraniuk (unpublished)].

1.4 Electronic Charge Reconstruction at an Interface

In surface physics, the process of a surface reconstruction, where the atoms on the surface rearrange themselves in response to the dangling bonds resulting from the interface with the vacuum, is well-known. The surface reconstruction of silicon was one of the first systems to be imaged with the scanning tunneling microscope [Binnig, *et al.* (1983)]. Much of the study of surfaces and how they interact with material deposited on the surfaces relies on understanding how the surface reconstructs itself.

In multilayered nanostructures, there are no open surfaces, and there is limited freedom for ions to rearrange their spatial locations in response to the interface with a different material (small relaxations of atoms near the interface certainly occur). But there is no reason why the chemical potential of the leads of the device needs to match the chemical potential of the barrier. This puts the barrier in an unstable situation, where some of the electrons are forced to either leave or enter the barrier from the leads (depending on the relation of the chemical potentials). Because the Coulomb interaction is long-ranged, the charge redistribution will be confined to the interface regions, with a healing length on the order of the Thomas-Fermi screening length [Thomas (1927);

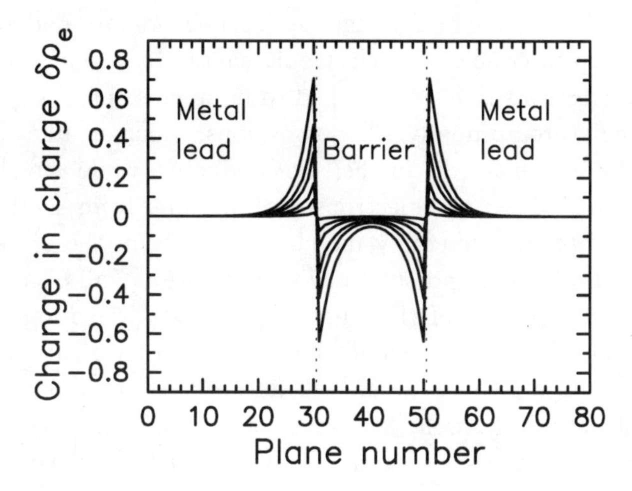

Fig. 1.13 Schematic of the electronic charge reconstruction at the interfaces of a multilayered nanostructure with a barrier that is 20 planes thick. The horizontal axis is the plane number (the barrier lies on planes numbered from 31 to 50—the interface is indicated with the dashed lines) and the vertical axis is the self-consistently calculated change in the charge density on each plane. The chemical potential of the barrier differs from that of the bulk metallic leads for each of the different curves. The screening length is chosen to be approximately 2.2 lattice spacings in both the metallic lead and the barrier. *Figure adapted with permission from* [Nikolić, Freericks and Miller (2002a)] (original figure © 2002 the American Physical Society).

Fermi (1928)] (usually less than an Angstrom in metals). The result is a screened-dipole layer at the interface, which creates an electric potential that causes scattering to electrons moving through the device and is plotted in Fig. 1.13 [Freericks, Nikolić and Miller (2002)]. One can see how charge spills from the barrier into the lead as the mismatch of the chemical potentials is increased. This effect is well known in the semiconductor community when a metal is placed in contact with a semiconductor creating a Schottky barrier [Schottky (1940)]. It is used to create a number of the different semiconductor-based devices.

The electric fields created by these screened dipole layers can be quite large. They do not cause current to flow, however, because they are exactly compensated by an opposite force due to the diffusion current arising from the change in the electron concentration. This is because the system has reached a static, equilibrium, rearrangement of the electronic charge.

One of the most interesting applications of interface charge reconstruction is the case of a metal-oxide-semiconductor field-effect transistor (MOS-

FET). In this device, one brings together a semiconductor and an insulator forming a sharp interface. The electronic charge reconstruction creates a thin layer of electrons that are trapped to lie in close proximity to the interface. If engineered properly, the dopant ions, which created the electron carriers in the first place, lie in the semiconductor, while the electrons lie in the insulator. Then the electrons are far away from scattering sites, and they can become incredibly mobile. It is within these systems that the quantum Hall effect and the fractional quantum Hall effect were both discovered. The creation of this "nearly free" two-dimensional electronic gas follows from the physics behind charge reconstruction at an interface.

Interface charge reconstruction will naturally occur in strongly correlated nanostructures as well, leading to even more interesting behavior when one of the materials is a strongly correlated insulator, since the charge depletion (or enhancement) can "dope" the insulator into a strongly correlated metal phase (or *vice versa* if the material is already a strongly correlated metal). These effects have been imaged in grain boundaries of high temperature superconductors, where the grain boundaries are known to be electrically active [Mannhart and Hilgenkamp]. A grain boundary occurs in the growth of a material where islands of different grains meet, and the temperature is too low for the system to anneal the crystallite boundaries out of the system. A TEM image of just such a grain boundary can be seen in the left panel of Fig. 1.14 [Browning, *et al.* (1993)]. This grain boundary has a large angle orientational mismatch, as is easily seen. Unfortunately, these grain boundaries have a significant deleterious effect on superconducting wires, as they create Josephson junction weak links between the grains, and the critical current of the weak link is much smaller than the maximal critical current of a bulk single crystal. This has proved to be the single largest hurdle to get over in making high temperature superconducting wires (of course, the presence of the grain boundaries can be employed to manufacture Josephson junctions, if desired).

The right panel of Fig. 1.14 [Browning, *et al.* (1993)] depicts the valence of the Copper atom as a function of the distance away from the grain boundary. Clearly the grain boundary is electrically active, and has a charge reconstruction. What is amazing is how far away from the grain boundary this charge rearrangement extends, which is likely due to the fact that the strongly correlated metal does not screen charge as efficiently as a more conventional metal. The charge distortion is reduced as the misorientation angle of the grain boundary is reduced; this is the underlying phenomenon that governs the reduction of critical current at a grain boundary.

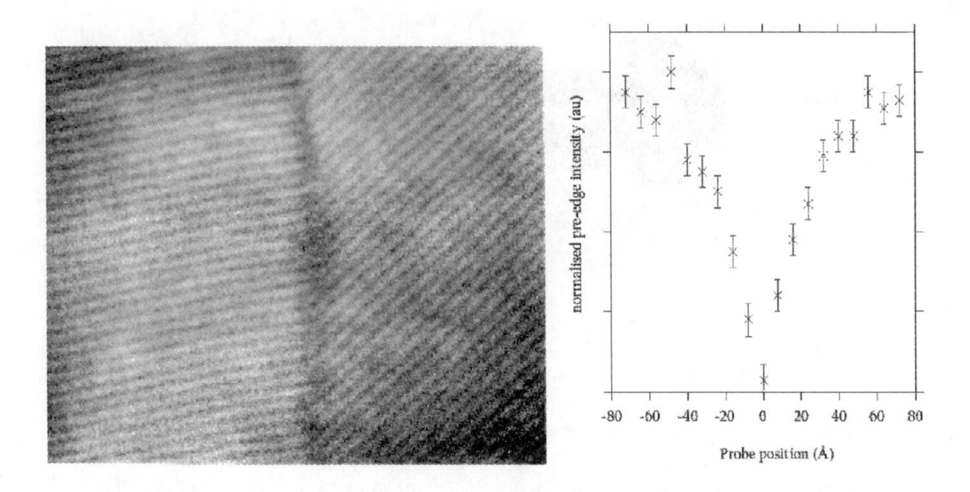

Fig. 1.14 Left panel: high-angle grain boundary in a high temperature superconductor. Right panel: charge profile around the grain boundary (Copper valence) as measured with electron energy loss spectroscopy (see Fig. 1.16 below). The probe position (horizontal axis) is relative to the center of the grain boundary. The vertical axis is proportional to the valence on the copper atom, which changes from a maximum of 2.6 at the top to a minimum of 1.0 at the bottom, as the probe is moved across the grain boundary. *Reprinted with permission from* [Browning *et al.* (1993)].

Since diffusion of chemical species is easier along grain boundaries than within the grains themselves, it was discovered that the critical current across a grain boundary could be enhanced by diffusing Calcium ions to the grain boundary location [Hammerl, *et al.* (2000)]. The Ca ions must be modifying the local charge reconstruction at the grain boundary to do this. An interesting way to improve the critical current density of a high-temperature superconducting tape is to grow multilayers of pure Yttrium-Barium-Copper-Oxide, and of Calcium-doped YBCO. Between the grain boundaries the current will be carried predominately in the pure YBCO, but at the grain boundaries, because the presence of Calcium reduces the charge reconstruction, the critical current density is not reduced as much as in the pure YBCO. A schematic of this multilayered device is shown in left panel of Fig. 1.15, and the improvement in the critical current is shown in the right panel. At this point, it is not clear whether this process can be used to make high temperature superconducting wires into a viable technology.

Another example is the artificially engineered band-insulator/strongly correlated insulator heterostructure made from $SrTiO_3$ (a band insula-

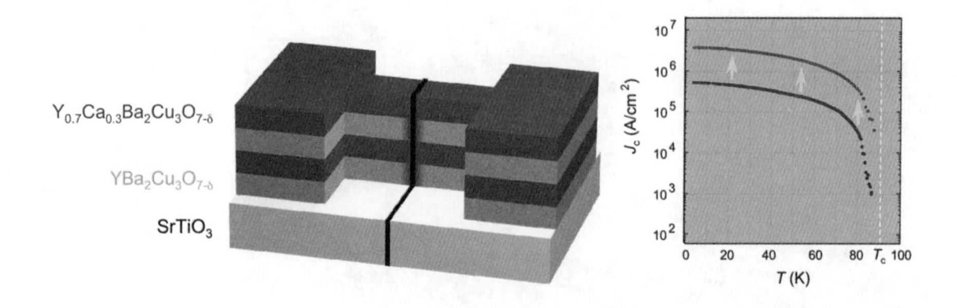

Fig. 1.15 Left panel: schematic of the growth of pure YBCO (yellow) and Ca doped YBCO (red) for increasing the critical current at the grain boundary. Note how the Calcium dopes preferentially into the grain boundary region (the grain boundary is the black line), presumably changing the electronic charge reconstruction. Right panel: enhancement of the critical current density due to Ca doping (increase from the red to the orange curve after doping). *Reprinted with permission from* [Mannhart (2005)].

tor that is nearly ferroelectric) and $LaTiO_3$ (a strongly correlated insulator) [Ohtomo, *et al.* (2002)]. The heterostructures of these materials are made using PLD, and varying the Sr or La content within the titanate background. The heterostructures are grown with nearly atomically flat precision and excellent control over the thicknesses of the different layers. A detailed analysis of the structure shows little interdiffusion of the species across the interface. What is surprising, is that the system has metallic conducting channels in the transverse direction (along the planes rather than perpendicular to the planes), which vary with the thickness of and the spacing of the $LaTiO_3$ layers within the $SrTiO_3$ matrix. Sophisticated experimental equipment is needed to image the charge redistribution in multilayered nanostructures, because one needs to have both sensitivity to the local charge, and an ability to achieve atomic resolution. One way that this is accomplished is by combining electron microscopy observations with electron energy loss spectra (EELS) as shown in Fig. 1.16. This is done with a dedicated scanning transmission electron microscope (STEM) that is equipped with an annular detector and an electron spectrometer. In the STEM, the optics are devoted to focusing the electron beam to a very fine probe (0.13 nm diameter), which is raster scanned over the sample. The transmitted electrons scattered at high angles are collected into an annular dark field detector which is used for the imaging. Since these electrons are primarily Rutherford scattered by the ion cores, the image intensity will be roughly proportional to the square of the atomic number. This is why this technique is called Z-contrast microscopy (for a review of the instrument see [Pennycook (2002)]). It is capable of producing incoherent images with atomic resolution and atomic specificity. Electrons traveling parallel to the optical axis (*i.e.* through the hole in the annulus) are collected into the EELS, so simultaneous EELS measurements can be obtained. These spectra can be employed to determine the local electronic charge, or the energies for the thresholds of different excitations, or the local chemical environment of a particular ion.

This imaging technique was used to measure the charge profile near the grain boundary, shown in Fig. 1.14 [Browning, *et al.* (1993)], and was used in the $SrTiO_3/LaTiO_3$ heterostructures [Ohtomo, *et al.* (2002)]. This imaging technique has also been applied to $YBa_2Cu_3O_{7-\delta}/La_{0.67}Ca_{0.33}MnO_3$ heterostructures [Varela, *et al.* (2003); Varela, *et al.* (2005)]. They find that the interfaces are nearly atomically flat, with essentially no interdiffusion of chemical species across the interface (determined by examining the EELS results). They also can use the STEM-EELS apparatus to map out the local charge density, which is plotted in Fig. 1.16. One can see how the charge screening length is much shorter in the LCMO material than in the YBCO, but the heterostructure is not thick enough for the LCMO material to heal its charge to the bulk value.

The phenomena described above has been termed electronic charge reconstruction [Okamoto and Millis (2004a); Okamoto and Millis (2004b)], due to its similarity with the well-known surface reconstruction. Okamoto and Millis analyzed the $SrTiO_3/LaTiO_3$ system [Ohtomo, *et al.* (2002)] using a hybrid density functional theory/many-body theory approach. The low-energy bands are modeled with a tight-binding scheme, and Coulomb interactions are introduced to describe the electron correlations. The many-body theory was analyzed in a static mean-field theory approach [Okamoto and Millis (2004a)] and in another approximate many-body physics method that can produce the MIT [Okamoto and Millis (2004b)]; both produced much insight into the physics behind this behavior. In particular, since the different systems are at different chemical potentials in the bulk, there is a localized charge transfer at the interfaces, which artificially dopes each of the insulators. This leads to metallic regions near the interfaces that can conduct electricity in the transverse (planar) directions. The results of their calculations are summarized in Fig. 1.17.

Electronic charge reconstruction is a phenomenon that naturally occurs at the interface of any two materials unless they happen to have exactly the same chemical potential (which is unlikely to occur in any real system at all

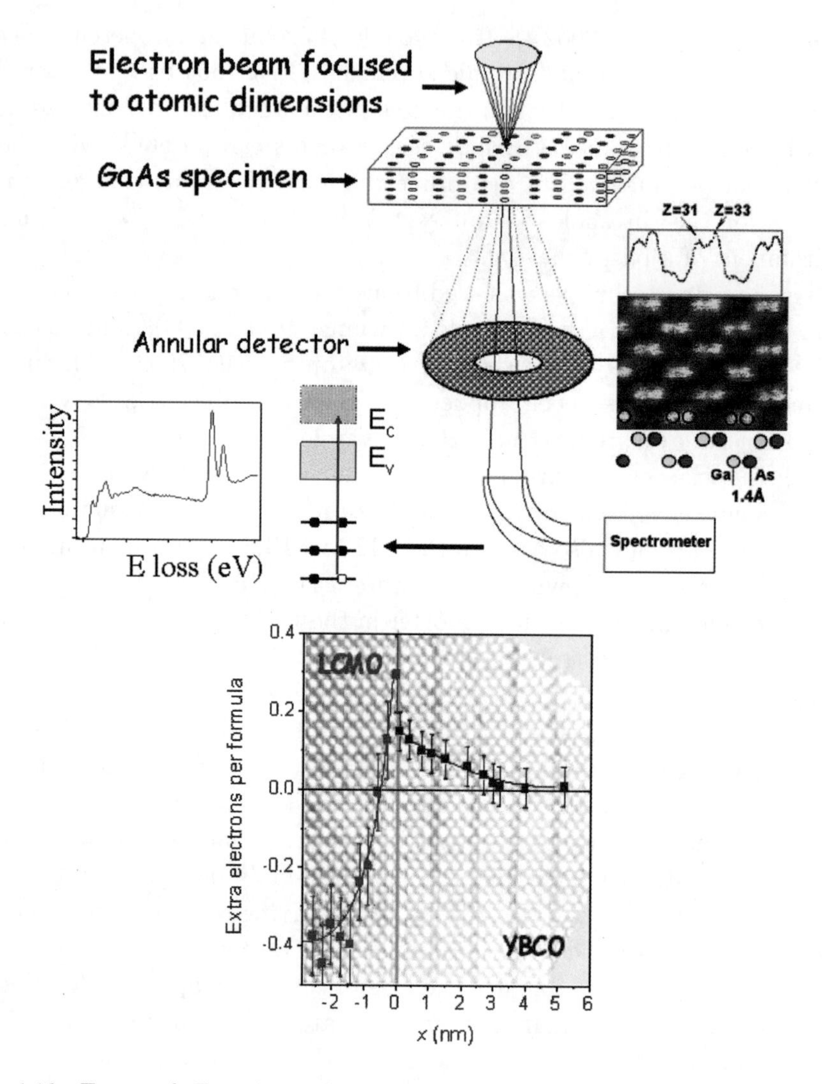

Fig. 1.16 Top panel: Experimental setup for a STEM-EELS measurement. This example shows a measurement on GaAs. A simultaneous measurement of the positions and types of the atoms and of the charge profile can be achieved. Bottom panel: Charge profile overlaid over the atomic positions of a $YBa_2Cu_3O_{7-\delta}/La_{0.67}Ca_{0.33}MnO_3$ heterostructure. *Top panel adapted with permission from* [Pennycook (2002)]. *Bottom panel reprinted with permission from* [Varela (2005)].

temperatures). We describe how to perform self-consistent calculations of electronic charge reconstruction in Chapter 3 and give additional numerical examples in Chapter 6.

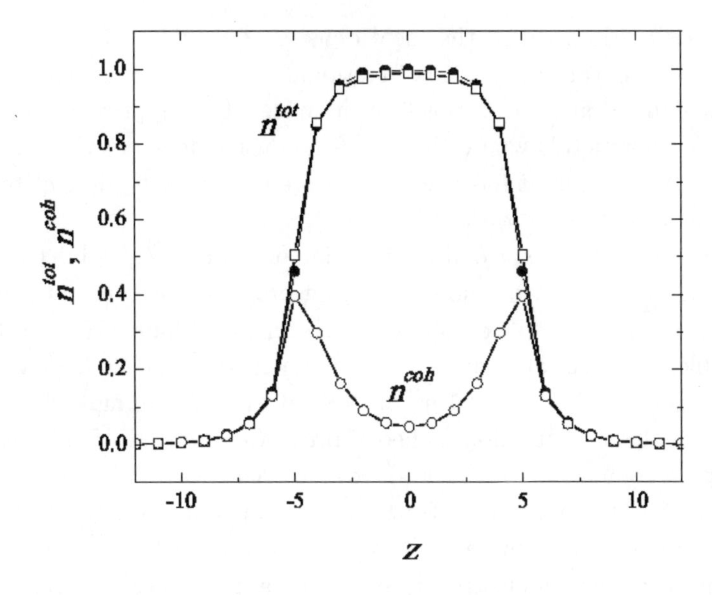

Fig. 1.17 Calculated local charge density near the interface of a $SrTiO_3/LaTiO_3$ heterostructure with six planes of $LaTiO_3$. The top curve shows the charge density as a function of distance in the inhomogeneous (longitudinal) direction. The bottom curve shows the charge density in the Fermi-liquid-like coherence peak near the electrochemical potential. It is clear from the curves, that the interface regions are conducting, with a thickness on the order of three atomic planes. *Reprinted with permission from* [Okamoto and Millis (2004b)] (© 2004 the American Physical Society).

1.5 Roadmap to Real-Materials Calculations

In this book, we concentrate on calculations for model Hamiltonians which usually include one itinerant electron band only. Model systems have been used for decades in many-body physics because they capture the important quantum-mechanical aspects of the problem, but are simpler than materials-specific calculations. Much can be learned about the many-body problem, and about strongly correlated nanostructures by examining these model systems.

But eventually we want to be able to handle real-materials problems in a "first principles" fashion. First principles calculations usually start from the density functional theory with the local density approximation or a generalized gradient expansion. Density functional theory is exact for the ground state energies of real materials if the exact exchange-correlation functional is known for the material [Hohenberg and Kohn (1964)]. The exchange-correlation functional is complicated, and not known in the general case.

The approximation, called the local density approximation, assumes that the functional is the same as the functional of a uniform electron gas with the same density at a given position in space. This approximation should be accurate for metals where the electron density does not change sharply through the material. It has been generalized to include gradient terms for the change of the density as well.

The common way that density functional theory is solved is to map the interacting problem onto a noninteracting problem with the same electron density (but in a complex potential) [Kohn and Sham (1965)]. Solving this problem numerically yields a band structure, which is believed to be similar to the true band structure of the material (assuming such a concept exists), but density functional theory provides no proof of this result. If we want to go beyond the mean-field-like treatment of the local density approximation, we first parameterize the density functional theory bands by a tight-binding model (where choosing the appropriate basis can be critical), and then add electron-electron interaction terms. These interactions can be solved with the techniques of dynamical mean-field theory under the assumption that the electronic self-energy is local (that is, independent of momentum). There has been much progress in solving for properties of strongly correlated materials in the bulk with this procedure, but it is a computationally intense project.

Plutonium is one of the most interesting materials from a solid-state physics context. It possesses numerous phases as functions of temperature and pressure, and it is generally believed that a number of these phase transitions arise from strong electron correlations. The $\alpha - \delta$ phase transition of Pu is interesting because it is accompanied by a 25% volume change, which is believed to be governed primarily by a change in character of the electrons from a band-like metal to a localized insulator. Since this transition can affect the stability of Pu when it is stored for long periods of time (Pu will be self-heated due to the nuclear radioactivity), it is of significant importance to understand its properties. This system was chosen as one of the first systems to apply the DFT+DMFT approach to [Savrasov, Kotliar and Abrahams (2001)]. The first calculations focused on the electronic structure, as modified by the strong electron interactions, and were then followed up by work on the phonons [Dai, *et al.* (2003)]. The phonon dispersions were later verified by inelastic X-ray scattering [Wong, *et al.* (2003); Wong, *et al.* (2005)] and summarized in a short review [Kotliar and Vollhardt (2004)] (see Fig. 1.18); this agreement between experiment and theory shows the power of these methods.

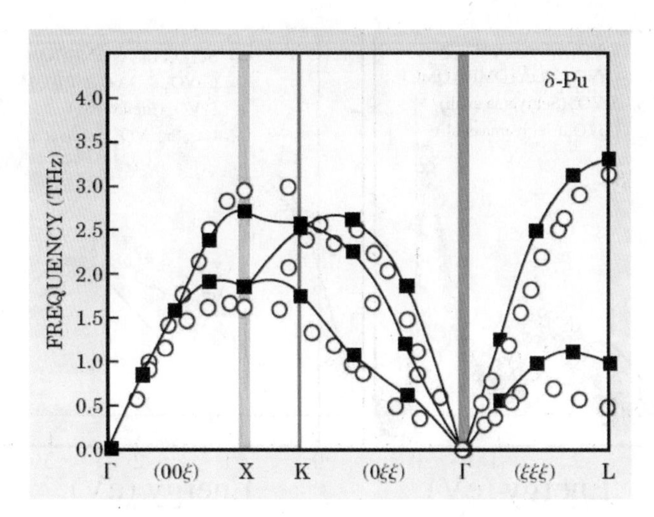

Fig. 1.18 Theoretically calculated phonon dispersion of δ-phase Pu (squares connected by full lines) which were later confirmed by inelastic X-ray scattering measurements (open red circles). *Figure reprinted with permission from* [Kotliar and Vollhardt (2004)] (©2004 American Institute of Physics) (*adapted with permission from* [Dai *et al.* (2003)], [Wong *et al.* (2003)] and [Wong *et al.* (2005)]).

The work of Vollhardt and collaborators has concentrated on examining transition-metal oxides. Initially they focused on V_2O_3 [Held, *et al.* (2001); Mo, *et al.* (2003); Keller, *et al.* (2004)], which is close to the Mott metal-insulator transition; more recent work has examined strongly correlated metals like $CaVO_3$ and $SrVO_3$ [Sekiyama, *et al.* (2004); Nekrasov, *et al.* (2005)]. In the left panel of Fig. 1.19, we show the spectral function (below the Fermi energy) obtained from a high energy (bulk) photoemission study. Photoemission is an experiment where high energy light is shone onto a clean surface of the material and it expels an electron (Einstein won his Nobel prize for this photoelectric effect). By varying the angle of incidence of the photon, one can map out the excitation spectra (multiplied by a Fermi factor) as a function of momentum. Here we show the local spectra, summed over all momenta. The agreement between experiment and theory is excellent, and the theory shows small differences between the two systems. The right panel is a comparison with X-ray absorption spectroscopy, since inverse photoemission data is not yet available (inverse photoemission corresponds to electrons shone onto a surface and light emitted; X-ray absorption spectroscopy measures the ease with which X-rays can be absorbed by forcing an electron to have a transition from a K-edge oxygen core state

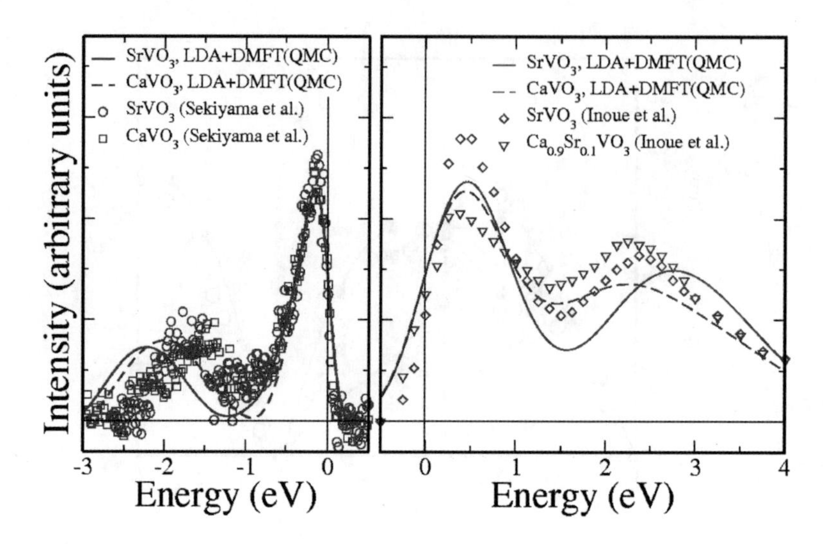

Fig. 1.19 Comparison of the parameter free LDA+DMFT(QMC) spectra of $SrVO_3$ (solid red line) and $CaVO_3$ (dashed blue line) with experimental data (symbols) below and above the Fermi energy. The left panel is high resolution photoemission spectroscopy for $SrVO_3$ (red circles) and $CaVO_3$ (blue squares) [Sekiyama *et al.* (2004)]. The right panel is 1s X-ray absorption spectroscopy for $SrVO_3$ (red diamonds) and $CaVO_3$ (blue triangles) [Inoue *et al.* (1994)]. The horizontal line is the experimental subtraction of the uniform background intensity. *Reprinted with permission from* [Nekrasov *et al.* (2005)] (© 2005 the American Physical Society).

to the Fermi energy of the correlated bands–hence they measure the DOS for the unoccupied states modified by atomic matrix elements, which are not expected to change the signal in this case).

The same strategy can be used for multilayered nanostructures. First a density functional theory is used to determine the band structure and from that an effective (inhomogeneous) tight-binding model is created. The interactions are introduced to represent the strong electron correlation effects. These are treated with inhomogeneous dynamical mean-field theory in order to fully solve the problem. So far, no one has attempted such a calculation. It requires a generalization of the techniques developed in this book along the lines of the progress made with first-principles calculations of bulk strongly correlated materials.

Chapter 2

Dynamical Mean-Field Theory in the Bulk

2.1 Models of Strongly Correlated Electrons

The most general Hamiltonian for matter is actually quite simple to write down. If we consider a collection of N nuclei, whose momenta, position, and mass are denoted by capital letters ($\mathbf{P}_i$, $\mathbf{R}_i$, and M_i), and a collection of N' electrons, whose momenta, position, and mass are denoted by lower case letters ($\mathbf{p}_i$, $\mathbf{r_i}$, and m_e), then the Hamiltonian involves just the sum of the kinetic energies of all of the particles and their mutual Coulomb interactions:

$$\mathcal{H} = \sum_{i=1}^{N} \frac{\hat{P}_i^2}{2M_i} + \sum_{i=1}^{N} \sum_{j=1, j\neq i}^{N} \frac{Z_i Z_j e^2}{2|\hat{\mathbf{R}}_i - \hat{\mathbf{R}}_j|} + \sum_{i=1}^{N'} \frac{\hat{p}_i^2}{2m_e} + \sum_{i=1}^{N'} \sum_{j=1, j\neq i}^{N'} \frac{e^2}{2|\hat{\mathbf{r}}_i - \hat{\mathbf{r}}_j|}$$
$$- \sum_{i=1}^{N} \sum_{j=1}^{N'} \frac{Z_i e^2}{|\hat{\mathbf{R}}_i - \hat{\mathbf{r}}_j|} \quad , \tag{2.1}$$

with Z_i the atomic number of the ith nucleus, and e the electric charge of an electron (the factors of two in the denominators of the nuclear and electronic potentials are to remove double counting); for a neutral system, we have $\sum_{i=1}^{N} Z_i = N'$. We use a hat to indicate a quantum-mechanical operator. A moment's reflection shows that the solution of the quantum-mechanical problem represented by this Hamiltonian will include all of the equilibrium properties of solids, liquids, and gases, and hence represents an enormously complex set of solutions as the parameters are varied over different nuclei.

Since the nuclear mass is so much larger than the electronic mass, the first approximation to be made is to take the limit $M_i \to \infty$, which means that the nuclei will be treated as classical particles. Next, since we are interested in solids, and since most solids condense in periodic lattice structures, we can fix the nuclear positions to the spatial locations of a lattice with periodic boundary conditions. We also assume that the core electrons are so tightly bound to the nuclei that they do not contribute to the low-energy dynamics, and thereby we remove the core electrons from the Hamiltonian, and change the nuclear charges to the ionic charges of the nuclei plus the core electrons (we still denote the number of conduction electrons by N', and the ionic charges by Z_i). Since the ion cores are fixed at their lattice sites, the potential energy from the ion cores is just a constant and can be ignored. Note that the first and third approximations amount to neglecting the effects of phonons or lattice vibrations on the physical processes we will investigate. The phonons do make important contributions to thermodynamic properties and transport properties, but they can be added back later if desired.

So the problem has been reduced to considering the interaction of electrons with static charges (the ion cores) located in a periodic arrangement on a lattice. If we ignore the electron-electron interaction term, then we have the fundamental problem of constructing the electronic band structure for electrons moving in a periodic potential. This is a noninteracting problem, whose ground state is found by forming a Slater determinant of the N' lowest energy states (due to the Pauli exclusion principle). The band structure, $\epsilon_n(\mathbf{k})$, are the energy eigenvalues of the reduced Hamiltonian at each wavevector $\mathbf{k}$ in the first Brillouin zone. The index n denotes the label for the different bands.

When we construct the ground state, we find that for band metals, many of the bands are either totally filled or totally empty, and only a small number of the bands are partially filled. If we are interested in the low-energy properties, we can restrict ourselves to consider those partially filled bands only. The band structure can be parameterized by a tight-binding scheme, where we consider the hopping of electrons from one lattice site (at site $\mathbf{R}_i$) to another lattice site (at site $\mathbf{R}_j$), with a strength t_{ij} called the hopping matrix. It is the hopping matrix that determines the connectivity of the lattice that the electrons are moving on. Now we can add the electron-electron interaction terms back into our Hamiltonian, and try to solve the resulting many-body problem.

This turns out to still be a rather complicated problem, so physicists have tried to simplify it even further to try to understand basic properties of electron correlations. The Hubbard model [Hubbard (1963)] is the simplest proposal in this scheme. It assumes that there is only one partially filled band, and it consists of s-wave orbitals, so the hopping matrix depends only on the distance to neighboring lattice sites. Finally, it assumes that the Coulomb interaction in the metal is screened, so it is sufficient to approximate it by an interaction U when two electrons occupy the same lattice site only. This Coulomb interaction is like the average electron-electron interaction in a helium atom, since the electrons on the same lattice site, really represent electrons restricted to lie in the same unit cell of the lattice; because of this, the Hubbard model cannot describe the formation of chemical bonds between ions, like the dimerization that happens if we have a real lattice of hydrogen atoms in Mott's metal-insulator transition problem. The Hubbard model is illustrated schematically in Fig. 2.1. In spite of the apparent simplicity of the Hubbard model, exact solutions exist only in one dimension [Lieb and Wu (1968)] and in infinite dimensions [Georges and Kotliar (1992); Jarrell (1992); Georges, *et al.* (1996); Kehrein (1998); Bulla (1999)].

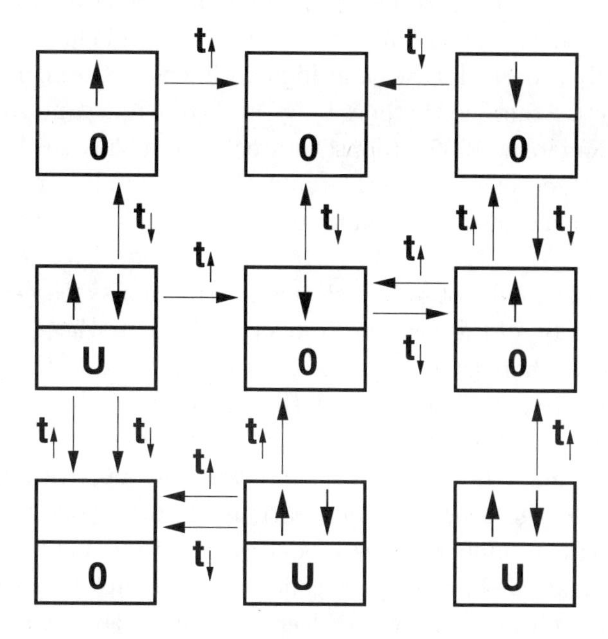

Fig. 2.1 Schematic diagram of the Hubbard model on a 3×3 square lattice. The boxes denote the lattice sites, and the arrows in the upper half of the box indicate the electrons (spin-up or spin-down) that occupy a given lattice site for the specific electron configuration shown. The allowed hoppings are indicated with the arrows between the boxes (some hoppings are forbidden for this configuration by the Pauli exclusion principle). The labels U and 0 in the lower half of the box denote the interaction energy for each lattice site in the current configuration.

The Hubbard model describes the interplay between delocalization effects coming from the kinetic energy and localization effects coming from the potential energy. This competition is most apparent in the d-shells of solids, even though the model assumes that the orbitals are symmetric. In the limit where U becomes large, no lattice site can be doubly occupied, which implies the system is an insulator if there is on average one electron per lattice site (called a half-filled band). On the other hand, if U is sufficiently small, one would expect that the band structure is only perturbed slightly from the noninteracting case, and the system will remain a metal. Hence there will be a critical value of U, where the system undergoes a metal-insulator transition. In one dimension, we have $U_c = 0$ [Lieb and Wu (1968)], but in infinite dimensions, U_c is approximately equal to the bandwidth [Bulla (1999)], so there is a nontrivial metal-insulator transition. The literature on this problem is immense, and a thorough and modern account can be found in Gebhard's book [Gebhard (1997)].

In addition, the Hubbard model possesses a number of interesting magnetic ground states. At half filling, the ground state is believed to be antiferromagnetically ordered for two and higher dimensions [Anderson (1959a)], and for large U, there is the possibility of ferromagnetism occurring near half filling [Nagaoka (1966)] (nonsaturated ferromagnetism has been seen with DMFT [Obermeier, Pruschke and Keller (1997)]); incommensurate magnetic order has also been found for intermediate U [Freericks and Jarrell (1995)].

The simplest model of strong electron correlations emerges from a further simplification of the Hubbard model to the case where the down-spin electrons do not hop. The up-spin electrons then move through a background of down-spin electrons and the quantum-mechanical problem is a simple bandstructure problem for any given configuration of the static particles. The many-body physics enters when we take an annealed thermodynamic average over all configurations of static particles that share the same particle number. This model was first introduced by Hubbard as the alloy-analogy solution (also called the Hubbard-III solution) [Hubbard (1965)]. It was rediscovered [Kennedy and Lieb (1986)] as a model for crystallization: the static particles were interpreted as ions, and they form a periodic arrangement when both the electrons and the ions are half filled on all lattices in two or higher dimensions. This periodic "crystallization" arises solely from the Pauli exclusion principle, and helps answer the fundamental question of solid-state physics—why do nearly all solids form periodic arrangements at low temperature? Another interpretation is that of a binary-alloy problem [Freericks and Falicov (1990)], where we interpret the presence of a static particle at site i as the presence of an A ion and the absence of a static particle at site i as the presence of a B ion—the Coulomb interaction U is then the difference in site energies for an electron on an A ion and an electron on a B ion. It is conventional now to view this simplified Hubbard model as the spinless Falicov-Kimball model [Falicov and Kimball (1969)]. The schematic of this model can be seen in Fig. 2.1 where we set $t_\downarrow = 0$.

The spinless Falicov-Kimball model has much known about its properties. In one dimension, the ground state is periodically ordered when the number of electrons equals the number of ions, and the interaction is attractive [Lemberger (1992)]. The phase diagram for different fillings is quite complex, and likely includes a devil's staircase [Gajek, J edrzejewski and Lemański (1996)]. For small-U, one also finds phase separation instead of the analog of a Peierl's distortion over some of the phase space [Freericks, Gruber and Macris (1999)]. In two dimensions, many similar results occur [Kennedy (1994); Kennedy (1998); Haller and Kennedy (2001)], and one also finds a large number of charge-stripe-like phases [Lemański, Freericks and Banach (2002); Lemański, Freericks and Banach (2004)]. If the electron concentration plus the ion concentration is less than one, and U is large and repulsive, then the system always phase separates into what is called the segregated phase [Freericks, Lieb and Ueltschi (2002a); Freericks, Lieb and Ueltschi (2002b)].

In the limit of large dimensions, DMFT has been employed to solve the spinless Falicov-Kimball model for a wide range of properties ranging from charge-density wave order to transport to Raman scattering. A comprehensive review serves as an introduction to these topics [Freericks and Zlatić (2003)].

The original Falicov-Kimball model [Falicov and Kimball (1969)] is more complex than the spinless version. It was originally introduced to describe the first-order jumps in the resistivity of a number of rare-earth and transition-metal compounds. It assumes that the static particles are electrons that are localized on the lattice (like f-electrons) and it assumes that both the localized and the conduction electrons are spin one-half. Since the localized electrons are strongly correlated, we usually set the localized-electron–localized-electron interaction to infinity, so we restrict the f-occupation to be less than or equal to one at each lattice site. The remaining Coulomb interaction is then repulsive, and acts between localized and conduction electrons when they sit on the same lattice site. If the localized electron level E_f lies above the electron chemical potential as $T \to 0$, the system will be metallic at low temperature, but as the temperature is raised, the high entropy of the localized electron states tends to transfer electrons from the conduction band to the localized band. This can then lead to a strongly scattering "insulating" phase, perhaps with a discontinuous jump in the resistivity. Hence, the thermodynamics of these models depends greatly on what interpretation is given for the localized particles, and how we perform the thermodynamic averaging (in this case we fix the total number of electrons not the separate numbers of conduction and localized electrons). A schematic of the spin-one-half Falicov-Kimball model is given in Fig. 2.2.

This version of the spin-one-half Falicov-Kimball model provides an alternative to the Mott-Hubbard picture of a metal-insulator transition. Here

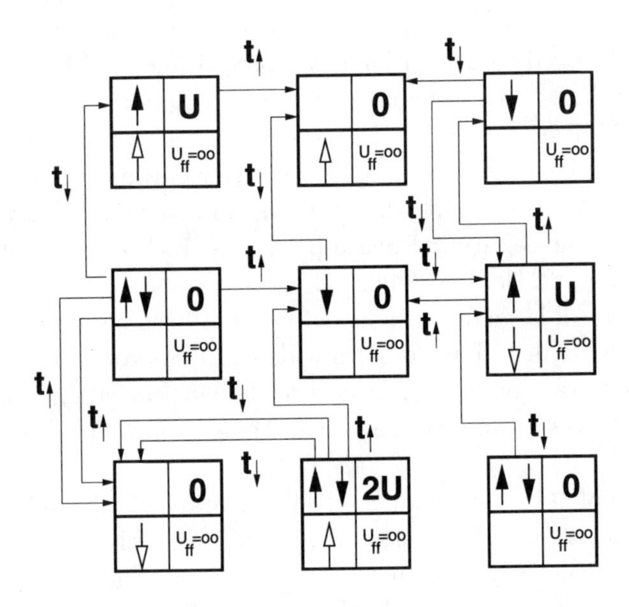

Fig. 2.2 Schematic diagram of the spin one-half Falicov-Kimball model on a 3×3 square lattice. A specific configuration of conduction (solid arrows) and localized (open arrows) electrons is given, along with the allowed hoppings of the conduction electrons. The boxes correspond to the given lattice sites. The upper right corner is the configuration of conduction electrons, the lower right corner is the configuration of the localized electrons. The upper left box is the on-site Coulomb interaction between conduction and localized electrons. The lower left corner is the localized-electron–localized-electron interaction which is taken to be infinitely large, forbidding double occupancy. We do not depict the localized electron site energy E_f.

we have two kinds of electrons: those that conduct electricity and those that do not; the metal-insulator transition occurs due to a change in occupancy of the different electrons, not to a change in the character of the electrons themselves, as in the Mott transition. In the Falicov-Kimball model, one can show that there are discontinuous metal-insulator transitions that occur as a function of T [Chung and Freericks (1998)].

The spin-one-half Falicov-Kimball model also has intermediate-valence phases, where the average occupancy of the localized electrons lies somewhere between zero and one as $T \to 0$ [Chung and Freericks (2000)]. This is called a classical intermediate valence state, because the occupancy of f-electrons on each lattice site is either zero or one, but the average over all lattice sites is noninteger. Hence, the system is inhomogeneous on the nanoscale, but appears uniform only when we perform an annealed average over all possible configurations of electrons.

There is one piece of physics that is left out of the spin-one-half Falicov-Kimball model—it is the possibility of hybridization of the localized electron levels with other electron levels. The f-electrons are so tightly bound to the ion cores that their direct overlap (with neighboring sites) is so small that it can usually be neglected, but the overlap of an f-orbital with an s, p, or d orbital on a neighboring lattice site may not be so small. Including the hybridization and a (finite) direct on-site Coulomb interaction between f-electrons (but neglecting the Falicov-Kimball-like Coulomb interaction between the f-electrons and conduction electrons), yields what is called the periodic Anderson model, since it is a periodic generalization of the single-impurity Anderson model [Anderson (1961)]. Once again, we usually make a number of further simplifications of the model—one takes both the conduction and the f-electron orbitals to have s-wave symmetry, and we assume the f-electrons have no degeneracy. Finally, it is often assumed that the hybridization runs not between neighboring sites, but is a direct hybridization between the localized and conduction electron at the same lattice site (this possibility is usually forbidden by symmetry unless the conduction electron band has some p-character to it). A schematic of the periodic Anderson model appears in Fig. 2.3. It is widely believed that the periodic Anderson model describes the physics behind the so-called heavy Fermion compounds, which have strong electron correlations, Kondo-like physics, and often display exotic magnetic and superconducting phases. It possesses a quantum intermediate-valence state, where the average occupation of the f-electrons is noninteger, and is uniform throughout the lattice.

Less is known about the solutions of the periodic Anderson model than the other two models introduced here. In one dimension, how-

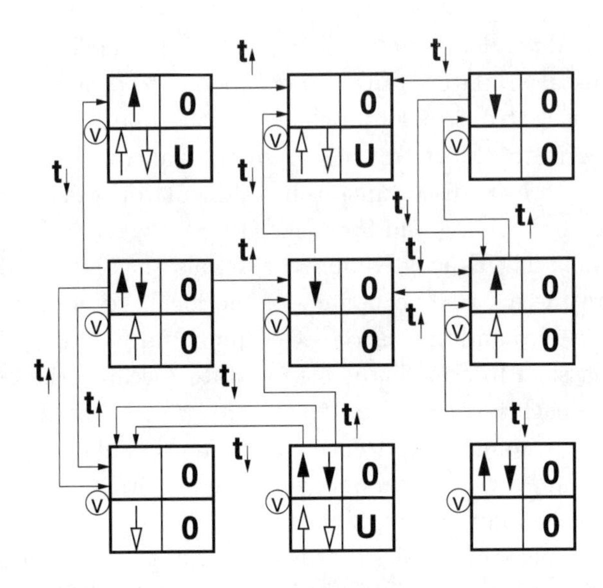

Fig. 2.3 Schematic diagram of the periodic Anderson model. As in the previous figures, the boxes denote the lattice sites, and the upper right corner is the conduction electrons on the site, while the lower right corner is the f-electrons on a site. There is no Falicov-Kimball interaction, so the upper left corner is always 0. The lower right corner denotes the f-f interaction, which is now finite. In addition there is a hybridization between the conduction electrons and the f-electrons on the same site (denoted V). The f-electron site energy is not depicted.

ever, the model can be solved by the density matrix renormalization group [Guerrero and Noack (1996)] where a number of different ferromagnetic and antiferromagnetic solutions are found. In infinite dimensions, the model can be solved with DMFT [Jarrell (1995); Tahvildar-Zadeh, Jarrell and Freericks (1998); Pruschke, Bulla and Jarrell (2000); Grenzebach, *et al.* (2006)] where one finds an enhanced Kondo scale for the symmetric case, which yields a Kondo insulating ground state, and one finds an interesting protracted formation of the quasiparticle peak in the density of states as a function of temperature, which is seen in the photoemission of heavy Fermion compounds.

In the most general case, we need to be able to include all of the relevant electronic bands that lie close to the chemical potential and all relevant Coulomb interactions and hybridizations. Doing so introduces a number of significant complications to the models which makes them harder to solve, but much effort has been devoted to solving these problems with DMFT techniques. This approach is currently being focused on bulk problems only.

2.2 Second Quantization

In the early days of quantum mechanics, Dirac [Dirac (1958)] invented an abstract method for solving the simple-harmonic oscillator problem. His technique is now commonly taught in undergraduate quantum mechanics courses, and it is assumed that the reader is familiar with it.

Dirac introduced the raising and lowering operators $a^\dagger$ and a, respectively. These operators satisfy the commutation relations: $[a^\dagger, a^\dagger] = 0$, $[a, a] = 0$ and $[a, a^\dagger] = 1$. The simple harmonic oscillator Hamiltonian can be written as

$$\mathcal{H}_{\text{sho}} = \hbar\omega_0 \left(a^\dagger a + \frac{1}{2} \right), \tag{2.2}$$

with ω_0 being the frequency of the oscillator. The raising operator increases the energy of an eigenstate by $\hbar\omega_0$ when it acts on an eigenstate, similarly the lowering operator decreases the energy by $\hbar\omega_0$ when acting on an eigenstate. Hence, there must be a lowest energy state $|0\rangle$ which satisfies $a|0\rangle = 0$. From this state, all excited states can be formed, and they are

$$|n\rangle = \frac{1}{\sqrt{n!}} (a^\dagger)^n |0\rangle, \tag{2.3}$$

with the numerical prefactor being the normalization constant; the energy of the $|n\rangle$ eigenstate is $\hbar\omega_0(n+1/2)$. Because there is a simple formula that relates the position and momentum operators to the raising and lowering operators, any matrix element of any function of position or momentum between any two eigenstates can be calculated using just algebraic manipulations.

The power of this abstract technique is obvious: the solution of many different problems is facilitated by these algebraic manipulations instead of requiring one to perform integrals over Hermite polynomials, and the like. But this approach is even more important than just providing efficiency to calculations. It has created a new way of approaching quantum-mechanical problems where one can abstractly relate operators to eigenstates. The economy of Eq. (2.3) for representing an eigenstate is one of the most important properties that will be generalized to Fermionic systems in the second quantization formalism.

The N-electron eigenstates are cumbersome to determine for a general many-body physics problem, even if they solve a relatively simple Schroedinger equation. To get a flavor of this complexity, we need to first construct a set of N-particle basis states that span the Hilbert space for the many-body Hamiltonian. These states are constructed from a complete set of single-particle states (they can be plane waves, Bloch functions, or any other convenient set of states; usually it is most convenient to use the single-particle eigenstates of the many-body Hamiltonian, but that is not necessary), labeled as $\{\psi_n(\mathbf{r})\}$ for $n = 1, 2, 3, \ldots$. In this notation, the symbol $\mathbf{r}$ denotes both spatial and spin coordinates. Then the N-electron basis functions can be written as Slater determinants

$$\Psi_{k_1,k_2,\ldots,k_N}(\mathbf{r_1},\mathbf{r_2},\ldots,\mathbf{r_N}) = \frac{1}{\sqrt{N!}}\sum_P(-1)^P\psi_{k_1}(P\mathbf{r_1})\psi_{k_2}(P\mathbf{r_2})\ldots\psi_{k_N}(P\mathbf{r_N}), \tag{2.4}$$

with the sum over all $N!$ permutations P of N objects [the symbol $(-1)^P$ denotes the parity of the permutation]. The state is labeled by the N different single-particle states $(k_1, k_2, \ldots, k_N)$ from which it is constructed. Any N-particle eigenstate can be written as a linear combination over these basis states. The wavefunction in Eq. (2.4) is antisymmetric under the interchange of any two particles

$$\Psi_{k_1,\ldots,k_N}(\mathbf{r_1},\mathbf{r_2},\ldots,\mathbf{r_i},\ldots,\mathbf{r_j}\ldots,\mathbf{r_N}) = -\Psi_{k_1,\ldots,k_N}(\mathbf{r_1},\mathbf{r_2},\ldots,\mathbf{r_j},\ldots,\mathbf{r_i}\ldots,\mathbf{r_N}). \tag{2.5}$$

Since the notation of the Slater determinant is cumbersome to deal with, a shorthand notation called the occupation-number representation was developed which denotes which wavefunctions explicitly appear in the Slater determinant (*i.e.*, the set of $\{k_i\}$ used). Since each single-particle state appears zero or one times in the Slater determinant, we can represent the N-electron basis states by vectors that consist of a string of zeros and ones, denoting whether the nth single-particle state ψ_n appears in the given Slater determinant (for an N-electron state, we must have exactly N terms that are equal to one). For example, the notation $|1, 0, 0, \ldots\rangle$ represents the state $\psi_1(\mathbf{r_1})$ and the notation $|0, 1, 1, 0, \ldots\rangle$ represents the state $[\psi_2(\mathbf{r_1})\psi_3(\mathbf{r_2}) - \psi_2(\mathbf{r_2})\psi_3(\mathbf{r_1})]/\sqrt{2}$. This notation is a very compact way to represent states for large numbers of particles.

In the spirit of Dirac, we introduce abstract operators called the electron creation operators $c_k^\dagger$ and the electron annihilation operators c_k which "create" an electron in state k or "destroy" and electron in state k, respectively:

$$c_k^\dagger|n_1, n_2, ..., n_k, ...\rangle = |n_1, n_2, ..., n_k + 1, ...\rangle,$$
$$c_k^\dagger|n_1, n_2, ..., n_k, ...\rangle = |n_1, n_2, ..., n_k - 1, ...\rangle. \tag{2.6}$$

Since the Pauli exclusion principle allows each state to be occupied at most one time, we must have $(c^\dagger)^2 = 0$ and $c^2 = 0$. Since the Slater determinant is antisymmetric under the interchange of any two k_i indices, it is easy to show that

$$c_k^\dagger c_{k'}^\dagger + c_{k'}^\dagger c_k^\dagger = 0 = \{c_k^\dagger, c_{k'}^\dagger\}. \tag{2.7}$$

We call this sum of the two operators in different orders the anticommutator of the two operators, which is denoted by the curly braces instead of the straight brackets for a commutator. Similarly, one can use the definitions to show

$$\{c_k, c_{k'}\} = 0, \quad \{c_k, c_{k'}^\dagger\} = 1 \tag{2.8}$$

too. Since $c_k|0_k\rangle = 0$, we have $c_k^\dagger c_k|0_k\rangle = 0$. Since $c_k|1_k\rangle = |0_k\rangle$ and $c_k^\dagger|0_k\rangle = |1_k\rangle$, we have $c_k^\dagger c_k|1_k\rangle = |1_k\rangle$. We call the operator $n_k = c_k^\dagger c_k$ the number operator for state k, because it counts the number of times (zero or one) that the state k appears in the given Slater determinant. It satisfies the following commutation relations with the creation and annihilation operators:

$$[n_k, c_k^\dagger] = c_k^\dagger, \quad [n_k, c_k] = -c_k, \tag{2.9}$$

which are similar to the relations for the raising and lowering operators of the simple harmonic oscillator. We call $\mathcal{N} = \sum_{k=1}^{\infty} n_k$ the total number operator, since it counts the total number of occupied states in the given Slater determinant.

We can use the second quantization formalism to represent the three different Hamiltonians that were depicted graphically in the previous section. We let $c_{i\sigma}^\dagger$ ($c_{i\sigma}$) denote the creation (annihilation) operator for a conduction electron at lattice site i with z-component of spin σ. We let $f_{i\sigma}^\dagger$ ($f_{i\sigma}$) denote the creation (annihilation) operator for a localized (or f) electron at site i with spin σ. Then the Hubbard model [Hubbard (1963)] Hamiltonian becomes

$$\mathcal{H}_{\text{Hub}} = -\sum_{ij\sigma} t_{ij}(c_{i\sigma}^\dagger c_{j\sigma} + c_{j\sigma}^\dagger c_{i\sigma}) + U\sum_i n_{i\uparrow}n_{i\downarrow}. \tag{2.10}$$

We use the standard convention of including a minus sign before the hopping matrix here. The screened Coulomb interaction term counts the number of double occupancies, and multiplies them by the strength U. The Falicov-Kimball model [Falicov and Kimball (1969)] Hamiltonian becomes

$$\mathcal{H}_{\rm FK} = -\sum_{ij\sigma} t_{ij}(c_{i\sigma}^\dagger c_{j\sigma} + c_{j\sigma}^\dagger c_{i\sigma}) + E_f \sum_{i\sigma} f_{i\sigma}^\dagger f_{i\sigma} + U\sum_{i\sigma\sigma'} c_{i\sigma}^\dagger c_{i\sigma} f_{i\sigma'}^\dagger f_{i\sigma'}, \tag{2.11}$$

for the spin-one-half case, where we need to project onto states where the f-electron occupation on each lattice site is less than or equal to 1. The spinless case corresponds to removing the spin labels. Finally, the periodic Anderson model [Anderson (1961)] Hamiltonian becomes

$$\begin{aligned}\mathcal{H}_{\rm pam} = &-\sum_{ij\sigma} t_{ij}(c_{i\sigma}^\dagger c_{j\sigma} + c_{j\sigma}^\dagger c_{i\sigma}) + E_f \sum_{i\sigma} f_{i\sigma}^\dagger f_{i\sigma} + U\sum_{i} f_{i\uparrow}^\dagger f_{i\uparrow} f_{i\downarrow}^\dagger f_{i\downarrow} \\ &+ V\sum_{i\sigma}(c_{i\sigma}^\dagger f_{i\sigma} + f_{j\sigma}^\dagger c_{i\sigma}),\end{aligned} \tag{2.12}$$

where we assumed that the hybridization is between the conduction electrons and the f-electrons on the same lattice site (the generalization to other situations is easy to write down).

We have been working with creation and annihilation operators in real space. It is sometimes convenient to also consider the Fourier transform to momentum space. We define the momentum-space creation and annihilation operators by

$$c_{\mathbf{k}\sigma}^\dagger = \frac{1}{\sqrt{\Lambda}}\sum_j c_{j\sigma}^\dagger e^{i\mathbf{k}\cdot\mathbf{R_j}}, \quad c_{\mathbf{k}\sigma} = \frac{1}{\sqrt{\Lambda}}\sum_j c_{j\sigma} e^{-i\mathbf{k}\cdot\mathbf{R_j}}, \tag{2.13}$$

with Λ being the number of lattice sites. It is a simple exercise to show that the creation and annihilation operators in momentum space satisfy the usual anticommutation relations, in particular, we have $\{c_{\mathbf{k}\sigma}, c_{\mathbf{k}'\sigma'}^\dagger\} = \delta_{\mathbf{k}\mathbf{k}'}\delta_{\sigma\sigma'}$ (readers not familiar with this should verify that it does hold). If the hopping matrix t_{ij} connects only nearest-neighbor sites (sites i and j where $\mathbf{R_i} - \mathbf{R_j} = \bar{\delta}$ with $\bar{\delta}$ a translation vector to a nearest-neighbor site) with strength t, then the kinetic energy operator for the conduction electrons becomes

$$-\sum_{ij\sigma} t_{ij} c_{i\sigma}^\dagger c_{j\sigma} = \sum_{\mathbf{k}\sigma} \epsilon_{\mathbf{k}} c_{\mathbf{k}\sigma}^\dagger c_{\mathbf{k}\sigma}, \tag{2.14}$$

with $\epsilon_{\mathbf{k}} = -t\sum_{\delta}\exp(i\mathbf{k}\cdot\delta)$ being the bandstructure. In a momentum-space representation, the Hubbard model Hamiltonian then becomes

$$\mathcal{H}_{\text{Hub}} = \sum_{\mathbf{k}\sigma} \epsilon_{\mathbf{k}} c^{\dagger}_{\mathbf{k}\sigma} c_{\mathbf{k}\sigma} + \frac{U}{\Lambda} \sum_{\mathbf{k}_1\mathbf{k}_2\mathbf{k}_3} c^{\dagger}_{\mathbf{k}_1\uparrow} c_{\mathbf{k}_2\uparrow} c^{\dagger}_{\mathbf{k}_3\downarrow} c_{\mathbf{k}_1-\mathbf{k}_2+\mathbf{k}_3\downarrow}, \tag{2.15}$$

with similar relations for the Falicov-Kimball and periodic Anderson models. It is interesting to note that the kinetic energy operator is diagonal in the momentum-space representation, while the screened Coulomb interaction is diagonal in the real-space representation. All of the complexities of the many-body problem arise from the competition that occurs between these two terms when we simultaneously diagonalize the Hamiltonian with $U \neq 0$.

On the other hand, in the case where $U = 0$, we can solve the Hubbard model for any number of electrons. The problem corresponds to a tight-binding bandstructure problem, which is diagonalized by making the Fourier transform to momentum space [see Eq. (2.15)]. The bandstructure $\epsilon_{\mathbf{k}}$ gives the Λ energy eigenvalues for the single-electron problem. Since the Hamiltonian is diagonal, the many-electron eigenstates correspond to Slater determinants in the momentum-space representation. The ground state is found by the "bathtub principle": choose the N eigenstates with the lowest energy eigenvalues (taking into account the spin degeneracy) to "fill the energy bathtub".

We end this section by presenting a nontrivial application of the second quantization formalism. The problem is the ground-state energy of the jellium model. In the jellium model, we start from the full solid-state Hamiltonian in Eq. (2.1) and we make the approximation that the ionic cores are spread out uniformly to produce a homogeneous positive background charge that cancels out the net negative charge of the conduction electrons, so there is no structure to the ion-electron interaction. We will express the Hamiltonian for jellium in a momentum basis. Since there is no lattice potential anymore, the electrons are free and the bandstructure becomes $\epsilon_{\mathbf{k}} = \hbar^2 k^2/2m$. When we evaluate the electron-electron interaction potential in the momentum-space basis, we find that it involves the Fourier transform of the Coulomb interaction. This Fourier transform is not well defined in general, so we need to evaluate in by a limiting procedure. We do this by replacing the Coulomb interaction with a screened Coulomb interaction $e^2/r \to e^2\exp(-\kappa r)/r$ and then we take the limit $\kappa \to 0$ at the end of the calculation. The Fourier transform is easiest to carry out in spherical coordinates, with the z-axis lying along the direction of the $\mathbf{k}$-vector. It becomes

$$\int_0^\infty dr\frac{e^2}{r}\exp(-\kappa r)\int_0^\pi d\theta r\exp[ikr\cos\theta]\int_0^{2\pi}d\phi r\sin\theta. \tag{2.16}$$

The integral over ϕ is trivial; the integral over θ is straightforward, and yields

$$\int_0^\infty dr 2\pi e^2 r\exp(-\kappa r)\frac{\exp(ikr)-\exp(-ikr)}{ikr}. \tag{2.17}$$

The integration over r becomes $4\pi e^2/(\kappa^2+k^2)$. Finally, taking the limit $\kappa\to 0$, gives the Fourier transform of the Coulomb interaction $4\pi e^2/k^2$. Note that the Fourier transform of the Coulomb interaction diverges as $k\to 0$. This divergence is canceled by the positive background charge, which only has a $k=0$ component. Hence the Coulomb potential for the jellium model is

$$V_{\text{jellium}}(k)=\begin{cases}0 \text{ if } \mathbf{k}=0\\ 4\pi e^2/k^2 \text{ if } \mathbf{k}\neq 0\end{cases}, \tag{2.18}$$

in momentum space.

The jellium Hamiltonian can then be written as

$$\mathcal{H}_{\text{jellium}}=\sum_{\mathbf{k}\sigma}\frac{\hbar^2k^2}{2m}c^\dagger_{\mathbf{k}\sigma}c_{\mathbf{k}\sigma}+\frac{4\pi e^2}{2V}\sum_{\mathbf{k}\mathbf{k}'\mathbf{q}\neq\mathbf{0}\sigma\sigma'}\frac{1}{q^2}c^\dagger_{\mathbf{k}+\mathbf{q}\sigma}c^\dagger_{\mathbf{k}'-\mathbf{q}\sigma'}c_{\mathbf{k}'\sigma'}c_{\mathbf{k}\sigma}. \tag{2.19}$$

where the factor of 2 in the denominator of the potential energy piece is to avoid double counting, the restriction to $\mathbf{q}\neq 0$ is because the positive background cancels the $\mathbf{q}=0$ term in the potential, and where V is the volume of the jellium "solid" (the normalization factor in the Fourier transform for the momentum-space operators is $1/\sqrt{V}$ in the continuum).

An estimate of the ground-state energy can be made in a number of different ways. If we think of the potential energy as a perturbation to the kinetic energy piece of the Hamiltonian, then the first-order perturbation theory says the energy shifts by the expectation value of the potential energy in the ground-state of the kinetic energy. This N-electron ground state is found from the bathtub principle by filling in the lowest N energy levels. In many-body physics, the first-order perturbation theory is also called the Hartree-Fock approximation. It can be viewed as a variational approximation to the true energy as well, so it will be an upper bound to the exact ground-state energy.

Since the band structure is $\epsilon_{\mathbf{k}} = \hbar^2 k^2/2m$, the surfaces of constant energy are spheres, and the bathtub principle says we fill in the electronic energy states up to the Fermi momentum k_F. If there are N states, then $\sum_{k<k_F\sigma} 1 = N$, or if we let $\rho_e = N/V$ be the density of electrons, we have

$$2V \int_0^{k_f} \frac{dk}{(2\pi)^3} k^2 4\pi = N, \tag{2.20}$$

with the factor of 2 coming from spin, the factor of 4π coming from the integral over solid angle, and the factor of $V/(2\pi)^3$ is the integration measure. The integral is easy to evaluate and gives

$$\rho_e = \frac{k_F^3}{3\pi^2}, \quad k_F = (3\pi^2 \rho_e)^{\frac{1}{3}}. \tag{2.21}$$

The ground-state wavefunction for the kinetic energy operator is

$$|\psi_{\mathrm{gs}}^0\rangle = \prod_{k<k_F\sigma} c_{\mathbf{k}\sigma}^\dagger |0\rangle. \tag{2.22}$$

Since the operator $c_{\mathbf{k}\sigma}^\dagger c_{\mathbf{k}\sigma}$ is the number operator, which counts the electron number in state $\mathbf{k}$ with spin σ, the expectation value of the kinetic energy $\hat{T}$ satisfies

$$\langle \psi_{\mathrm{gs}}^0 | \hat{T} | \psi_{\mathrm{gs}}^0 \rangle = 2 \sum_{k<k_F} \frac{\hbar^2 k^2}{2m} = \frac{\hbar^2}{2m} \frac{k_F^5}{5\pi^2} V, \tag{2.23}$$

because the number of electrons is 1 for $k < k_F$ and 0 for $k > k_F$.

The expectation value of the potential-energy operator $\hat{V}$ is more complicated. To begin, we need to work out the operator expectation value

$$\langle \psi_{\mathrm{gs}}^0 | c_{\mathbf{k}+\mathbf{q}\sigma}^\dagger c_{\mathbf{k}'-\mathbf{q}\sigma'}^\dagger c_{\mathbf{k}'\sigma'} c_{\mathbf{k}\sigma} | \psi_{\mathrm{gs}}^0 \rangle, \tag{2.24}$$

in the kinetic-energy ground state. The operator average has $2N + 2$ creation and annihilation operators which act on the vacuum state to the left and to the right. Each creation operator must be paired with an annihilation operator for the same state in order for the average not to vanish. This means the momentum vectors of the operators in Eq. (2.24) must lie below the Fermi wavevector. In addition, since the operators coming from the wavefunction are already paired, we need to pair each creation and annihilation operator in the operator average. Hence there are two possibilities where the expectation value does not vanish: (i) when $\mathbf{q} = 0$ and (ii) when

$\sigma = \sigma'$ and $\mathbf{k}+\mathbf{q} = \mathbf{k}'$. The average in the first case is equal to 1 and in the second case is equal to -1. It can be written in the following fashion

$$\begin{aligned}\langle\psi^0_{\rm gs}|c^\dagger_{\mathbf{k}+\mathbf{q}\sigma}c^\dagger_{\mathbf{k}'-\mathbf{q}\sigma'}c_{\mathbf{k}'\sigma'}c_{\mathbf{k}\sigma}|\psi^0_{\rm gs}\rangle &= \langle\psi^0_{\rm gs}|c^\dagger_{\mathbf{k}+\mathbf{q}\sigma}c_{\mathbf{k}\sigma}|\psi^0_{\rm gs}\rangle\langle\psi^0_{\rm gs}|c^\dagger_{\mathbf{k}'-\mathbf{q}\sigma'}c_{\mathbf{k}'\sigma'}|\psi^0_{\rm gs}\rangle\\ &- \langle\psi^0_{\rm gs}|c^\dagger_{\mathbf{k}+\mathbf{q}\sigma}c_{\mathbf{k}'\sigma'}|\psi^0_{\rm gs}\rangle\langle\psi^0_{\rm gs}|c^\dagger_{\mathbf{k}'-\mathbf{q}\sigma'}c_{\mathbf{k}\sigma}|\psi^0_{\rm gs}\rangle,\end{aligned} \tag{2.25}$$

where the expectation value of the four-Fermion operator is written as the product of the expectation value of two two-Fermion operators (paired in each possible way). This result is quite general and is called Wick's theorem, although we do not prove it here. Using the results of Eq. (2.25), allows us to evaluate the expectation value of the potential energy (see Problem A.1). The result is

$$\langle\psi^0_{\rm gs}|\hat{V}|\psi^0_{\rm gs}\rangle = -e^2\frac{k_F^4}{4\pi^3}V. \tag{2.26}$$

The standard way to express the total energy per electron is with respect to the radius r of a sphere that contains one electron, $r_s = r/a_0$ with $a_0 = \hbar^2/me^2 = 0.0529$ nm the Bohr radius. The parameter r_s can be expressed in terms of the Fermi wavevector as $r_s = (9\pi/4)^{1/3}1/k_Fa_0$ yielding the final result

$$\frac{E_{\text{jellium HF}}}{N} = \left[\frac{2.210}{r_s^2} - \frac{0.916}{r_s}\right]\text{Ry}, \tag{2.27}$$

in Rydbergs (Ry $= e^2/2a_0 = 13.6$ eV). The higher terms in the jellium energy are much more complicated to derive and were a significant project in the 1950s [Pines (1953); Gell-Mann and Brueckner (1957)].

2.3 Imaginary Time Green's Functions

In many-body physics, the technique of Green's functions has been developed to solve problems. This method is used, instead of trying to solve directly for the many-body wavefunctions (via a Schroedinger equation), because it turns out to be much simpler. This is because the Green's function satisfies a simpler differential equation (and a boundary condition), which allows it to be more readily solved. You will see the details of how this works in the next few sections.

The starting point is to define the partition function as in statistical mechanics

$$\mathcal{Z} = \mathrm{Tr}\{e^{-\beta(\mathcal{H}-\mu\mathcal{N})}\} = \sum_m e^{-\beta(E_m-\mu N_m)}, \tag{2.28}$$

with the trace over all normalized many-body wavefunctions $|m\rangle$, which satisfy $\mathcal{H}|m\rangle = E_m|m\rangle$ and $\mathcal{N}|m\rangle = N_m|m\rangle$ (we suppress the spin indices for simplicity here except where they are necessary; they should be able to be added back by the reader if needed). The symbol $\beta = 1/T$ is the inverse of the temperature. Then the Green's function (in real space) is defined by

$$\begin{aligned} G_{ij}(\tau,\tau') &= [-\theta(\tau-\tau')\mathrm{Tr}\{e^{-\beta(\mathcal{H}-\mu\mathcal{N})}c_i(\tau)c_j^\dagger(\tau')\} \\ &\quad + \theta(\tau'-\tau)\mathrm{Tr}\{e^{-\beta(\mathcal{H}-\mu\mathcal{N})}c_j^\dagger(\tau')c_i(\tau)\}]/\mathcal{Z} \\ &= -\mathrm{Tr}\{e^{-\beta(\mathcal{H}-\mu\mathcal{N})}\mathcal{T}_\tau c_i(\tau)c_j^\dagger(\tau')\}/\mathcal{Z}, \\ &= -\langle \mathcal{T}_\tau c_i(\tau)c_j^\dagger(\tau')\rangle, \end{aligned} \tag{2.29}$$

where the last two lines define the time-ordering operator $\mathcal{T}_\tau$ which orders earlier times to the right taking into account the Fermionic sign when operators are interchanged, and the operator average is defined by $\langle\mathcal{O}\rangle = \mathrm{Tr}\{\exp[-\beta(\mathcal{H}-\mu\mathcal{N})]\mathcal{O}\}/\mathcal{Z}$. The symbol $\theta(\tau-\tau')$ is the unit step function, which equals 1 for $\tau > \tau'$ and equals 0 for $\tau < \tau'$. Similarly, the momentum-space Green's function is defined by

$$G_{\mathbf{k}}(\tau,\tau') = -\langle \mathcal{T}_\tau c_{\mathbf{k}}(\tau)c_{\mathbf{k}}^\dagger(\tau')\rangle, \tag{2.30}$$

where we assume the system is translationally invariant (*i.e.*, G_{ij} depends only on $\mathbf{R_i} - \mathbf{R_j}$). The translational invariance guarantees that the momentum-space Green's function is diagonal in the momentum indices, *i.e.*, $-\langle \mathcal{T}_\tau c_{\mathbf{k}}(\tau)c_{\mathbf{k}'}^\dagger(\tau')\rangle = 0$ when $\mathbf{k} \neq \mathbf{k}'$.

The Green's function in Eq. (2.29) can be simplified by representing it with the Lehmann representation. A complete set of states is introduced in between the two Fermionic operators and the trace is written out explicitly over all possible states. The result is

$$\begin{aligned} G_{ij}(\tau,\tau') &= \frac{1}{\mathcal{Z}}\sum_{mn}[-\theta(\tau-\tau')e^{-\beta(E_m-\mu N_m)} + \theta(\tau'-\tau)e^{-\beta(E_n-\mu N_n)}] \\ &\quad \times e^{(\tau-\tau')(E_m-\mu N_m-E_n+\mu N_n)}\langle m|c_i|n\rangle\langle n|c_j^\dagger|m\rangle, \end{aligned} \tag{2.31}$$

where the symbol $\langle n|c_j^\dagger|m\rangle$ is a matrix element with respect to different many-body wavefunctions, and should not be confused with the $\langle\mathcal{O}\rangle$ notation. The first thing to note is that Eq. (2.31) shows that the Green's function is a function of $\tau-\tau'$ only. This property is called time-translation invariance, and it occurs because the system is in equilibrium, so it appears the same for any origin of time. In addition, we see that as long as $|\tau-\tau'|<\beta$, the Green's function is well defined, because the statistical factor guarantees the convergence of the summations. Writing the Lehmann representation for the momentum-space Green's function shows it is also a function of $\tau-\tau'$ too. The Green's function depends, in a complicated fashion, on the energy eigenvalues and the matrix elements of the exact wavefunctions of the many-body problem. It turns out that these particular combinations are the combinations that allow us to perform many calculations, which is why the Green's functions are so useful. Note that the fact that the Green's function is time-translation invariant can also be shown directly from Eq. (2.29) by using the cyclic property of the trace ($\mathrm{Tr}AB=\mathrm{Tr}BA$) and the fact that $\mathcal{H}-\mu\mathcal{N}$ commutes with itself.

The imaginary-time Green's functions can be employed to calculate a wide range of different static properties of the many-body system. The simplest such property is the average electron filling, which satisfies

$$\rho_e=\frac{1}{\Lambda}\sum_i\langle n_i\rangle=\frac{1}{\Lambda}\sum_i G_{ii}(0^-)=G_{\mathrm{loc}}(0^-), \tag{2.32}$$

where the symbol 0^- means that we take the limit $\tau'\to\tau$ with $\tau'>\tau$. The last equality holds in a homogeneous system, where the local Green's function does not change from one site to another (which holds on any periodic lattice that has not gone into an ordered phase which breaks the lattice translational symmetry). One can also determine the average kinetic energy, which satisfies

$$\langle\hat{T}\rangle=-\sum_{ij}t_{ij}G_{ij}(0^-). \tag{2.33}$$

Similar averages are determined in momentum space. The filling of electrons in different regions of the Brillouin zone is just $\langle n_{\mathbf{k}}\rangle=G_{\mathbf{k}}(0^-)$ and the average kinetic energy is $\langle\hat{T}\rangle=\sum_{\mathbf{k}}\epsilon_{\mathbf{k}}G_{\mathbf{k}}(0^-)$. A little thought shows that trying to calculate these averages at nonzero temperature from the individual wavefunctions would entail much more work than doing so from the Green's function.

Our next step is to show how one can actually calculate the Green's function itself. The basic idea is that the Green's function satisfies a differential equation with a boundary condition. If we can solve the differential equation, then we can determine the Green's functions. In general, the differential equation is not easily solved, and often one develops perturbative methods to approximately solve it. In later sections we will show how the DMFT approach actually allows us to solve for a wide class of Green's functions.

Since we know that the Green's function depends on one time variable τ, we need to differentiate with respect to τ in order to find the differential equation that it satisfies. This is called the equation of motion (EOM) technique. The imaginary time derivative of Eq. (2.29) is

$$\begin{aligned}\frac{\partial}{\partial\tau}G_{ij}(\tau) &= -\delta(\tau)\langle c_i(\tau)c_j^\dagger(0)+c_j^\dagger(0)c_i(\tau)\rangle \\ &\quad -\theta(\tau)\langle[\mathcal{H}-\mu\mathcal{N},c_i(\tau)]c_j^\dagger(0)\rangle+\theta(\tau)\langle c_j^\dagger(0)[\mathcal{H}-\mu\mathcal{N},c_i(\tau)]\rangle,\end{aligned} \tag{2.34}$$

where $\delta(\tau)=\partial_\tau\theta(\tau)$ is the Dirac delta function. The following manipulations were used in deriving Eq. (2.34):

$$\begin{aligned}\frac{\partial}{\partial\tau}c_i(\tau) &= \frac{\partial}{\partial\tau}\left[e^{(\mathcal{H}-\mu\mathcal{N})\tau}c_i(0)e^{-(\mathcal{H}-\mu\mathcal{N})\tau}\right] \\ &= (\mathcal{H}-\mu\mathcal{N})e^{(\mathcal{H}-\mu\mathcal{N})\tau}c_i(0)e^{-(\mathcal{H}-\mu\mathcal{N})\tau} \\ &\quad -e^{(\mathcal{H}-\mu\mathcal{N})\tau}c_i(0)e^{-(\mathcal{H}-\mu\mathcal{N})\tau}(\mathcal{H}-\mu\mathcal{N})=[\mathcal{H}-\mu\mathcal{N},c_i(\tau)] \\ &= e^{(\mathcal{H}-\mu\mathcal{N})\tau}[\mathcal{H}-\mu\mathcal{N},c_i(0)]e^{-(\mathcal{H}-\mu\mathcal{N})\tau},\end{aligned} \tag{2.35}$$

where the last line is the simplest for calculations. Using the facts that $\{c_i(0),c_j^\dagger(0)\}=\delta_{ij}$, $[\mathcal{N},c_i(0)]=-c_i(0)$ and $[\hat{T},c_i(0)]=\sum_\delta t_{ii+\delta}c_{i+\delta}(0)$ yields the final differential equation for the Green's function in real space

$$\begin{aligned}&(-\partial_\tau+\mu)G_{ij}(\tau)+\sum_\delta t_{ii+\delta}G_{i+\delta j}(\tau) \\ &-\theta(\tau)\langle e^{(\mathcal{H}-\mu\mathcal{N})\tau}[\hat{V},c_i(0)]e^{-(\mathcal{H}-\mu\mathcal{N})\tau}c_j^\dagger(0)\rangle \\ &+\theta(-\tau)\langle c_j^\dagger(0)e^{(\mathcal{H}-\mu\mathcal{N})\tau}[\hat{V},c_i(0)]e^{-(\mathcal{H}-\mu\mathcal{N})\tau}\rangle=\delta_{ij}\delta(\tau).\end{aligned} \tag{2.36}$$

In general, the commutator $[\hat{V},c_i(0)]$ involves at least three Fermion operators, so the terms with the θ functions in Eq. (2.36) involve new Green's functions, called two-particle Green's functions, but we don't define them here. For the Hubbard model, the EOM in Eq. (2.36) becomes

$$\begin{aligned}&(-\partial_\tau+\mu)G_{ij\sigma}(\tau)+\sum_\delta t_{ii+\delta}G_{i+\delta j\sigma}(\tau)+\theta(\tau)U\langle n_{i-\sigma}(\tau)c_{i\sigma}(\tau)c_{j\sigma}^\dagger(0)\rangle \\ &-\theta(-\tau)U\langle c_{j\sigma}^\dagger(0)n_{i-\sigma}(\tau)c_{i\sigma}(\tau)\rangle=\delta_{ij}\delta(\tau),\end{aligned} \tag{2.37}$$

with the spin variables restored. The EOM can be employed to determine the average potential energy $\langle\hat{V}\rangle$. We illustrate this for the Hubbard model, but it can be worked out for any model. Starting from Eq. (2.37), take $i = j$, $\sigma =\uparrow$, $\tau \to 0^-$, and sum over i to yield

$$\langle\hat{V}_{\text{Hub}}\rangle = U\sum_i\langle n_{i\uparrow}n_{i\downarrow}\rangle = \sum_i[(-\partial_\tau + \mu)G_{ii\uparrow}(\tau) - \delta(\tau)]\Big|_{\tau\to 0^-} - \frac{1}{2}\langle\hat{T}\rangle, \tag{2.38}$$

and the expectation value for the energy follows from $\langle\mathcal{H}\rangle = \langle\hat{T}\rangle + \langle\hat{V}\rangle$.

In momentum space, the EOM is similar, but somewhat simpler, because the kinetic energy part is now diagonal; hence the analog of Eq. (2.36) is

$$\begin{aligned}&(-\partial_\tau + \mu - \epsilon_{\mathbf{k}})G_{\mathbf{k}}(\tau) - \theta(\tau)\langle e^{(\mathcal{H}-\mu\mathcal{N})\tau}[\hat{V}, c_{\mathbf{k}}(0)]e^{-(\mathcal{H}-\mu\mathcal{N})\tau}c^\dagger_{\mathbf{k}}(0)\rangle \\ &+\theta(-\tau)\langle c^\dagger_{\mathbf{k}}(0)e^{(\mathcal{H}-\mu\mathcal{N})\tau}[\hat{V}, c_{\mathbf{k}}(0)]e^{-(\mathcal{H}-\mu\mathcal{N})\tau}\rangle = \delta(\tau),\end{aligned} \tag{2.39}$$

but often the commutator of the annihilation operator with the potential energy is more complicated than in real space.

In addition to the differential equation, we need a boundary condition to determine a unique solution. Since the EOM is a first-order differential equation, we require only one boundary condition. Going back to the Lehmann representation [Eq. (2.31)], with $-\beta < \tau < 0$ and $\tau' = 0$, it is easy to show that $G_{ij}(\tau + \beta) = -G_{ij}(\tau)$ by direct substitution. Hence the Green's function is antiperiodic with period β [the antiperiodicity can also be shown by using invariance of the trace starting from Eq. (2.29)]. This antiperiodicity can be used to extend the definition of the Green's function to all τ by taking the antiperiodic extension (in which case we also need to extend the definition of the delta function by the same antiperiodic extension). Then, we can expand the Green's function in a Fourier series—the only nonzero Fourier components occur at the so-called Fermionic Matsubara frequencies $i\omega_n = i\pi T(2n + 1)$ with n an integer. Hence we have

$$G_{ij}(\tau) = T\sum_n e^{-i\omega_n\tau}G_{ij}(i\omega_n) \tag{2.40}$$

with

$$G_{ij}(i\omega_n) = \int_0^\beta d\tau e^{i\omega_n\tau}G_{ij}(\tau). \tag{2.41}$$

We illustrate a solution for the Green's function in the case of noninteracting band electrons, where the potential-energy operator $\hat{V}$ vanishes.

It is easiest to work in momentum space, so we start from Eq. (2.39) and substitute in the momentum-space analog of Eq. (2.40). Using the fact that the delta function can be written as $\delta(\tau) = T\sum_n \exp[-i\omega_n\tau]$ and that the time-derivative can be evaluated by switching the order of the derivative and the summation, we find an exact solution for the Matsubara frequency Green's function

$$G_{\mathbf{k}}^{\text{nonint}}(i\omega_n) = \frac{1}{i\omega_n + \mu - \epsilon_{\mathbf{k}}}. \tag{2.42}$$

If we substitute these results into Eq. (2.40) and evaluate the summations using identities 1.445.1 and 1.445.2 in [Gradshteyn and Ryzhik (1980)], we find the noninteracting Green's function is

$$G_{\mathbf{k}}^{\text{nonint}}(\tau) = -\text{sgn}(\tau)\left[\frac{\sinh(\epsilon_{\mathbf{k}}-\mu)(\beta-|\tau|)}{\sinh\beta(\epsilon_{\mathbf{k}}-\mu)} - \frac{1}{2}\frac{\sinh(\epsilon_{\mathbf{k}}-\mu)(\frac{\beta}{2}-|\tau|)}{\sinh\frac{1}{2}\beta(\epsilon_{\mathbf{k}}-\mu)}\right]$$
$$-\frac{\cosh(\epsilon_{\mathbf{k}}-\mu)(\beta-|\tau|)}{\sinh\beta(\epsilon_{\mathbf{k}}-\mu)} + \frac{1}{2}\frac{\cosh(\epsilon_{\mathbf{k}}-\mu)(\frac{\beta}{2}-|\tau|)}{\sinh\frac{1}{2}\beta(\epsilon_{\mathbf{k}}-\mu)} \tag{2.43}$$

[it is important for the reader to verify Eq. (2.43)]. Note that in the limit $\tau \to 0^{\pm}$, we find $G_{\mathbf{k}}^{\text{nonint}}(0^+) = f(\epsilon_{\mathbf{k}} - \mu) - 1$ and $G_{\mathbf{k}}^{\text{nonint}}(0^-) = f(\epsilon_{\mathbf{k}} - \mu)$, which is what we expect for a noninteracting system at finite temperature. Furthermore, one can verify that if $-\beta < \tau < 0$, then $G_{\mathbf{k}}^{\text{nonint}}(\tau) = -G_{\mathbf{k}}^{\text{nonint}}(\beta + \tau)$. Finally, a direct substitution into the differential equation in Eq. (2.39) with $\hat{V} = 0$, shows that it solves the differential equation. Hence this is the noninteracting Green's function. One can also calculate $\langle\hat{T}\rangle$, but we won't do so here. As an example, we plot $G_{\text{loc}}(\tau) = \sum_{\mathbf{k}} G_{\mathbf{k}}(\tau)/\Lambda = G_{ii}(\tau)$ in Fig. 2.4 for a variety of different cases, to illustrate how G can vary as the interactions are turned on (techniques needed to produce this plot are developed in the next two sections).

In general, for an interacting system, the Green's function varies from the noninteracting solution. We can summarize the deviations from the noninteracting system by introducing the so-called self-energy, which includes all of the effects of the interactions. We do this by modifying Eq. (2.42) to

$$G_{\mathbf{k}}(i\omega_n) = \frac{1}{i\omega_n + \mu - \Sigma_{\mathbf{k}}(i\omega_n) - \epsilon_{\mathbf{k}}}, \tag{2.44}$$

with $\Sigma_{\mathbf{k}}(i\omega_n)$ being the momentum and frequency-dependent self-energy. An alternative way to express Eq. (2.44) is through the noninteracting

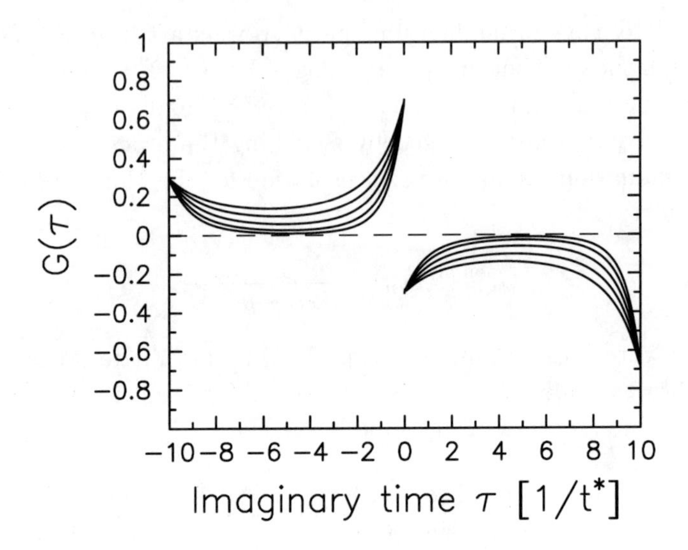

Fig. 2.4 Local Green's function as a function of imaginary time for $T = 0.1$ in the spinless Falicov-Kimball on the hypercubic lattice with $d \to \infty$. We choose $\rho_e = 0.7$, $w_1 = 0.3$, and vary U to include $U = 0$, 1.0 1.5 2.0 and 2.5 (top to bottom for $-\beta < \tau < 0$). The metal-insulator transition occurs near $U = 1.5$. Notice how the Green's functions for different U values agree at $\tau = \pm\beta$ and $\tau = 0$, but differ the most near $\pm\beta/2$. Indeed, the Green's function becomes exponentially small near $\pm\beta/2$ in the insulating phase.

Green's functions and the Dyson equation

$$\begin{aligned} G_{\mathbf{k}}(i\omega_n) &= G_{\mathbf{k}}^{\text{nonint}}(i\omega_n) + G_{\mathbf{k}}^{\text{nonint}}(i\omega_n)\Sigma_{\mathbf{k}}(i\omega_n)G_{\mathbf{k}}(i\omega_n), \\ G_{\mathbf{k}}(i\omega_n)^{-1} &= G_{\mathbf{k}}^{\text{nonint}}(i\omega_n)^{-1} - \Sigma_{\mathbf{k}}(i\omega_n). \end{aligned} \tag{2.45}$$

Since a product of Fourier transforms will be Fourier transformed into a series of convolutions in imaginary time, the Dyson equation involves a number of imaginary-time integrals in the time domain. It turns out, that in the limit of large spatial dimensions, the self-energy becomes local, *i.e.*, it is independent of momentum, which allows a large number of strongly correlated problems to be solved exactly.

Finally, we end this section by showing how one can calculate different expectation values using the Matsubara frequency Green's functions. To begin, the average filling is $\langle n_i \rangle = T\sum_n G_{ii}(i\omega_n)$, but one must use caution in evaluating this summation, because it requires a regularization scheme to converge (see Problem A.6). Similarly, the kinetic energy satisfies $\langle \hat{T} \rangle = T\sum_n \sum_{\mathbf{k}} \epsilon_{\mathbf{k}} G_{\mathbf{k}}(i\omega_n)$, and the convergence of the summation must be handled carefully here as well. Determining the potential energy is somewhat more complicated, because one needs to evaluate a derivative with respect to imaginary time; the derivation is straightforward, and yields for the Hubbard model

$$\langle \hat{V}_{\text{Hub}} \rangle = T \sum_n \sum_{\mathbf{k}} \frac{\Sigma_{\mathbf{k}\uparrow}(i\omega_n)}{i\omega_n + \mu - \Sigma_{\mathbf{k}\uparrow}(i\omega_n) - \epsilon_{\mathbf{k}}} = \Lambda T \sum_n \Sigma_\uparrow(i\omega_n) G_{\text{loc}\uparrow}(i\omega_n), \tag{2.46}$$

where the last equality holds if the self-energy is local.

2.4 Real Time Green's Functions

In the last section, we developed a complete theory for the imaginary time Green's functions. They can be used to calculate a large array of static properties of strongly correlated systems. But we often are interested in dynamical properties as well, such as the many-body density of states (seen in a photoemission experiment), *dc* charge and heat transport, optical conductivity, and so on. Dynamical properties require a real-time formalism. This cannot be achieved by simply replacing τ by $-it$, because such a procedure requires us to properly know the functional form of the function of τ and verify that the substitution maintains the relevant analyticity properties. In cases where we only have numerical values for $G(\tau)$, such a procedure is impossible.

Instead, we can proceed formally by using the Lehmann representation in Eq. (2.31) and Fourier transforming with respect to τ (setting $\tau' = 0$) as in Eq. (2.41). The result yields

$$G_{ij}(i\omega_p) = \frac{1}{\mathcal{Z}} \sum_{mn} \frac{e^{-\beta(E_m - \mu N_m)} + e^{-\beta(E_n - \mu N_n)}}{i\omega_p + E_m - \mu N_m - E_n + \mu N_n} \langle m|c_i|n\rangle \langle n|c_j^\dagger|m\rangle, \tag{2.47}$$

where we used the fact that $\exp[-i\beta\omega_p] = -1$. It is easy to show that in order for the matrix elements in Eq. (2.47) not to vanish, we must have $N_n = N_m + 1$, then all of the Matsubara frequency dependence is in the term $1/[i\omega_p + \mu + E_m - E_n]$ which will be summed over all states labeled by m and n. If we analytically continue by replacing $i\omega_p$ by a complex variable z, then we can easily learn two important facts: (i) if $|z| \to \infty$, then $G_{ii}(z) \to 1/|z|$ (since $\{c_i, c_i^\dagger\} = 1$), and (ii) the term $1/[z+\mu+E_m-E_n]$ has a singularity only if $\text{Im}(z) = 0$ and $\text{Re}(z) + \mu + E_m - E_n = 0$. Hence, if we consider the set of Matsubara frequencies with $\omega_p > 0$, the analytic continuation of Eq. (2.47) [by substituting $i\omega_p \to z$] is analytic in the upper half plane. Similarly, if we consider the set of Matsubara frequencies with

$\omega_p < 0$, the analytic continuation of Eq. (2.47) [by substituting $i\omega_p \to z$] is analytic in the lower half plane. If we take the limit where z approaches the real axis from each respective domain, then we can write $z = \omega + i\delta$ for the upper half plane, and $z = \omega - i\delta$ for the lower half plane, with ω being real and $\delta \to 0^+$. The Lehmann representation of these two real frequency Green's functions is

$$G_{ij}(\omega \pm i\delta) = \frac{1}{\mathcal{Z}} \sum_{mn} \frac{e^{-\beta(E_m - \mu N_m)} + e^{-\beta(E_n - \mu N_n)}}{\omega \pm i\delta + \mu + E_m - E_n} \langle m|c_i|n\rangle \langle n|c_j^\dagger|m\rangle. \tag{2.48}$$

If the two Green's functions defined above are different, then there is a branch cut along the real axis for those frequencies, otherwise, the Green's functions are analytic across the real axis. In the general case, the branch cut extends along the real axis for the regions of frequency corresponding to the differences in the energies of the many-body states; *i.e.*, for the single-particle excitation energies. As we know from band theory for an infinite system, these regions will correspond to finite length pieces along the real axis. Since these two Green's functions are complex conjugates of one another, we learn that if they are purely real, then there is no branch cut, and that the branch cut exists whenever they have a nonzero imaginary part.

It is convenient to summarize the discontinuity of the imaginary part at the branch cut by defining the local density of states via

$$A_{ii}(\omega) = -\frac{G_{ii}(\omega + i\delta) - G_{ii}(\omega - i\delta)}{2\pi i}. \tag{2.49}$$

If we recall the identity

$$\frac{1}{\omega \pm i\delta} = \frac{\mathrm{P}}{\omega} \mp i\pi\delta(\omega), \tag{2.50}$$

where the symbol P denotes the principle value, and means that when it is substituted into an integral, the integration routine must be performed in a symmetric fashion about the singularity at $\omega = 0$, so that it can be well defined. Then, substituting Eq.(2.50) into Eq. (2.49) gives

$$A_{ii}(\omega) = \frac{1}{\mathcal{Z}} \sum_{mn} [e^{-\beta(E_m - \mu N_m)} + e^{-\beta(E_n - \mu N_n)}] |\langle m|c_i|n\rangle|^2 \delta(\omega + \mu + E_m - E_n). \tag{2.51}$$

This shows that the local DOS is always a real-valued function and that it is nonnegative $A_{ii}(\omega) \geq 0$, since each term in Eq. (2.51) is nonnegative. If we integrate Eq. (2.51) over all frequency, we find

$$\int d\omega A_{ii}(\omega) = \frac{1}{\mathcal{Z}}\sum_{mn}[e^{-\beta(E_m-\mu N_m)} + e^{-\beta(E_n-\mu N_n)}]\langle m|c_i|n\rangle\langle n|c_i^\dagger|m\rangle$$
$$= \langle c_i c_i^\dagger + c_i^\dagger c_i\rangle = 1, \tag{2.52}$$

which is an important sum rule. We can repeat the above steps for the momentum-space Green's functions, to arrive at

$$G_{\mathbf{k}}(\omega \pm i\delta) = \frac{1}{\mathcal{Z}}\sum_{mn}\frac{e^{-\beta(E_m-\mu N_m)} + e^{-\beta(E_n-\mu N_n)}}{\omega \pm i\delta + \mu + E_m - E_n}|\langle m|c_{\mathbf{k}}|n\rangle|^2. \tag{2.53}$$

Then, we define the spectral function via

$$A_{\mathbf{k}}(\omega) = -\frac{G_{\mathbf{k}}(\omega + i\delta) - G_{\mathbf{k}}(\omega - i\delta)}{2\pi i}, \tag{2.54}$$

and find that $A_{\mathbf{k}}(\omega) \geq 0$, $\int d\omega A_{\mathbf{k}}(\omega) = 1$, and $A_{ii}(\omega) = \sum_{\mathbf{k}} A_{\mathbf{k}}(\omega)/\Lambda$.

The local DOS, or the spectral function, can both be used to provide another general integral form of the Green's function, that is valid in either the upper half plane or the lower half plane. To derive this relation, we must first go back to the Cauchy formula for an analytic function, which says

$$F(z) = -\frac{1}{2\pi i}\int_C dz' \frac{F(z')}{z - z'}, \tag{2.55}$$

which follows from the residue theorem for any closed contour C which encloses the point z where the function is to be evaluated. If $F(z)$ decays like $1/|z|$ for large $|z|$, and if $F(z)$ is analytic in the upper half plane, we can choose the contour to run from $-\infty$ to ∞ infinitesimally above the real axis, and then return along a radial arc that has a radius $R \to \infty$ in the upper half plane. Since the function $F(z)$ decays for large $|z|$, the contribution along the circular arc vanishes. Then if we let $z \to \omega + i\delta$ in Eq. (2.55) and we recall the identity in Eq. (2.50), we get

$$F(\omega + i\delta) = -\frac{1}{2\pi i}\mathrm{P}\int_{-\infty}^{\infty} d\omega' \frac{F(\omega')}{\omega - \omega'} + \frac{1}{2}F(\omega + i\delta). \tag{2.56}$$

Taking the real part gives $\mathrm{Re}F(\omega + i\delta) = -\mathrm{P}\int d\omega' \mathrm{Im}F(\omega')/[(\omega - \omega')\pi]$ which is commonly called the Kramers-Kronig relation (see Problem A.14 for further examples of how to apply the Kramers-Kronig relation). This result for the real part shows that the following integral identity (called the spectral formula) holds for the Green's function

$$G_{ii}(z) = \int d\omega' \frac{A_{ii}(\omega')}{z - \omega'}, \tag{2.57}$$

which we explicitly wrote down for the local Green's function.

The first step that we need to take to check that Eq. (2.57) is valid, is to reproduce the Matsubara-frequency form of the Lehmann representation in Eq. (2.47) when $z = i\omega_p$. Substituting Eq. (2.51) into Eq. (2.57), and performing the integration over ω' immediately produces Eq. (2.47). The next step is to examine the integral form as $|z| \to \infty$. In this case, we find $G_{ii}(z) \to 1/|z|$ because of the integral sum rule for the DOS in Eq. (2.52). Hence, Eq. (2.57) is an analytic function representation (when restricted to either the upper or lower half plane) for the Green's function that decays to zero for large arguments, and it produces the correct values at all Matsubara frequencies. There is a theorem [Baym and Mermin (1961)], which states that the only function that can satisfy these three properties is the unique analytic continuation of the Matsubara-frequency Green's functions. This theorem is similar in spirit to the conventional unique analytic-continuation theorems, which require the two analytic functions to agree over a continuous domain, but here it is limited to having them agree over a countably infinite set of discrete points, which is why the additional condition, that the function decays for large argument, is needed for the theorem. A similar set of results holds for the spectral function $A_{\mathbf{k}}(\omega)$, but we won't repeat them here.

Our analysis on analytic continuation has shown that the Green's functions in the upper or the lower half plane can be determined from the DOS or the spectral function. Hence, solving for the relevant $A(\omega)$ can be employed to find all Green's functions. In the next two sections, we will show how this function can be calculated within the DMFT approach. In order to complete the formalism, we need to determine how the analytic continuation of the Green's function relates to the real-time Green's functions. This requires us to perform a Fourier transformation with respect to frequency. If we start from the Lehman representation in Eq. (2.48), then we need to compute the Fourier transform

$$\frac{1}{2\pi} \int d\omega \frac{e^{-i\omega t}}{\omega \pm i\delta + \mu + E_m - E_n} = \mp i\theta(\pm t) e^{i(\mu + E_m - E_n)t}, \tag{2.58}$$

which follows from the residue theorem: if $t < 0$ we close the contour integral in the upper half plane and if $t > 0$ we close in the lower half plane and the simple pole in the integrand occurs at $z = -\mu - E_m + E_n \mp i\delta$. The Fourier transform of the Green's function then becomes

$$G_{ij}(t) = \frac{\mp i\theta(\pm t)}{\mathcal{Z}} \sum_{mn} [e^{-\beta(E_m - \mu N_m)} + e^{-\beta(E_n - \mu N_n)}] e^{i(E_m - \mu N_m - E_n + \mu N_n)t}$$

$$\times \langle m|c_i|n\rangle\langle n|c_j^\dagger|m\rangle$$

$$= \mp i\theta(\pm t)\langle c_i(t)c_j^\dagger(0) + c_j^\dagger(0)c_i(t)\rangle = \mp i\theta(\pm t)\langle\{c_i(t), c_j^\dagger(0)\}\rangle, \quad (2.59)$$

where we have included the time-dependence of an operator $\mathcal{O}$ via $\mathcal{O}(t) = \exp[it(\mathcal{H} - \mu\mathcal{N})]\mathcal{O}\exp[-it(\mathcal{H} - \mu\mathcal{N})]$. So the real-time representation of the Green's function that is analytic in the upper half plane is $G^R(t) = -i\theta(t)\langle\{c_i(t), c_j^\dagger(0)\}\rangle$, which is called the retarded Green's function and the real-time representation of the Green's function that is analytic in the lower half plane is $G^A(t) = i\theta(-t)\langle\{c_i(t), c_j^\dagger(0)\}\rangle$, which is called the advanced Green's function (the names come from the nature of the θ functions). *Note that the time-ordered Green's function on the imaginary-time axis analytically continues to either the retarded or advanced Green's function on the real-time axis!*

We can derive an EOM for the advanced and retarded Green's functions in real time as well. In real space, we have

$$\left(i\frac{\partial}{\partial t} + \mu\right)G_{ij}^{R,A}(t) + \sum_\delta t_{ii+\delta}G_{i+\delta j}^{R,A}(t) \quad (2.60)$$

$$\mp i\theta(\pm t)\langle\{e^{i(\mathcal{H}-\mu\mathcal{N})t}[\hat{V}, c_i(0)]e^{-i(\mathcal{H}-\mu\mathcal{N})t}, c_j^\dagger(0)\}\rangle = \delta_{ij}\delta(t),$$

with a similar equation for the momentum Green's functions. The top sign is for the retarded Green's function and the bottom sign is for the advanced Green's function.

In the case of noninteracting electrons, where $\hat{V} = 0$, the analysis simplifies. We have an explicit expression for the eigenstates and eigenvalues: the n-electron eigenstates are formed by products of n distinct creation operators acting on the vacuum state. The eigenvalue is the sum of the n eigenvalues of the bandstructure associated with those creation operators. We can evaluate the retarded Green's function in momentum space from Eq. (2.53) by choosing the top sign. The matrix element tells us that the state $|m\rangle$ does not have the state labeled by $\mathbf{k}$ in it, and $E_n - E_m = \epsilon_{\mathbf{k}}$ and

$N_n - N_m = 1$. Furthermore, for each state $|m\rangle$ there is only one state $|n\rangle$ that contributes to the summation. Putting this all together yields

$$G_{\mathbf{k}}^{R\ \mathrm{nonint}}(\omega) = \frac{1}{\mathcal{Z}} \sum_{m:\ c_{\mathbf{k}}^{\dagger}|m\rangle \neq 0} \frac{e^{-\beta(E_m - \mu N_m)}[1 + e^{-\beta(\epsilon_{\mathbf{k}} - \mu)}]}{\omega + \mu - \epsilon_{\mathbf{k}} + i\delta}. \tag{2.61}$$

The only m-dependence is now in the summation over the exponential factors. Writing the partition function as

$$\begin{aligned} \mathcal{Z} &= \sum_{m:\ c_{\mathbf{k}}^{\dagger}|m\rangle \neq 0} e^{-\beta(E_m - \mu N_m)} + \sum_{m:\ c_{\mathbf{k}}^{\dagger}|m\rangle = 0} e^{-\beta(E_m - \mu N_m)} \\ &= \sum_{m:\ c_{\mathbf{k}}^{\dagger}|m\rangle \neq 0} e^{-\beta(E_m - \mu N_m)}[1 + e^{-\beta(\epsilon_{\mathbf{k}} - \mu)}], \end{aligned} \tag{2.62}$$

by noting that states that satisfy $c_{\mathbf{k}}^{\dagger}|m\rangle = 0$ can be written as $c_{\mathbf{k}}^{\dagger}|n\rangle$ with $c_{\mathbf{k}}^{\dagger}|n\rangle \neq 0$, and the energy of the $|m\rangle$ state is $\epsilon_{\mathbf{k}}$ larger than the energy of the $|n\rangle$ state. Plugging Eq. (2.62) into Eq. (2.61) produces the noninteracting Green's function and spectral function

$$G_{\mathbf{k}}^{R\ \mathrm{nonint}}(\omega) = \frac{1}{\omega + \mu - \epsilon_{\mathbf{k}} + i\delta}, \quad A_{\mathbf{k}}^{\mathrm{nonint}}(\omega) = \delta(\omega + \mu - \epsilon_{\mathbf{k}}). \tag{2.63}$$

Summing over momentum gives the local Green's function and the DOS

$$G_{ii}^{R\ \mathrm{nonint}}(\omega) = \frac{1}{\Lambda} \sum_{\mathbf{k}} \frac{1}{\omega + \mu - \epsilon_{\mathbf{k}} + i\delta}, \quad A_{ii}^{\mathrm{nonint}}(\omega) = \frac{1}{\Lambda} \sum_{\mathbf{k}} \delta(\omega + \mu - \epsilon_{\mathbf{k}}). \tag{2.64}$$

Performing these summations on hypercubic lattices is an exercise in Problem A.2. Fourier transforming to real time yields

$$\begin{aligned} G_{\mathbf{k}}^{R\ \mathrm{nonint}}(t) &= -i\theta(t) e^{-i(\epsilon_{\mathbf{k}} - \mu)t}, \\ G_{ii}^{R\ \mathrm{nonint}}(t) &= -i\theta(t) \int d\epsilon A_{ii}^{\mathrm{nonint}}(\epsilon) e^{-i(\epsilon - \mu)t}. \end{aligned} \tag{2.65}$$

We illustrate the solution for the noninteracting Green's functions for the case of a Bethe lattice, which is to be thought of as the limit of a Cayley tree that has no boundary (a Cayley tree of coordination 3 is illustrated in Fig. 2.5). A Cayley tree is a strange structure that actually has nearly all lattice sites lying on the surface and is sensitive to surface properties; the Bethe lattice, however, is free from those problems because it has no surface (see [Thorpe (1981)] for a clear review). The noninteracting Green's function for a Bethe lattice with coordination Z can be solved in the spirit

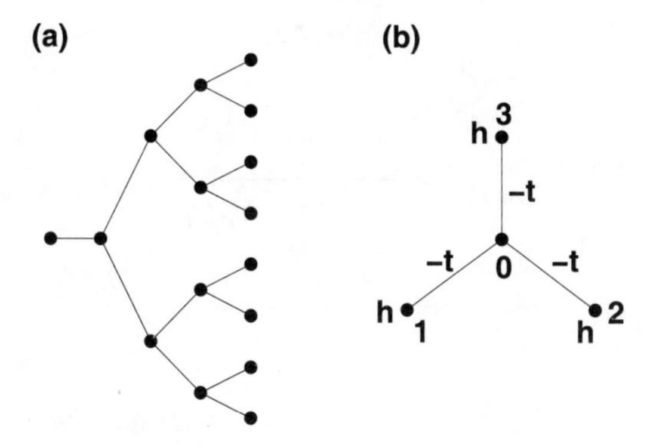

Fig. 2.5 (a) Schematic of a coordination $Z = 3$ Cayley tree. Note how the number of sites grows on each column. (b) Schematic of the Bethe-lattice cluster used to calculate the noninteracting Green's function from the Hamiltonian in Eq. (2.66).

of the Bethe approximation [Bethe (1935)]: we examine a single site 0 and its cluster of nearest neighbors $(1-Z)$ by introducing an effective field h on each of the neighboring sites, which represents the effect of all other sites on the lattice (see Fig. 2.5); our discussion follows that of [Thorpe (1981)]. The Hamiltonian of the cluster is

$$\mathcal{H}_{\text{cluster}} = -t\sum_{i=1}^{Z}(c_i^\dagger c_0 + c_0^\dagger c_i) + h\sum_{i=1}^{Z} c_i^\dagger c_i - \mu\sum_{i=0}^{Z} c_i^\dagger c_i. \tag{2.66}$$

Taking the Fourier transform of Eq. (2.60) for the case where $\hat{V} = 0$, yields

$$(\omega + \mu)G_{ij}^{R\ \text{nonint}}(\omega) + t\sum_{\delta} G_{i+\delta j}^{R\ \text{nonint}}(\omega) = \delta_{ij}. \tag{2.67}$$

Applying a generalization of this (to include the h-field) to the Bethe cluster gives

$$\begin{aligned}
&(\omega + \mu)G_{00}^{R\ \text{nonint}}(\omega) + t\sum_{i=1}^{Z} G_{i0}^{R\ \text{nonint}}(\omega) = 1, \\
&(\omega + \mu - h)G_{i0}^{R\ \text{nonint}}(\omega) + tG_{00}^{R\ \text{nonint}}(\omega) = 0 \\
&(\omega + \mu - h)G_{ii}^{R\ \text{nonint}}(\omega) + tG_{i0}^{R\ \text{nonint}}(\omega) = 1.
\end{aligned} \tag{2.68}$$

Using the facts that $G_{i0} = G_{10}$ and $G_{ii} = G_{11}$ for all $1 \le i \le Z$, allows these equations to be solved by

$$G_{00}^{R\ \text{nonint}}(\omega) = \frac{1}{\omega + \mu - \frac{Zt^2}{\omega+\mu-h}},$$
$$G_{11}^{R\ \text{nonint}}(\omega) = \frac{\left[\omega + \mu - \frac{(Z-1)t^2}{\omega+\mu-h}\right]\frac{1}{\omega+\mu-h}}{\omega + \mu - \frac{Zt^2}{\omega+\mu-h}}. \tag{2.69}$$

Now we use the fact that the Bethe lattice is homogeneous, to set $G_{00} = G_{11}$, which yields

$$h = \frac{\omega + \mu}{2} \pm \frac{1}{2}\sqrt{(\omega + \mu)^2 - 4(Z-1)t^2}. \tag{2.70}$$

Finally, plugging the value for h back into Eq. (2.68), we get

$$A_{\text{loc}}(\omega) = \frac{Z\sqrt{4(Z-1)t^2 - (\omega + \mu)^2}}{2\pi[(Zt)^2 - (\omega + \mu)^2]}, \tag{2.71}$$

for the local DOS of the Bethe lattice, when $|\omega + \mu| < 2\sqrt{Z-1}t$.

We end this section with some important definitions, and with some generalizations of imaginary axis formulas. The momentum-dependent self-energy is defined by

$$G_{\mathbf{k}}(\omega \pm i\delta) = \frac{G_{\mathbf{k}}^{\text{nonint}}(\omega \pm i\delta)}{1 - G_{\mathbf{k}}^{\text{nonint}}(\omega \pm i\delta)\Sigma_{\mathbf{k}}(\omega \pm i\delta)}, \tag{2.72}$$

which follows from the Dyson equation

$$G_{\mathbf{k}}(\omega \pm i\delta) = G_{\mathbf{k}}^{\text{nonint}}(\omega \pm i\delta) + G_{\mathbf{k}}^{\text{nonint}}(\omega \pm i\delta)\Sigma_{\mathbf{k}}(\omega \pm i\delta)G_{\mathbf{k}}(\omega \pm i\delta). \tag{2.73}$$

In cases where the self-energy does not depend on momentum, the summation over momentum to determine the local Green's function is simple, and involves an integral over the noninteracting DOS, since the momentum-dependent Green's function depends on $\mathbf{k}$ only through $\epsilon_{\mathbf{k}}$.

The analog of Eqs. (2.32), (2.33), and (2.46) are $\rho_e = \int d\omega f(\omega) A_{\text{loc}}(\omega)$, $\langle n_{\mathbf{k}} \rangle = \int d\omega f(\omega) A_{\mathbf{k}}(\omega)$, $\langle \hat{T} \rangle = -\int d\omega \text{Im}\{[\omega + \mu - \Sigma(\omega)]G_{\text{loc}}(\omega)\}/\pi$, and $\langle \hat{V} \rangle = \int d\omega \text{Im}\{[\mu - \Sigma(\omega)]G_{\text{loc}}(\omega)\}/\pi$. Techniques used to derive these results are developed in Problem A.4.

2.5 The Limit $d \to \infty$ and the Mapping onto a Time-Dependent Impurity Problem

There is a long history of examining complicated problems in quantum mechanics or statistical mechanics in unphysically large dimensions to find simpler behavior. In this way, for example, we have learned about how critical exponents in the statistical mechanics of phase transitions assume mean-field-theory values above some upper critical dimension. The application of these ideas to interacting electrons began in 1989 [Metzner and Vollhardt (1989)]. The original work of Metzner and Vollhardt showed that a particular scaling of the nearest-neighbor hopping with spatial dimension d would lead to a finite kinetic energy, which competes with the potential energy, when we minimize the total energy of the system. Their idea has its roots in the vast literature on alloy theory from the 1960s and 1970s, which showed that the inverse of the number of nearest neighbors $1/Z$ was the small parameter that governed the convergence of the coherent potential approximation [Schwartz and Siggia (1972)].

Metzner and Vollhardt chose to scale the nearest-neighbor hopping as

$$t = \frac{t^*}{2\sqrt{d}}, \tag{2.74}$$

on a simple hypercubic lattice in d-dimensions. This choice leads to a finite average kinetic energy for the noninteracting bands. Indeed, the band structure $\epsilon_{\mathbf{k}} = -t^* \lim_{d\to\infty} \sum_{i=1}^{d} \cos(\mathbf{k}_i)/\sqrt{d}$ can be viewed as the sum of a collection of d "random" numbers distributed from -1 to 1. When we take the limit $d \to \infty$, the sum of the random numbers will grow like $\sqrt{d}$ (from the random-walk problem), so dividing by $\sqrt{d}$ will yield a finite limit as $d \to \infty$. The central limit theorem then says that the distribution of energies will be a Gaussian distribution. Hence the noninteracting DOS is a Gaussian with an infinite bandwidth. But the average kinetic energy for any finite filling is finite and can be expressed as

$$\langle \hat{T} \rangle_{\text{nonint}} = \frac{\Lambda}{t^* \sqrt{\pi}} \int_{-\infty}^{E_F} d\epsilon \; \epsilon \; e^{-\epsilon^2/t^{*2}} = -\frac{\Lambda t^* e^{-E_F^2/t^{*2}}}{2\sqrt{\pi}} \tag{2.75}$$

with E_F the Fermi energy at $T = 0$.

The noninteracting Fermi surface is defined by the surface in the Brillouin zone where $\epsilon_{\mathbf{k}} = E_F$. It is common in many theories to ignore the lattice potential, and describe the Fermi surface as a sphere due to the quadratic dispersion $\epsilon_{\mathbf{k}} = \hbar^2 \mathbf{k} \cdot \mathbf{k}/2m$. But in the limit of large dimensions, the Fermi surface is never spherical. The reason why is that we always measure volumes relative to a unit cube. In infinite dimensions, the diagonal of the cube has an infinite length (even though the length along each axis is finite). Hence, any finite radius sphere will occupy a set of measure zero inside the unit cube, so they correspond to a vanishing density of the electrons. As the dimension increases, the Fermi surfaces become more and more "porcupine-like" and less and less spherical. In fact, the infinite bandwidth of the noninteracting DOS arises from rare regions in the Brillouin zone where the band energy can be very large (think of points like the zone center, or the end of the zone diagonal). These regions have very small DOS though.

The main simplification of the $d \to \infty$ limit is that the self-energy becomes local, *i.e.*, $\Sigma_{\mathbf{k}}(z)$ has no $\mathbf{k}$ dependence. Unfortunately, there is no simple way to show this without examining either perturbative expansions, or path-integral methods, both of which are beyond the scope of this book. But, we can motivate the idea heuristically. A local self-energy means that we neglect spatially correlated pieces of the self-energy like Σ_{ij} with $i \neq j$. If we examine the perturbative expansion for the self-energy, we see that the lowest-order nontrivial contribution for the case when i and j are nearest neighbors is proportional to $t^3 2d$, since it involves three propagations from site i to site j and there are $2d$ total nearest neighbors. Using the scaling of the hopping in Eq. (2.74), we find that $\Sigma_{ij} \propto \lim_{d\to\infty} 1/\sqrt{d} \to 0$. A complete proof of this idea from a perturbative expansion for the Hubbard model can be found in [Metzner (1991)], or from a path-integral approach using the so-called cavity method in [Georges, *et al.* (1996)] or [Freericks and Zlatić (2003)].

If the self-energy for the lattice-based many-body problem is local, then that problem can be mapped onto an impurity-based many-body problem, but in a time-dependent local field, which is called the dynamical mean field. Once again, a proof of this idea requires either perturbation theory or path-integral techniques, but the physical principle is easy to describe. Since the self-energy is local, it should be able to be represented as the self-energy of a purely local problem. But it isn't clear just how one can find the correct Hamiltonian for the impurity problem to produce the self-energy for the lattice. The idea, is that one adds an additional time-dependent field to the impurity evolution, and adjusts that field until the self-energy of the impurity is identical to the self-energy of the lattice. Since we do not know *a priori* what the self-energy of the lattice is, we must solve this problem self-consistently and hope that the self-consistent solution is unique. The only issue that we must accept is that by varying the time-dependent field over all possible functional forms, we have enough freedom to describe every possible form that the local self-energy can take on the lattice. This time-dependent field mimics the hopping of the electrons on the lattice, because it allows electrons to hop onto and to hop off of the impurity as a function of time. Since the potential-energy piece of the Hamiltonian is local, we use that piece for the impurity problem with the identical values of the interaction strengths as used on the lattice.

We illustrate these ideas with the concrete example of a spinless electron. This problem can be solved analytically for any possible time-dependent field. In cases where there is an interaction between electrons with different spins, the problem becomes more complicated and is mapped onto a generalized single-impurity Anderson model, which can be solved numerically, but cannot be solved analytically. We discuss that case only briefly when we describe the numerical renormalization group method for DMFT.

We consider the problem of a single impurity electron evolving in a nontrivial time-dependent field. The Hamiltonian of the impurity is taken to be

$$\mathcal{H}_{\text{imp}} - \mu\mathcal{N} = -\mu c^\dagger c. \tag{2.76}$$

Next we introduce a general time-dependent evolution operator in the interaction representation on the imaginary-time axis

$$S(\lambda) = \mathcal{T}_\tau \exp\left[-\int_0^\beta d\tau \int_0^\beta d\tau' \lambda(\tau,\tau') c^\dagger(\tau) c(\tau')\right], \tag{2.77}$$

where the time-dependence of the operators is with respect to the impurity Hamiltonian in Eq. (2.76). The impurity partition function is defined to be

$$\mathcal{Z}_{\text{imp}}(\lambda,\mu) = \text{Tr}_c\left\{\mathcal{T}_\tau e^{-\beta(\mathcal{H}_{\text{imp}}-\mu\mathcal{N})} S(\lambda)\right\}, \tag{2.78}$$

with the trace taken over the two Fermionic states with zero or one electron.

This operator trace is not simple to calculate. The procedure is to first examine the functional derivative for the change in the partition function due to a small change in the field λ

$$\delta\mathcal{Z}_{\text{imp}}(\lambda,\mu) = \text{Tr}_c\left\{\mathcal{T}_\tau e^{-\beta(\mathcal{H}_{\text{imp}}-\mu\mathcal{N})} \delta S(\lambda)\right\}. \tag{2.79}$$

To find the change in the evolution operator S, we use the calculus of variations

$$\delta S(\lambda) = -\mathcal{T}_\tau \left[S(\lambda) \int_0^\beta d\tau \int_0^\beta d\tau' \delta\lambda(\tau,\tau') c^\dagger(\tau) c(\tau') \right]. \tag{2.80}$$

Substituting Eq. (2.80) into Eq. (2.79) and noting that the time ordering can be collected into one time-ordering operation, yields

$$\delta \mathcal{Z}_{\text{imp}}(\lambda,\mu) = -\mathcal{Z}_{\text{imp}}(\lambda,\mu) \int_0^\beta d\tau \int_0^\beta d\tau' \delta\lambda(\tau,\tau') G_{\text{imp}}(\tau',\tau), \tag{2.81}$$

where the impurity Green's function is defined in Eq. (2.29) with the spatial indices dropped, and we need to note the order of the time arguments in Eq. (2.81). Hence, the impurity Green's function can be written as a functional derivative of the impurity partition function with respect to the dynamical mean field as

$$G_{\text{imp}}(\tau,\tau') = -\frac{\delta \ln \mathcal{Z}_{\text{imp}}(\lambda,\mu)}{\delta\lambda(\tau',\tau)}. \tag{2.82}$$

Since the Green's function depends only on the difference of its time arguments, we can infer that we can restrict the dynamical mean field to depend only on the time difference, and then we can Fourier transform this result to yield

$$G_{\text{imp}}(i\omega_n) = -\frac{\partial \ln \mathcal{Z}_{\text{imp}}(\lambda,\mu)}{\partial\lambda(i\omega_n)}. \tag{2.83}$$

The next step is to determine the impurity Green's function by solving the EOM that it satisfies. This EOM is more complicated than the ones we derived before, because of the presence of the evolution operator. The complication arises from the fact that the evolution operator is defined as a time-ordered product, and we have to order with respect to both time arguments in the double integral of the exponent. It simplifies, however, when we discover that if we plan to differentiate with respect to the imaginary time τ, then we need only write the time-ordered product in the schematic form

$$\mathcal{T}_\tau S(\lambda) c(\tau) c^\dagger(\tau') = \left[\mathcal{T}_\tau \bar{S}(\lambda)\right] c(\tau) \left[\mathcal{T}_\tau \bar{\bar{S}}(\lambda)\right] c^\dagger(\tau'), \tag{2.84}$$

with

$$\bar{S}(\lambda) = \exp\left[-\int_{\tau}^{\beta} d\tau'' \int_{0}^{\beta} d\tau''' \lambda(\tau'' - \tau''') c^\dagger(\tau'') c(\tau''')\right], \tag{2.85}$$

and

$$\bar{\bar{S}}(\lambda) = \exp\left[-\int_{0}^{\tau} d\tau'' \int_{0}^{\beta} d\tau''' \lambda(\tau'' - \tau''') c^\dagger(\tau'') c(\tau''')\right], \tag{2.86}$$

because the τ''' terms do not enter the derivative since $c^2(\tau') = 0$, implying they give no contribution. Explicitly evaluating the derivative with respect to τ finally gives the EOM for the impurity Green's function

$$\left(-\frac{\partial}{\partial\tau} + \mu\right) G_{\mathrm{imp}}(\tau - \tau') - \int_0^\beta d\tau'' \lambda(\tau - \tau'') G_{\mathrm{imp}}(\tau'' - \tau') = \delta(\tau - \tau'). \tag{2.87}$$

If we think of Eq. (2.87) as a continuous operator equation over a functional space, then it can be written schematically as $G^{-1}G = 1$ where multiplication of the functional operators is accomplished by an integral over τ. Then we learn that

$$G^{-1}(\tau, \tau') = \left(-\frac{\partial}{\partial\tau} + \mu\right)\delta(\tau - \tau') - \lambda(\tau - \tau'), \tag{2.88}$$

in the imaginary-time domain. This operator can be diagonalized by performing a double Fourier transform via the integrals $\int_0^\beta d\tau \exp[i\omega_n \tau]$ and $T\int_0^\beta d\tau' \exp[-i\omega_{n'}\tau']$. The end result is

$$G^{-1}(i\omega_n, i\omega_{n'}) = (i\omega_n + \mu - \lambda_n)\delta_{nn'}, \tag{2.89}$$

where we used the notation $\lambda_n = \lambda(i\omega_n)$ for the dynamical mean field. The matrix operator is diagonal, so we can invert it to find

$$G_{\mathrm{imp}}(i\omega_n) = \frac{1}{i\omega_n + \mu - \lambda_n} = -\frac{\partial \mathcal{Z}_{\mathrm{imp}}(\lambda, \mu)}{\partial \lambda_n}. \tag{2.90}$$

This differential equation can be easily integrated to yield $\mathcal{Z}_{\mathrm{imp}}(\lambda, \mu) = C\prod_{n=-\infty}^{\infty}(i\omega_n + \mu - \lambda_n)$ with C a numerical constant independent of λ. Using the fact that the partition function for vanishing dynamical mean field can be determined directly as $\mathcal{Z}_{\mathrm{imp}}(\lambda = 0, \mu) = 1 + e^{\beta\mu}$, allows us to find the constant C, or to learn that

$$\mathcal{Z}_{\mathrm{imp}}(\lambda, \mu) = 2e^{\beta\mu/2} \prod_{n=-\infty}^{\infty} \frac{i\omega_n + \mu - \lambda_n}{i\omega_n}, \tag{2.91}$$

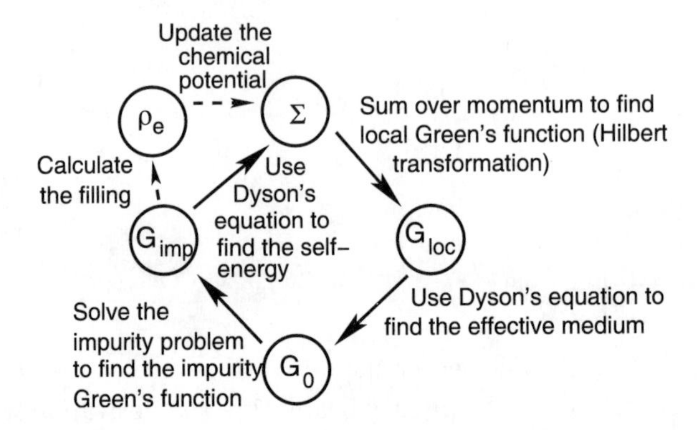

Fig. 2.6 Flow diagram for the DMFT algorithm. Starting with an initial self-energy (usually chosen to be zero, unless something is known about the system *a priori*), we use the Hilbert transformation to calculate the local Green's function. Next, the effective medium for the impurity is extracted from the current self-energy and the local Green's function via Dyson's equation. The impurity problem is solved determining the impurity Green's function from the effective medium. If we want to work with a fixed filling of electrons, then the electron filling is calculated, and the chemical potential is updated to get closer to the target filling; the update step is sometimes performed only every ten or so iterations rather than every iteration, to give the system time to adjust to the new chemical potential. Finally, Dyson's equation is used to find the new self-energy from the impurity Green's function and the effective medium. The algorithm is iterated until converged. Once the chemical potential is determined from an imaginary-axis calculation, it is fixed for the real-axis calculation.

which is proved in Problem A.9. Hence we can analytically solve for the partition function and Green's functions of the impurity problem for a spinless electron in a time-dependent field. The case of electrons which have an interaction between the spins can often only be solved numerically using more sophisticated techniques.

We end this section with a description of how we adjust the dynamical mean field to give us the solution to the lattice problem with a local self-energy. The local Green's function of the lattice can be found by summing the momentum-dependent Green's function over all momenta $G_{ii}(z) = \sum_{\mathbf{k}} G_{\mathbf{k}}(z)$. Since the self-energy is local, and has no momentum dependence, the sum can be expressed as a one-dimensional integral over the noninteracting DOS (which is called a Hilbert transform)

$$G_{ii}(z) = \int d\epsilon\ \rho(\epsilon) \frac{1}{z + \mu - \Sigma(z) - \epsilon}, \tag{2.92}$$

with $\rho(\epsilon)$ the noninteracting DOS and z denoting a variable in the complex plane. Since we need to equate this local Green's function with the impurity

Green's function, we need to find a way to extract the dynamical mean field. This is normally done by defining the effective medium $G_0(z) = 1/[z+\mu-\lambda(z)]$ which is found from Dyson's equation

$$G(z)=\frac{1}{G_0(z)^{-1}-\Sigma(z)},\quad G_0(z)=\frac{1}{G^{-1}(z)+\Sigma(z)},\tag{2.93}$$

where we have written it in two equivalent forms and suppressed the subscripts for the local Green's function or the impurity Green's function. The second form allows us to extract $\lambda(z)$ from the local Green's function of the lattice and the self-energy. The former allows us to extract the new self-energy from the impurity Green's function and the effective medium. Hence, the DMFT algorithm is as follows (see Fig. 2.6): (i) start with a guess for the self-energy (which is often chosen to be zero); (ii) use Eq. (2.92) to determine the local Green's function from the current self-energy; (iii) use Eq. (2.93) to extract the effective medium and the dynamical mean field; (iv) solve the impurity problem in the presence of the evolution operator for the given dynamical mean field to determine the impurity Green's function; (v) extract the new guess for the self-energy by extracting it from the Dyson equation in Eq. (2.93); and (vi) iterate steps (ii) through (v) until the results stop changing. This algorithm, originally proposed by Jarrell [Jarrell (1992)], solves for the dynamical mean field, self-energy, and Green's functions via an iterative technique. There is no guarantee that this method will converge in general, or that there is only one solution to these equations, but it is found in practice, that the convergence can be quite rapid, and that the solutions are often unique. Note that in the spinless case, it is obvious that there is a one-to-one correspondence between any set of local Green's functions and the corresponding dynamical mean field.

2.6 Impurity Problem Solvers

The only step of the DMFT algorithm that remains to be worked out is to construct an impurity problem solver, that will determine the impurity Green's function for the given dynamical mean field or effective medium. Finding efficient and accurate impurity problem solvers is the hardest part of the DMFT approach. We first illustrate an analytic solution for the spinless Falicov-Kimball model, then discuss the ideas behind the numerical renormalization group (NRG) approach, and end with a discussion of other types of solvers (perturbative, quantum Monte Carlo, and so on).

The spinless Falicov-Kimball model has an impurity Hamiltonian that satisfies

$$\mathcal{H}_{\text{imp}}^{\text{FK}} - \mu\mathcal{N} = -\mu c^\dagger c + E_f f^\dagger f + U c^\dagger c f^\dagger f. \tag{2.94}$$

The impurity partition function becomes

$$\mathcal{Z}_{\text{imp}}^{\text{FK}}(\lambda) = \text{Tr}_c \text{Tr}_f \mathcal{T}_\tau \left[e^{-\beta(\mathcal{H}_{\text{imp}}^{\text{FK}} - \mu\mathcal{N})} S(\lambda) \right], \tag{2.95}$$

with the evolution operator $S(\lambda)$ given by Eq. (2.77). Since $n_f = f^\dagger f$ commutes both with the evolution operator and the Hamiltonian, the trace over the f-particles is easy, because it separates into the sum of two simple terms. The first has $n_f = 0$, where $\mathcal{H}_{\text{imp}}^{\text{FK}} - \mu\mathcal{N} = -\mu c^\dagger c$, and the second has $n_f = 1$, where $\mathcal{H}_{\text{imp}}^{\text{FK}} - \mu\mathcal{N} = E_f + (U - \mu)c^\dagger c$. The first case is identical to the spinless case solved in the previous subsection, the second requires us to shift $\mu \to \mu - U$ and to include the extra factor $\exp(-\beta E_f)$. The net result is

$$\mathcal{Z}_{\text{imp}}^{\text{FK}}(\lambda) = \mathcal{Z}_{\text{imp}}(\lambda, \mu) + e^{-\beta E_f} \mathcal{Z}_{\text{imp}}(\lambda, \mu - U), \tag{2.96}$$

where we use the result for the impurity partition function in Eq. (2.91), and the second term corresponds to shifting $\mu \to \mu - U$ in Eq. (2.91).

The Green's function for the conduction electrons still satisfies Eq. (2.83), so we can immediately learn that

$$G(i\omega_n) = \frac{\mathcal{Z}_{\text{imp}}(\lambda)}{\mathcal{Z}_{\text{imp}}^{\text{FK}}(\lambda)} \frac{1}{i\omega_n + \mu - \lambda_n} + \frac{e^{-\beta E_f} \mathcal{Z}_{\text{imp}}(\lambda + U)}{\mathcal{Z}_{\text{imp}}^{\text{FK}}(\lambda)} \frac{1}{i\omega_n + \mu - \lambda_n - U}, \tag{2.97}$$

by taking the derivative with respect to λ_n. If we make the following definitions

$$w_0 = \frac{\mathcal{Z}_{\text{imp}}(\lambda)}{\mathcal{Z}_{\text{imp}}^{\text{FK}}(\lambda, \mu)}, \quad w_1 = \frac{e^{-\beta E_f} \mathcal{Z}_{\text{imp}}(\lambda, \mu - U)}{\mathcal{Z}_{\text{imp}}^{\text{FK}}(\lambda)}, \tag{2.98}$$

then we can write the impurity Green's function for the spinless Falicov-Kimball model as

$$\begin{aligned} G(i\omega_n) &= \frac{w_0}{i\omega_n + \mu - \lambda_n} + \frac{w_1}{i\omega_n + \mu - \lambda_n - U} \\ &= w_0 G_0(i\omega_n) + \frac{w_1}{G_0^{-1}(i\omega_n) - U}, \end{aligned} \tag{2.99}$$

where we have expressed the Green's function in terms of the effective medium in the lower line. This form for the Green's function is quite simple, and is identical to that used in the coherent-potential approximation. The symbol w_1 is the filling of the f-electrons and w_0 is the concentration of lattice sites with no f-electrons. This follows from the fact that $w_0 = 1 - w_1$ and that

$$\langle f^\dagger f \rangle = \frac{1}{\mathcal{Z}_{\rm imp}^{\rm FK}(\lambda)} {\rm Tr}_c {\rm Tr}_f \mathcal{T}_\tau \left[e^{-\beta(\mathcal{H}_{\rm imp}^{\rm FK} - \mu \mathcal{N})} S(\lambda) f^\dagger f \right] = \frac{e^{-\beta E_f} \mathcal{Z}_{\rm imp}(\lambda, \mu - U)}{\mathcal{Z}_{\rm imp}^{\rm FK}(\lambda)}. \tag{2.100}$$

Now we will investigate the self-energy of the impurity in more detail. The self-energy is extracted from the relation $\Sigma_n = G_0^{-1}(i\omega_n) - G_n^{-1}$. This is the standard way to find the impurity self-energy which will be equated to the local self-energy of the lattice in the DMFT approach. But for the spinless Falicov-Kimball model, we can determine an interesting relation between the self-energy and the Green's function [Brandt and Mielsch (1989)]. We substitute $G_0^{-1}(i\omega_n) = G_n^{-1} + \Sigma_n$ into the last line of Eq. (2.99) and multiply both sides by $G_n(G_n^{-1} + \Sigma_n)(G_n^{-1} + \Sigma_n - U)$. Simplifying the terms yields a quadratic equation for the self-energy

$$\Sigma_n^2 G_n^2 + \Sigma_n [G_n - U G_n^2] - w_1 G_n U = 0, \tag{2.101}$$

which is solved by

$$\Sigma_n = -\frac{1}{2G_n} + \frac{U}{2} \pm \frac{1}{2G_n} \sqrt{1 - 2(1 - 2w_1) U G_n + U^2 G_n^2}, \tag{2.102}$$

where the sign is determined by analyticity if the imaginary part of the self-energy is nonzero, and by continuity if the self-energy is real. For a given value of w_1, this is an explicit equation for the self-energy in terms of the local Green's function G_n! It can be employed to find an alternative method to solve the DMFT equations than the iterative approach. This will be discussed in the next section.

Unfortunately, the analytic approach given here for the spinless Falicov-Kimball model will only work for other so-called static models like the higher-spin variants of the Falicov-Kimball model, or the static Holstein model. For other types of correlated systems, like the Hubbard model or the periodic Anderson model, new techniques must be developed. This is because the partition function for the impurity will involve a trace over the spin up and the spin down particles, which will each have their own evolution operators and dynamical mean fields. If there is now an interaction between these particles, then we can no longer evaluate the trace over both particles as we did before, because the number operators of each of the particles are not conserved by the evolution operators. Trying to solve this problem by introducing an EOM as before does not help, because the coupled EOMs can no longer be solved.

It turns out that this problem can generically be mapped onto a single impurity Anderson model, but with a nontrivial conduction density of states and a nontrivial time-dependent hybridization function. Nevertheless, the NRG approach originally developed by [Wilson (1975)] can be applied to this problem, and it can be solved with a numerically exact procedure that is most accurate near the Fermi energy.

This procedure is quite technical and involves a number of subtle steps to be taken. In the following, the basic ideas and equations for the NRG process will be developed, but the presentation here will be insufficient to allow the reader to create a full DMFT-NRG code directly from this discussion. Instead, it serves as an introduction to the subject that interested readers can follow up with the original literature and create their own computer codes after digesting the additional technical and computational issues required for the procedure.

We will focus on the Hubbard model for this exposition. The generalization to the PAM is not too complicated, and is fully covered in the scientific literature [Pruschke, Bulla and Jarrell (2000)]. We start with an examination of the partition function for the impurity Hubbard model in the presence of a dynamical mean field for both the up spin and the down spin electrons

$$\mathcal{Z}_{\rm imp}^{\rm Hubb}(\lambda_\uparrow,\lambda_\downarrow) = {\rm Tr}_c\left\{\mathcal{T}_\tau e^{-\beta(\mathcal{H}_{\rm imp}^{\rm Hubb}-\mu\mathcal{N})}S(\lambda_\uparrow)S(\lambda_\downarrow)\right\}, \tag{2.103}$$

with the impurity Hamiltonian for the Hubbard model [Hubbard (1963)] being

$$\mathcal{H}_{\rm imp}^{\rm Hubb} - \mu\mathcal{N} = -\mu(c_\uparrow^\dagger c_\uparrow + c_\downarrow^\dagger c_\downarrow) + Uc_\uparrow^\dagger c_\uparrow c_\downarrow^\dagger c_\downarrow\,, \tag{2.104}$$

and the evolution operators for the spin-up and the spin-down electrons given by Eq. (2.77) with the generalization to include a spin index on the field λ and on the electron creation and annihilation operators; in the absence of a magnetic field, we always have $\lambda_\uparrow = \lambda_\downarrow$. Unlike the Falicov-Kimball model case described above, in the Hubbard model case, we cannot write down an analytic expression for the partition function because the time-dependence of the up spin particles depends on the occupation of the down-spin particles (from the commutator of the U-term) and *vice versa.* So the trace of the time-ordered product in Eq. (2.103) over the four Fermionic states is complicated to evaluate.

In the limit where $U = 0$, however, the local Green's function can be written down, and it satisfies

$$G_\sigma^{R\ \text{nonint}}(\omega) = \frac{1}{\omega + \mu - \lambda_\sigma(\omega)}, \tag{2.105}$$

for either spin. The first step of the NRG calculation is to map the impurity problem in a time-dependent field onto a Hamiltonian on a finite one-dimensional chain, that reproduces a discrete approximation to the $U = 0$ Green's function of the impurity when evaluated on the chain. The physical picture is that the impurity electrons interact at the same time via the Coulomb interaction, so we need to know the field at $t = 0$, which corresponds to an average of $\lambda(\omega)$ over all ω, and they interact with the low-energy excitations (small ω) at low temperature. So we introduce a set of fictitious chain electron operators, with the goal of using the couplings of those electrons to themselves, and to the impurity, to set up a discretized version of the time-dependent field that the impurity electron is evolving in. Because the λ field has nontrivial frequency dependence, this mapping is complicated [Sakai and Kuramoto (1994); Chen and Jayaprakash (1995); Bulla, Pruschke and Hewson (1997); Gozales-Buxton and Ingersent (1998); Bulla, Hewson and Pruschke (1998)]. We describe the procedure for how to do this next. It is conventional to define the spectral function of the retarded dynamical mean field by $\Delta(\omega) = -\text{Im}\lambda^R(\omega)/\pi$ and to drop the spin index from λ.

The idea of Ken Wilson was to construct a frequency grid for the discretized Δ field on a logarithmic scale, rather than on a linear scale [Wilson (1975); Krishna-Murthy, Wilkins and Wilson (1980a); Krishna-Murthy, Wilkins and Wilson (1980b)]. We begin by choosing a maximum and a minimum frequency denoted by $\pm E$, and define a set of frequency points on the grid via $\omega_n^\pm = \pm E\Lambda^{-n}$ with Λ a numerical constant larger than 1, and typically chosen to be in the range of 1.5 to 3. We let $a_{n\sigma}^\dagger$ ($a_{n\sigma}$) denote a creation (destruction) operator for a fictitious electron associated with the frequencies near ω_n^+ and $b_{n\sigma}^\dagger$ ($b_{n\sigma}$) denote a creation (destruction) operator for a fictitious electron associated with the frequencies near ω_n^-. Next, we define a set of positive real numbers via

$$(\gamma_n^+)^2 = \int_{\omega_{n+1}^+}^{\omega_n^+} d\omega \Delta(\omega), \quad (\gamma_n^-)^2 = \int_{\omega_n^-}^{\omega_{n+1}^-} d\omega \Delta(\omega), \tag{2.106}$$

which denote the weight of the Δ field in the respective logarithmic frequency intervals, and

$$\xi_n^+ = \int_{\omega_{n+1}^+}^{\omega_n^+} d\omega\omega\Delta(\omega)/(\gamma_n^+)^2, \quad \xi_n^- = \int_{\omega_n^-}^{\omega_{n+1}^-} d\omega\omega\Delta(\omega)/(\gamma_n^-)^2, \tag{2.107}$$

which denote the weighted average of the frequency in each interval. The total weight of the Δ field is denoted by V^2:

$$V^2 = \int_{-E}^{E} d\omega\Delta(\omega) = \sum_n \left[(\gamma_n^+)^2 + (\gamma_n^-)^2\right]. \tag{2.108}$$

See Fig. 2.7 for a picture of the discretization process.

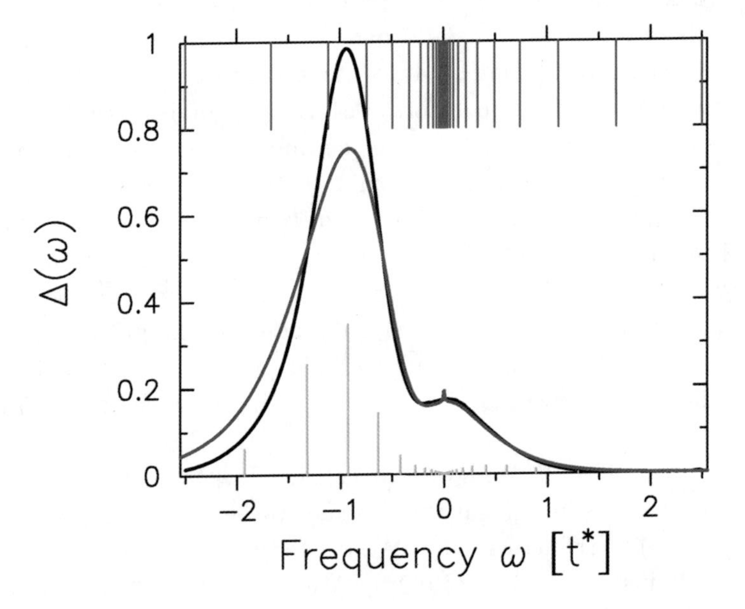

Fig. 2.7 Logarithmic frequency grid and discretized approximation to the Δ field used in the NRG approach. The solid black curve is the original $\Delta(\omega)$, the red lines are the discrete frequencies $\omega_n = 2.5 \times (1.5)^{-n}$, and we choose 19 positive and 19 negative frequencies, the green lines are the weights $(\gamma_n^{\pm})^2$ (described by the height of the line) and the positions $\xi_n^{\pm}$ of the delta function peaks for the discrete approximation to $\Delta(\omega)$, and the blue curve is the smoothed function using the logarithmic broadening with $b = 0.4$ (see below). Note how the blue curve is quite accurate for frequencies near the Fermi energy $\omega = 0$ (except for a "glitch" at the lowest frequencies) and becomes less accurate at higher frequencies. These inaccuracies are reduced by lowering Λ below 1.5, and increasing N beyond 19.

The chain Hamiltonian is now constructed via a Lanczos-like procedure in two steps. We first write the chain Hamiltonian in three terms: the impurity term, a hybridization term of the impurity electron with the "$t = 0$" field operator, and a diagonal energy term representing the number of electrons created at each discretized frequency with the corresponding weight associated with the Δ field. This chain Hamiltonian is

$$\tilde{\mathcal{H}}_{\text{chain}} = \mathcal{H}_{\text{imp}}^{\text{Hubb}}(U=0) + V\sum_{\sigma}(c_{\sigma}^{\dagger}\tilde{c}_{0\sigma} + \tilde{c}_{0\sigma}^{\dagger}c_{\sigma}) + \sum_{n\sigma}[\xi_n^{+}a_{n\sigma}^{\dagger}a_{n\sigma} + \xi_n^{-}b_{n\sigma}^{\dagger}b_{n\sigma}]. \tag{2.109}$$

The operator $\tilde{c}_{0\sigma}$ is $\sum_n(\gamma_n^{+}a_{n\sigma} + \gamma_n^{-}b_{n\sigma})/V$ which is "normalized" in the sense that $\{\tilde{c}_{0\sigma}^{\dagger}, \tilde{c}_{0\sigma}\} = 1$. One can check that the Green's function of Eq. (2.109) is equal to

$$G_{\text{imp}}(\omega) = \frac{1}{\omega + \mu - \sum_n\left[\frac{(\gamma_n^{+})^2}{\omega - \xi_n^{+}} + \frac{(\gamma_n^{-})^2}{\omega - \xi_n^{-}}\right] + i\delta} \tag{2.110}$$

by writing the Hamiltonian explicitly in terms of the a and b operators, and solving for the impurity Green's function via an EOM (see Problem A.11). Comparing with Eq. (2.105), shows that the NRG makes a discretization of the spectral function for the dynamical mean field via

$$-\frac{1}{\pi}\text{Im}\lambda^{R}(\omega) = \Delta(\omega) = \sum_n\left[(\gamma_n^{+})^2\delta(\omega - \xi_n^{+}) + (\gamma_n^{-})^2\delta(\omega - \xi_n^{-})\right], \tag{2.111}$$

but the delta functions lie at $\xi_n^{\pm}$ rather than $\omega_n^{\pm}$, although these quantities are quite close to one another.

The form for the Hamiltonian [in Eq. (2.109)] is not yet useful for calculations, because it is not in a tridiagonal form. We tridiagonalize it via a Lanczos-like procedure. First we take the state $\tilde{c}_{0\sigma}^{\dagger}|0\rangle$ and operate on it with the chain Hamiltonian $\tilde{\mathcal{H}}_{\text{chain}}$ to create a new state $\tilde{\mathcal{H}}_{\text{chain}}\tilde{c}_{0\sigma}^{\dagger}|0\rangle$. We find the overlap of this new state with the old one, giving us a local site energy at site 0,

$$\epsilon_0 = \langle 0|\tilde{c}_{0\sigma}\tilde{\mathcal{H}}_{\text{chain}}\tilde{c}_{0\sigma}^{\dagger}|0\rangle, \tag{2.112}$$

and the leftover piece defines the next state on the chain $\tilde{c}_{1\sigma}|0\rangle$ and the "hopping" λ_1 between site 0 and site 1 on the chain

$$\lambda_1\tilde{c}_{1\sigma}^{\dagger}|0\rangle = \tilde{\mathcal{H}}_{\text{chain}}\tilde{c}_{0\sigma}^{\dagger}|0\rangle - \epsilon_0\tilde{c}_{0\sigma}^{\dagger}|0\rangle, \tag{2.113}$$

with the normalization condition $\{\tilde{c}_{1\sigma}^{\dagger}, c_{1\sigma}\} = 1$. Since $\tilde{\mathcal{H}}_{\text{chain}}$ is Hermitian, it is easy to show that the hopping from site 1 to site 0 is also equal to

λ_1. A direct calculation shows that $\epsilon_0 = \langle\omega\rangle$ is the average value of the frequency over the field $\Delta(\omega)$ and that $\lambda_1^2 = \langle\omega^2\rangle - \langle\omega\rangle^2$ is the quadratic fluctuation over the field. Proceeding in this fashion, one constructs all of the site energies and hoppings on the chain, and puts the Hamiltonian into its final form (see Fig. 2.8)

$$\mathcal{H}_{\text{chain}} = \mathcal{H}_{\text{imp}}^{\text{Hubb}} + V\sum_{\sigma}(c_\sigma^\dagger \tilde{c}_{0\sigma} + \tilde{c}_{0\sigma}^\dagger c_\sigma) + \sum_{n\sigma}[\epsilon_n \tilde{c}_{n\sigma}^\dagger \tilde{c}_{n\sigma} + \lambda_n \tilde{c}_{n+1\sigma}^\dagger \tilde{c}_{n\sigma} + \text{h.c.}], \tag{2.114}$$

with each Fermionic operator satisfying the conventional anticommutation relations. The explicit computation is difficult. In some cases it can be performed analytically for simple fields, in other cases it must be done numerically. Since the elements of the chain decay as we move farther out along the chain, special care (such as using arbitrary precision arithmetic [Gozales-Buxton and Ingersent (1998); Bulla, Hewson and Pruschke (1998)] or carefully analyzing and stabilizing the recursions [Chen and Jayaprakash (1995)]) is needed to accurately determine the chain parameters (see Problem A.10).

Note that we did not specify that $U = 0$ in Eq. (2.114). This is because the actual Hamiltonian for the NRG calculation must restore the many-body interactions on the impurity site (denoted by a square box in Fig. 2.8) because the rest of the chain represents the time evolution in the dynamical mean field. The next step in the NRG approach is to find all the eigenvalues and eigenstates of the chain Hamiltonian. In a typical calculation, one takes between 30 and 100 sites in the chain. The impurity Hamiltonian has only four states corresponding to zero, one, or two electrons at the

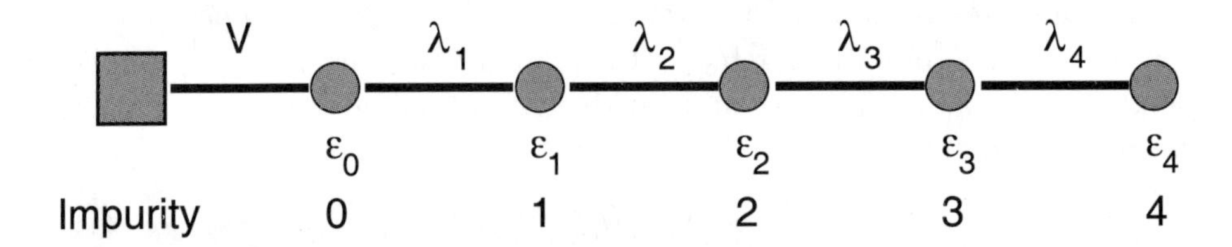

Fig. 2.8 Mapping of the Hubbard impurity in a time-dependent field onto a discrete Hamiltonian on a finite chain. The square box denotes the impurity site, where the impurity Hamiltonian lies, and the circles denote the "lattice" sites of the fictitious Fermions, determined by the Lanczos-like mapping, with diagonal and off-diagonal elements ϵ_n and λ_n respectively. The off-diagonal element V connects the chain to the impurity site and *vice versa*. This case has a total of $N = 5$ sites in the chain. Note that all of the many-body interactions take place only on the impurity site.

impurity. Each site of the chain introduces another 4 states, so the many-body Hilbert space has dimension 4^{L+1} when we have included the Lth site of the chain. As L increases, this number rapidly becomes too large to diagonalize the entire Hamiltonian matrix. The strategy of the NRG is to diagonalize the full Hamiltonian for the first few steps of the iteration, and then, once the Hamiltonian contains more than about 1000 states, we keep only the lowest 1000 states for the next iteration, implying we will need to diagonalize a 4000×4000 matrix as we add each subsequent site to the chain and then truncate to the lowest 1000 states. In reality, one takes into account a number of useful quantum numbers like the particle number, the z-component of spin, the total spin, and so on. Then each symmetry block is much smaller, but it is more complicated to construct each block with definite symmetry at the each step of the NRG iteration (see Problem A.12).

Once we have included every site in the finite chain, we will have a series of eigenvalues and eigenfunctions associated with each of the steps in the iteration. Since the couplings along the chain decrease, it turns out that the energy levels come closer and closer to the low-lying excitations near the Fermi level. We can use these states and energy levels to calculate things like the imaginary part of the retarded Green's function via the Lehmann representation. This requires us to know the matrix elements of the creation and annihilation operators between the low-lying states that we have calculated [Bulla, Pruschke and Hewson (1997)]. It requires some care to update these matrix elements as each new chain site is added, but it can be done. The final result is, however, a set of discrete delta functions for imaginary part of the retarded Green's function. They are constructed as a composite from the energy levels and weights (from the matrix elements) of a series of calculations for different numbers of sites in the chain. The delta functions are then broadened via a set of peaks broadened on a logarithmic scale,

$$\delta(\omega - E_n) \to \frac{e^{-b^2/4}}{b|E_n|\sqrt{\pi}} \exp\left[-\frac{(\ln[\omega/E_n])^2}{b^2}\right], \tag{2.115}$$

which are employed to represent the final density of states back on the linear frequency scale (usually we take $0.3 < b < 0.6$ and we set the broadening to zero if $\omega/E_n \leq 0$). Since the information about the higher-energy states is coarse, and because they arise in a composite fashion from a series of calculations at different chain lengths, the DOS may not exactly satisfy the requisite sum rules, but it usually is quite accurate at low energies (although the broadening can produce some errors here too), and typically has pointwise errors on the order of a few percent. The real part of the Green's function is finally generated via a Kramers-Kronig analysis.

To complete the DMFT loop, we need to extract the self-energy, compute the local Green's function from the Hilbert transform, and then determine the new effective medium and dynamical mean field. It turns out that calculating the self-energy via $\Sigma = G_0^{-1} - G^{-1}$ is not as accurate as extracting it from an EOM-based approach using the four-Fermion expectation value resulting from the commutator with $\hat{V}$. The details of how this works is quite technical, and can be found in the original literature [Bulla, Hewson and Pruschke (1998)].

There are other impurity solvers that can be used for the DMFT approach as well. The two most common ones employed involve a perturbative solution, or employ quantum Monte Carlo methods. The perturbative solvers rapidly become quite complex as the order of the perturbation theory is increased. They also often suffer from a breakdown in the intermediate and strong coupling regimes. They are never exact approaches. The quantum Monte Carlo methods can work only on the imaginary axis, so they are best employed to calculate static properties of the given systems under study. The quantum Monte Carlo approach involves two steps. First the imaginary time interval is discretized and then the four-Fermion interaction terms (like the U terms in the Hubbard model) are decoupled by introducing so-called Hirsch-Hubbard-Stratanovitch fields [Stratonovitch (1957); Hubbard (1959); Hirsch (1983)]. The resulting Fermionic piece of the action is now quadratic in the Fermions, so the path integral can be performed and it involves a Feynman determinant of a discretized matrix that represents the action. The Monte Carlo piece then sums over configurations of the Hirsch-Hubbard-Stratanovitch fields in imaginary time selecting the configurations that tend to be most important [Hirsch and Fye (1986)]. These methods suffer from discretization error (which is well understood) and statistical error (which can often be controlled). As a final step, a maximum-entropy analytic continuation can be performed on the quantum Monte Carlo data to determine the dynamical properties [Jarrell and Gubernatis (1996)]. This approach is numerically exact, but it can be expensive in computer time and is insensitive to sharp, narrow features in the DOS. Often it yields the best solutions for complicated many-body problems. We won't discuss these alternative methods further.

2.7 Computational Algorithms

Regardless of the choice for the impurity solver, the vast majority of DMFT problems are solved by employing an iterative algorithm [Jarrell (1992)]. Iterative methods for the solution of nonlinear equations can lead to a number of numerical issues that need to be properly controlled. There may be more than one fixed-point solution to the equations, the iterations may iterate away from rather than toward the fixed point, or the iterations may converge slowly. We discuss these computational issues here.

In most cases, the DMFT algorithm is first solved on the imaginary axis, yielding the Matsubara frequency Green's functions and self energies. This is done because imaginary-axis methods are often faster and more accurate than real-axis methods. When we initialize a numerical solution for DMFT, we must input the basic parameters of the Hamiltonian, including the noninteracting DOS, the temperature, and the potential energy parameters. We often are interested in performing calculations with a fixed density of particles. But we cannot input this density into the calculation, instead we input a chemical potential. For a given value of μ, with all other parameters fixed, we can iterate the DMFT equations to convergence, and then compute the electron density. If this is not equal to our target density, then we adjust the chemical potential and repeat. Since we are changing only one parameter in this process, a one-dimensional root finder can be used. The most efficient such root finder is Brent's false-position plus inverse quadratic substitution root-finder [Brent (1973)], which approaches the speed and accuracy of a Newton's method approach, with the safety of a false-position algorithm, and no need to calculate the derivative of the filling with respect to μ, which is complicated to do in practice. It does require that the final chemical potential be "bracketed" by the two initial guesses for the chemical potential, so we need to have an educated guess for where the chemical potential will be. This is not difficult to achieve. Once the chemical potential is known, we can switch to a real-axis code to find the dynamical properties, without needing to adjust the chemical potential anymore. Note that this strategy is not feasible for problems solved with the NRG approach, because there is no simple way to employ the NRG approach to calculate the Matsubara-frequency Green's functions without calculating the real-axis Green's functions first.

This scheme can be sped up, if we adjust the chemical potential within the iterative DMFT algorithm, so that we iterate to the DMFT solutions at the correct filling. There are two ways to achieve this goal. First, we can estimate the value of the compressibility, which is defined from the derivative of the filling with respect to μ, and use a Newton's method to adjust the chemical potential every 10 to 20 iterations (the compressibility is a two-particle susceptibility that we do not treat in this book). We don't adjust at each iteration because we want to give the system some time to close in on the DMFT solutions with the new value of μ before we adjust it again. In the second case, we simply shift the chemical potential upwards or downwards (depending on whether the current filling is too low or too high) by a small amount, proportional to the deviation from the target filling, and we reduce the maximal allowed shift of μ as the iterations increase.

The iterative solutions can sometimes fall into a limit cycle as they approach the fixed point. Rather than moving in a straight-line toward the solution, they orbit around the fixed point, spiraling inwards very slowly. The problem of limit cycles can be reduced, or even eliminated by using a weighted average scheme in the calculations. Since we need to store the old value of the self-energy to test for convergence of the algorithm, we construct the new lattice self-energy as a weighted average of the old lattice self-energy and the new impurity self-energy

$$\Sigma_{\text{new lattice}} = \alpha\Sigma_{\text{old lattice}} + (1-\alpha)\Sigma_{\text{new impurity}}. \tag{2.116}$$

The damping factor α is usually taken to lie between 0 (no averaging) and 1 (no updating). For the Falicov-Kimball model solutions, it is often convenient to choose $\alpha = 0$ for the first 50 iterations, then increase $\alpha = 0.5$ for the next 200 iterations, then increase $\alpha = 0.75$ for the next 500 iterations, then increase to $\alpha = 0.9$ for the next 2000 iterations, and so on. One needs to increase the number of iterations as α is increased, because the progress of the iterations goes more slowly for larger α. We usually want at least 6 to 8 digits of accuracy for the self-energy before we declare it converged. This criterion is relaxed with NRG or quantum Monte Carlo impurity solvers since they are so costly to run.

There are occasionally situations where there is more than one solution to these iterative equations. Fortunately this is rare, but it occurs, for example, in the Mott transition in the Hubbard model at finite temperature. When such a situation occurs, the multiple fixed points can be attractive or repelling. If they are unstable, then it is unlikely they will be found within the iterative scheme, and this is always an unfortunate uncertainty with this numerical technique. If they are stable, and hence attractive, one needs to start the iteration with self-energies sufficiently close to the self-energies of the corresponding fixed points in order to be able to find them. This is actually how their existence was discovered in the Hubbard model [Georges, *et al.* (1996)]. The only way to determine which fixed point is the physical fixed point is to calculate the free energy associated with each of them—the lowest free energy will be the thermodynamically stable solution.

When we are solving for the chemical potential, we need to calculate the Green's function at every Matsubara frequency, because they are all needed to determine the electron filling. But when we have the converged value for μ, and we want to calculate the real-axis results, then for the Falicov-Kimball model we can perform the iteration separately for each desired frequency value ω. This can greatly speed up the computational time, because most frequencies converge quite rapidly, and they will not need to be recalculated after they have converged. The algorithm often becomes inaccurate near the band edges in the insulating phase. There are a number of problems in the Appendix which deal with these issues. For other models, there might be a coupling between different frequencies in solving the impurity problem. If this is the case, then the results should be calculated for every frequency in each iteration.

Note that in the NRG calculations, there are a number of places where the numerics can be challenging: (i) in constructing the chain parameters for a given size chain; (ii) in accurately determining the self-energy from the NRG for use in the DMFT iterative algorithm; and (iii) in the convergence of the DMFT algorithm. This is made worse by there being a coarser knowledge of energy eigenvalues far from the chemical potential and by the need to artificially broaden the delta functions to get smooth functions of frequency. Nevertheless, it remains one of the most accurate approaches available for dynamical properties of strongly correlated systems, and it is particularly good for cases where a sharp peak develops near or at the chemical potential, as often occurs in Fermi-liquid systems.

There is one other method that can be employed to solve the DMFT equations for the spinless Falicov-Kimball model [Brandt and Mielsch (1989)]. Since we know the direct relation between the self-energy and the Green's function [from Eq. (2.102)], we can substitute it into the Hilbert transform and get a transcendental equation for G_n [or equivalently for $G(\omega)$ since the quadratic equation holds on the real axis too]. If we solve this transcendental equation using a root-finder in the complex plane, making sure to avoid the trivial root $G = 0$, we have an alternative way to solve the DMFT algorithm (Müller's square-root algorithm is a good root finder for the complex plane [Müller (1956)]). This approach is nice, because it does not require any iterations, but it appears to only be possible for the Falicov-Kimball model, since the relation between Σ and G is not known for other models. There also is a challenge in being able to choose the correct sign for the square root, since it can change as a function of where the variable z lies in the complex plane.

2.8 Linear-Response *dc*-Transport in the Bulk

Determining the transport properties of a material, namely the response of the electrons in the material to the application of an external electromagnetic field, or a temperature gradient, is the most useful way to classify materials properties. Often the *dc* conductivity is used to describe the material as metallic, insulating, or semiconducting. But the thermal transport, and the effects of a magnetic field (say in the Hall effect), are also important properties to measure. From a theoretical perspective, we would like to calculate the relevant transport coefficients in the linear-response regime, where the effect of the perturbing fields is taken into account to first-order only. Since many materials display linear behavior over a wide range of fields available in a laboratory, this approximation is usually quite adequate.

Our starting point is to determine the relevant current operators associated with particular transport mechanisms. We will then employ a Kubo-formula analysis to extract the corresponding transport coefficients. Our focus will be on *dc* effects, and we will not investigate any magnetic field effects here.

Suppose an electric field $\mathbf{E}(\mathbf{r},t)$ (that can vary with position and time) is applied to a material. This field will cause the electron density $\rho_e(\mathbf{r},t)$ to vary with position and time as well. The equation of continuity for the number current $\mathbf{j}$ says that

$$\frac{\partial \rho_e(\mathbf{r},t)}{\partial t} + \nabla \cdot \mathbf{j}(\mathbf{r},t) = 0. \tag{2.117}$$

If we introduce the definition of the electric polarization $\mathbf{P}$ by $\mathbf{P}(\mathbf{r},t) = \mathbf{r}\rho_e(\mathbf{r},t)$, take the partial derivative with respect to time, and integrate over all space, then we can use Eq. (2.117) and integration by parts to show that $\mathbf{j}(\mathbf{r},t) = \partial \mathbf{P}(\mathbf{r},t)/\partial t$. In quantum mechanics, we are interested in determining operators, where time derivatives are replaced by commutators with the Hamiltonian, so the polarization and the number current operator are related via $\mathbf{j}(\mathbf{r}) = i[\mathcal{H}, \mathbf{P}(\mathbf{r})]$. On a lattice, the position variables are replaced by site indices, so that $\mathbf{P}_j(t) = \mathbf{R}_j c_j^\dagger c_j$ and

$$\mathbf{j}_j = i[\mathcal{H}, P_j]. \tag{2.118}$$

Since the potential-energy operator is normally a function of the number operators, it does not contribute to the number current, and the only contribution to the current operator comes from the commutator with the kinetic energy. Performing the commutator yields two forms for the total number current operator $\mathbf{j} = \sum_j \mathbf{j}_j$ (in real space and momentum space)

$$\mathbf{j} = -it \sum_i \sum_\delta c_i^\dagger c_{i+\delta} \bar{\delta} = \sum_{\mathbf{k}} \mathbf{v}_{\mathbf{k}} c_{\mathbf{k}}^\dagger c_{\mathbf{k}}, \tag{2.119}$$

where the symbol $\bar{\delta}$ denotes the nearest-neighbor translation vector, $i+\delta$ is the corresponding nearest-neighbor to site i, and $\mathbf{v}_{\mathbf{k}}$ is the electron velocity $\mathbf{v}_{\mathbf{k}} = \nabla\epsilon(\mathbf{k})$. Note that in cases where the electrons have spin, Eq. (2.119) needs to be modified to include a sum over the spin indices. The charge-current operator, is then written as $\mathbf{j}_c = -|e|\mathbf{j}$, where we make explicit that the charge of the electrons is negative.

The heat-current operator is more complicated to determine. As electrons move through the lattice, they carry energy and heat with them. The heat current is just the energy current minus the chemical potential times the number current, because the heat is always measured relative to the chemical potential (in other words, an electron that has energy larger than μ carries heat, while one with energy less than μ carries coldness). It is easy to guess the kinetic-energy contribution to the energy current — one simply adds a factor of $\epsilon_{\mathbf{k}}$ to the last term in Eq. (2.119), but we need to account for the potential-energy contributions as well. The procedure for how to do this is described thoroughly in Chapter 1 of [Mahan (1990)]. We need to generalize the polarization operator to an "energy polarization" operator, which requires us to break the Hamiltonian up into pieces, each associated with a given lattice site. This can be easily done for the potential energy if the interactions are local, but there is no well-defined procedure for the kinetic-energy piece. The convention is to break each hopping term in half, associating half with the site in each nearest-neighbor pair. In this fashion, we write $\mathcal{H} = \sum_i \mathcal{H}_i$, with

$$\mathcal{H}_i = -\frac{t}{2}\sum_\delta (c_i^\dagger c_{i+\delta} + c_{i+\delta}^\dagger c_i) + \hat{V}_i, \tag{2.120}$$

where the local potential energy is, for example, just $\hat{V}_i = E_f f_i^\dagger f_i + U c_i^\dagger c_i f_i^\dagger f_i$ for the Falicov-Kimball model. The energy current is then defined via $\mathbf{j}_E = i[\mathcal{H}, \sum_j \mathbf{R}_j \mathcal{H}_j]$. Because the energy polarization operator is not just a function of the electron number operators anymore, the commutators will change for different models, due to the changing potential energy. Evaluating the commutators can be complicated as well. For the Falicov-Kimball model we find

$$\mathbf{j}_Q = \mathbf{j}_E - \mu \mathbf{j} = \sum_{\mathbf{k}} (\epsilon_{\mathbf{k}} - \mu) \mathbf{v}_{\mathbf{k}} c_k^\dagger c_k + \frac{1}{2} \sum_{\mathbf{k}\mathbf{k}'} (\mathbf{v}_{\mathbf{k}} + \mathbf{v}_{\mathbf{k}'}) \hat{W}(\mathbf{k} - \mathbf{k}') c_{\mathbf{k}}^\dagger c_{\mathbf{k}'}, \quad (2.121)$$

where $\hat{W}(\mathbf{q}) = \sum_j \exp(-i\mathbf{q} \cdot \mathbf{R}_j) f_j^\dagger f_j / V$.

We use the Kubo-Greenwood formula [Kubo (1957); Greenwood (1958)] to evaluate the transport coefficients. We will derive it in general, but apply it directly to the electrical conductivity. We work in the linear-response regime, taking the limit that the external field is infinitesimal. We imagine modifying our time-independent Hamiltonian by adding a time-dependent perturbation $\mathcal{H}'(t)$. The perturbation is turned on adiabatically as $t \to -\infty$. The expectation value for the particle-current operator as a function of time is

$$\langle \mathbf{j}(t) \rangle = \frac{1}{\mathcal{Z}} \mathrm{Tr} \left\{ e^{-\beta(\mathcal{H} - \mu \mathcal{N})} \hat{U}^\dagger(t, -\infty) \mathbf{j} \hat{U}(t, -\infty) \right\}, \quad (2.122)$$

where we introduced the time evolution operator $\hat{U}(t, -\infty)$, familiar from time-dependent problems in quantum mechanics, which determines the time evolution of all operators. The evolution operator satisfies an EOM

$$i\partial_t \hat{U}(t, -\infty) = [\mathcal{H} - \mu \mathcal{N} + \mathcal{H}'(t)] \hat{U}(t, -\infty), \quad (2.123)$$

which is solved by a time-ordered product

$$\hat{U}(t, -\infty) = \mathcal{T}_t e^{-i \int_{-\infty}^t d\bar{t} [\mathcal{H} - \mu \mathcal{N} + \mathcal{H}'(\bar{t})]}, \quad (2.124)$$

which requires the time ordering because the operator $\mathcal{H}'$ may not commute with itself or $\mathcal{H} - \mu \mathcal{N}$ for different times. Unfortunately, it is difficult to expand Eq. (2.122) in a power series in the perturbation because of the time ordering. We need to re-organize the expression, so the time ordering involves only the $\mathcal{H}'$ operator. This is done by introducing the interaction representation, defining the interaction representation evolution operator via

$$\hat{U}_I(t, -\infty) = e^{i(\mathcal{H} - \mu \mathcal{N})t} \hat{U}(t, -\infty). \quad (2.125)$$

It is easy to show that $i\partial_t \hat{U}_I(t,-\infty) = \mathcal{H}'_I(t)\hat{U}_I(t,-\infty)$ with the interaction representation of the perturbation being $\mathcal{H}'_I(t) = \exp[i(\mathcal{H} - \mu\mathcal{N})t]\mathcal{H}'(t)\exp[-i(\mathcal{H}-\mu\mathcal{N})t]$. Hence $\hat{U}_I(t,-\infty) = \mathcal{T}_t \exp[-i\int_{-\infty}^{t} d\bar{t}\mathcal{H}'_I(\bar{t})]$. Next, we move the rightmost factor $\hat{U}$ in Eq. (2.122) to the left by the cyclic property of the trace, and we replace $\hat{U}$ by $\hat{U}_I$ throughout. All of the dependence on the perturbation now lies in the term $\hat{U}_I \exp[-\beta(\mathcal{H}-\mu\mathcal{N})]\hat{U}_I^{\dagger}$, and we need to expand this to first order. Substituting in the Taylor series expansion for the time-ordered product in $\hat{U}_I$ then yields

$$\hat{U}_I(t,-\infty)e^{-\beta(\mathcal{H}-\mu\mathcal{N})}\hat{U}_I^{\dagger}(t,-\infty) = e^{-\beta(\mathcal{H}-\mu\mathcal{N})} - i\int_{-\infty}^{t} d\bar{t}[\mathcal{H}'_I(\bar{t}), e^{-\beta(\mathcal{H}-\mu\mathcal{N})}] + \mathcal{O}(\mathcal{H}'^2). \tag{2.126}$$

Now we can evaluate the linear-response current. It satisfies

$$\langle \mathbf{j}(t)\rangle = \langle \mathbf{j}\rangle - i\int_{-\infty}^{t} d\bar{t}\mathrm{Tr}\left\{e^{-i(\mathcal{H}-\mu\mathcal{N})t}[\mathcal{H}'_I(\bar{t}), e^{-\beta(\mathcal{H}-\mu\mathcal{N})}]e^{i(\mathcal{H}-\mu\mathcal{N})t}\mathbf{j}\right\}. \tag{2.127}$$

Using the invariance of the trace, and remembering the definition of time-dependent operators in the interaction representation, we get the final Kubo-Greenwood result

$$\langle \mathbf{j}(t)\rangle = \langle \mathbf{j}\rangle - i\int_{-\infty}^{t} d\bar{t}\langle[\mathbf{j}_I(t), \mathcal{H}'_I(\bar{t})]\rangle. \tag{2.128}$$

Note that since we did not use any special properties about the current operator, or the perturbation, the above relation is a completely general result. We have succeeded in representing the linear-response current with expectation values and time-evolutions that take place wholly within equilibrium. Knowing this, we can drop the I subscripts for the interaction picture from the remainder of our formulas since we know the time evolution is always with respect to $\mathcal{H} - \mu\mathcal{N}$.

The Kubo formula is normally written in the form of a current-current correlation function. This can be seen when we examine the effects of an electric field written in the gauge where the scalar potential vanishes, so that $\mathbf{E}_i(t) = -\partial_t \mathbf{A}_i(t)$ (we set the speed of light $c = 1$). The perturbation to the Hamiltonian is then $\mathcal{H}'(t) = |e|\sum_i \mathbf{j}_i \cdot \mathbf{A}_i(t)$ involving the sum of the negative of the dot product of the charge current operator with the vector potential over all sites of the lattice. We will examine only uniform fields here, so we assume $\mathbf{A}_i = \mathbf{A}$ is independent of i. Substituting this result into

Eq. (2.128) and noting that the expectation value of the current vanishes in equilibrium gives

$$\langle -|e|\mathbf{j}_a(t)\rangle = ie^2 \sum_b \int_{-\infty}^{t} d\bar{t} \langle [\mathbf{j}_a(t), \mathbf{j}_b(\bar{t})] \rangle \mathbf{A}_b(\bar{t}). \tag{2.129}$$

The symbols a and b refer to the spatial indices of the corresponding vectors (they are not the lattice sites).

Since we have time-translation invariance when we are in equilibrium, we can take the Fourier transform of Eq. (2.129), and note that $\mathbf{A}(\omega) = -i\mathbf{E}(\omega)/\omega$, to yield

$$\langle -|e|\mathbf{j}_a(\omega)\rangle = ie^2 \sum_b \frac{\Pi_{ab}^{jj\text{ret}}(\omega)}{\omega} \mathbf{E}_b(\omega). \tag{2.130}$$

We introduced the symbol $\Pi_{ab}^{jj\text{ret}}(\omega) = -i \int dt \exp[i\omega t]\theta(t)\langle [\mathbf{j}_a(t), \mathbf{j}_b(0)]\rangle$, the retarded current-current correlation function. This is almost in the form where we can extract the conductivity. The only remaining issue that we need to take into account is that when we add an electric field to our system, we have to replace the momentum by the Peierl's substitution $\mathbf{p} \to \mathbf{p} + |e|\mathbf{A}$. When we are on a lattice, the shift is $\mathbf{k} \to \mathbf{k} + |e|\mathbf{A}$ in the velocity $\mathbf{v}_\mathbf{k}$. This produces an additional term to the charge current in linear response, namely the term $-ie^2\langle \hat{T}_a\rangle \mathbf{E}_a/\omega$ to the response (where $\hat{T}_a$ is the $-\cos \mathbf{k}_a$ piece of the kinetic energy operator). Putting this together with the Kubo result gives

$$\langle -|e|\mathbf{j}_a(\omega)\rangle = ie^2 \sum_b \frac{\Pi_{ab}^{jj\text{ret}}(\omega) - \langle \hat{T}_a\rangle \delta_{ab}}{\omega} \mathbf{E}_b(\omega), \tag{2.131}$$

or, since the conductivity satisfies $-|e|\mathbf{j}_a(\omega) = \sum_b \sigma_{ab}(\omega)\mathbf{E}_b(\omega)$,

$$\sigma_{ab}(\omega) = ie^2 \frac{\Pi_{ab}^{jj\text{ret}}(\omega)}{\omega} - ie^2 \frac{\langle \hat{T}_a\rangle \delta_{ab}}{\omega}. \tag{2.132}$$

The real part of Eq. (2.132) is called the optical conductivity, which is normally what a theorist will calculate.

It is not easy to evaluate the conductivity by directly calculating Eq. (2.129) and then taking the Fourier transform. Instead, we proceed as we did with the Green's functions by first evaluating the response function on the imaginary axis, and then analytically continuing to the real axis.

This is done because the continuation of the imaginary-axis response function is to the retarded response function, and all calculations are straightforward, although they are lengthy.

We start with the appropriate generalization of the retarded current-current correlation function to the imaginary-time axis (one can verify that this is the correct form by examining the Lehman representation, and then converting from the imaginary to the real axis, similar to what we did with the Green's functions):

$$\Pi_{ab}^{jj}(\tau) = \frac{1}{\mathcal{Z}}\mathrm{Tr}\left[e^{-\beta(\mathcal{H}-\mu\mathcal{N})}\mathcal{T}_\tau \mathbf{j}_a(\tau)\mathbf{j}_b(0)\right] - \left(\frac{1}{\mathcal{Z}}\mathrm{Tr}\left[e^{-\beta(\mathcal{H}-\mu\mathcal{N})}\mathbf{j}_a(0)\right]\right) \times \left(\frac{1}{\mathcal{Z}}\mathrm{Tr}\left[e^{-\beta(\mathcal{H}-\mu\mathcal{N})}\mathbf{j}_b(0)\right]\right), \tag{2.133}$$

and the expectation values of the current vanish in equilibrium, so the last term is zero. This correlation function is periodic when $\tau \to \tau + \beta$ because the current operators involve two Fermionic operators, and there is no sign change when the current operators are interchanged by the time-ordering operation. The Matsubara frequency Fourier series corresponds to the Bosonic frequencies $i\nu_l = 2i\pi T l$ and the Fourier coefficients are

$$\Pi_{ab}^{jj}(i\nu_l) = \int_0^\beta d\tau e^{i\nu_l\tau}\Pi_{ab}^{jj}(\tau). \tag{2.134}$$

We calculate the correlation function by adding a field to the Hamiltonian, and taking a derivative of the appropriate expectation value with respect to the field. In this case, we add a field $-\mathbf{j}\cdot\mathbf{A}(\tau)$ with $\mathbf{A}(\tau) = T\sum_l \mathbf{A}_l \exp(-i\nu_l\tau)$; note, that the field $\mathbf{A}$ is not necessarily the vector potential here, but is a fictitious vector-valued function employed in the calculations. Since the added field does not commute with the Hamiltonian, we need to express it as a time-ordered product using the identity

$$e^{-\int_0^\beta d\tau[\mathcal{H}-\mu\mathcal{N}-\mathbf{j}\cdot\mathbf{A}(\tau)]} = e^{-\beta(\mathcal{H}-\mu\mathcal{N})}\mathcal{T}_\tau e^{\int_0^\beta d\tau \mathbf{j}(\tau)\cdot\mathbf{A}(\tau)}, \tag{2.135}$$

with $\mathbf{j}(\tau) = \exp[\tau(\mathcal{H}-\mu\mathcal{N})]\mathbf{j}\exp[-\tau(\mathcal{H}-\mu\mathcal{N})]$. Then we can take derivatives, to find that $\Pi_{ab}^{jj}(i\nu_l) = \partial\langle\mathbf{j}_b(0)\rangle/\partial\mathbf{A}_a(-i\nu_l)/T$ when $\mathbf{A}\to 0$. Using Green's functions to evaluate the expectation value gives

$$\Pi_{ab}^{jj}(i\nu_l) = \sum_{\mathbf{k}}\sum_{nm} \mathbf{v}_{\mathbf{k}b}\frac{\partial G_{\mathbf{k}}(i\omega_n, i\omega_m)}{\partial \mathbf{A}_a(-i\nu_l)}, \tag{2.136}$$

where we must use the double Fourier transform of the Green's function, since the $\mathbf{j} \cdot \mathbf{A}$ field makes the Green's function lose it's time-translation invariance so that

$$G_{\mathbf{k}}(i\omega_n, i\omega_m) = T \int_0^\beta d\tau \int_0^\beta d\tau' e^{i\omega_n \tau} G_{\mathbf{k}}(\tau, \tau') e^{-i\omega_m \tau'}; \tag{2.137}$$

after the derivative is taken, we can evaluate the Green's functions in equilibrium, where $G_{\mathbf{k}}(i\omega_n, i\omega_m) \propto \delta_{nm}$. The derivative is computed by using the identity

$$G_{\mathbf{k}}(i\omega_n, i\omega_m) = \sum_{m'n'} G_{\mathbf{k}}(i\omega_n, i\omega_{m'}) G_{\mathbf{k}}^{-1}(i\omega_{m'}, i\omega_{n'}) G_{\mathbf{k}}(i\omega_{n'}, i\omega_m), \tag{2.138}$$

and noting that any Green's function that does not have a derivative acting on it can be replaced by it's (diagonal) equilibrium value. This yields

$$\Pi_{ab}^{jj}(i\nu_l) = -\sum_{\mathbf{k}} \sum_{nm} \mathbf{v}_{\mathbf{k}b} G_{\mathbf{k}}(i\omega_n) \frac{\partial G_{\mathbf{k}}^{-1}(i\omega_n, i\omega_m)}{\partial \mathbf{A}_a(-i\nu_l)} G_{\mathbf{k}}(i\omega_m). \tag{2.139}$$

The derivative of the inverse of the Green's function is easier to find than the derivative of the Green's function itself. To do so, we must first examine the EOM for the Green's function, which is

$$\int_0^\beta d\tau'' [\{-\partial_\tau + \mu - \epsilon_{\mathbf{k}} + \mathbf{v}_{\mathbf{k}} \cdot \mathbf{A}(\tau)\} \delta(\tau - \tau'') - \Sigma(\tau, \tau'')] G_{\mathbf{k}}(\tau'', \tau') = \delta(\tau - \tau'), \tag{2.140}$$

and determines the inverse operator in the square brackets (note, the self-energy depends on two times now as well). Next, we take the double Fourier transform of the inverse operator, to find

$$\sum_{m'} [(i\omega_n + \mu - \epsilon_{\mathbf{k}}) \delta_{nm'} - \Sigma_{nm'} + T \sum_l \mathbf{v}_{\mathbf{k}} \cdot \mathbf{A}_l \delta_{nm'+l}] G_{\mathbf{k}}(i\omega_{m'}, i\omega_m) = \delta_{mn}. \tag{2.141}$$

The term inside the square brackets is $G_{\mathbf{k}}^{-1}(i\omega_n, i\omega_{m'})$. There are two terms that have a nonzero derivative with respect to $\mathbf{A}_b(-i\nu_l)$: the self-energy and the $\mathbf{A}$ field. Performing the derivatives gives

$$\begin{aligned} \Pi_{ab}^{jj}(i\nu_l) = &-T \sum_{\mathbf{k}} \sum_n \mathbf{v}_{\mathbf{k}a} \mathbf{v}_{\mathbf{k}b} G_{\mathbf{k}}(i\omega_n) G_{\mathbf{k}}(i\omega_{n+l}) \\ &+ \sum_{\mathbf{k}} \sum_{nm} \mathbf{v}_{\mathbf{k}b} G_{\mathbf{k}}(i\omega_n) \frac{\partial \Sigma_{nm}}{\partial \mathbf{A}_a(-i\nu_l)} G_{\mathbf{k}}(i\omega_m). \end{aligned} \tag{2.142}$$

The second term on the right hand side is called the vertex correction. The self-energy is a functional of the local Green's function, so we can use the chain rule to relate the derivative with respect to $\mathbf{A}$ to a derivative with respect to the (two-frequency) local Green's function. The derivative becomes

$$\frac{\partial \Sigma_{nm}}{\partial \mathbf{A}_a(-i\nu_l)} = \sum_{n'm'} \frac{\partial \Sigma_{nm}}{\partial G_{n'm'}} \frac{\partial G_{n'm'}}{\partial \mathbf{A}_a(-i\nu_l)}. \tag{2.143}$$

Since the derivative of the self-energy with respect to the local Green's function is independent of momentum, and since $\sum_{\mathbf{k}} \mathbf{v}_{\mathbf{k}} G_{\mathbf{k}}(i\omega_n) G_{\mathbf{k}}(i\omega_m) = 0$, because the velocity operator is an odd function of $\mathbf{k}$ and the Green's functions are even (since they depend only on $\epsilon_{\mathbf{k}}$), we have that the vertex-correction term in Eq. (2.142) vanishes [Khurana (1990)]. The current-current correlation function is equal to the bare correlation function with no vertex corrections in DMFT!

So the last task we have is to perform the analytic continuation from the imaginary frequency axis to the real frequency axis. This is done by using a variant of Cauchy's theorem [Mahan (1990)]. We first write the summation over Matsubara frequencies as an integral over the contour C

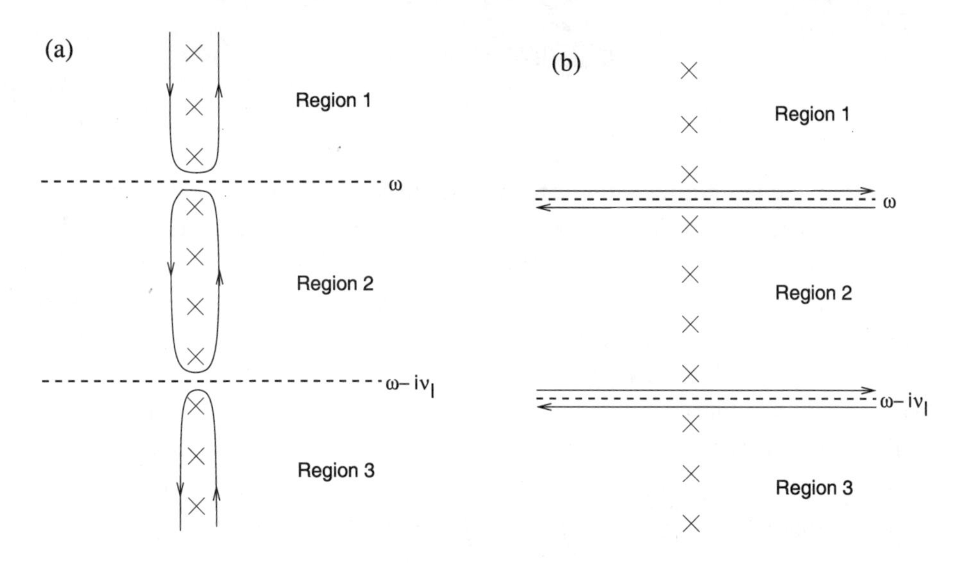

Fig. 2.9 Contours used in the analytic continuation: (a) three contours needed for the Matsubara frequency summation and (b) contours deformed to four lines parallel to the real axis. The Green's functions involve either $G(z)$ or $G(z+i\nu_l)$. The analytic functions are G^R in region 1 and G^A in regions 2 and 3 for the former, and G^R in regions 1 and 2 and G^A in region 3 for the latter.

shown in Fig. 2.9(a) which has contributions at the poles of the Fermi distribution which lie at the Fermionic Matsubara frequencies (the residue of the pole is $-T$). The contours are then deformed to lines parallel to the real axis, with the Green's functions evaluated with either retarded or advanced functions depending on the argument, and the regions of analyticity (since the functions are all analytic, there are no additional poles). The result is

$$\begin{aligned}
&\Pi_{ab}^{jj}(i\nu_l) = \\
&\frac{1}{2\pi i}\int_{-\infty}^{\infty} d\omega f(\omega)\sum_{\mathbf{k}} \mathbf{v}_{\mathbf{k}a}\mathbf{v}_{\mathbf{k}b}[G_{\mathbf{k}}^R(\omega) - G_{\mathbf{k}}^A(\omega)]G_{\mathbf{k}}^R(\omega + i\nu_l) \\
&+ \frac{1}{2\pi i}\int_{-\infty}^{\infty} d\omega f(\omega - i\nu_l)\sum_{\mathbf{k}} \mathbf{v}_{\mathbf{k}a}\mathbf{v}_{\mathbf{k}b}G_{\mathbf{k}}^A(\omega - i\nu_l)[G_{\mathbf{k}}^R(\omega) - G_{\mathbf{k}}^A(\omega)].
\end{aligned} \tag{2.144}$$

The analytic continuation (which is unique) is performed by first rewriting $f(\omega - i\nu_l) = f(\omega)$, then taking $i\nu_l \to \nu + i\delta$ and shifting the integration variable $\omega \to \omega + \nu$ in the second integral. Then, using the definition for σ_{ab}, we finally arrive at

$$\begin{aligned}
\sigma_{ab}(\nu) = \frac{e^2}{2\pi\nu}\int_{-\infty}^{\infty} d\omega \sum_{\mathbf{k}\sigma} \mathbf{v}_{\mathbf{k}a}\mathbf{v}_{\mathbf{k}b}\mathrm{Re}\Big\{ & f(\omega)G_{\mathbf{k}\sigma}(\omega)G_{\mathbf{k}\sigma}(\omega + \nu) \\
& - f(\omega + \nu)G_{\mathbf{k}\sigma}^*(\omega)G_{\mathbf{k}\sigma}^*(\omega + \nu) \\
& - [f(\omega) - f(\omega + \nu)]G_{\mathbf{k}\sigma}^*(\omega)G_{\mathbf{k}\sigma}(\omega + \nu)\Big\}.
\end{aligned} \tag{2.145}$$

We can perform the summation over $\mathbf{k}$ directly. Because $\epsilon_{\mathbf{k}}$ is an even function of $\mathbf{k}$ and $\mathbf{v}_{\mathbf{k}}$ is odd, we must have $a = b$. The average of $\sin^2 \mathbf{k}_\alpha$ times a function of $\epsilon_{\mathbf{k}}$ over the Brillouin zone turns out to be equal to the average of $\cos^2 \mathbf{k}_a$ times the same function of $\epsilon_{\mathbf{k}}$ (which can be related to the average of $\epsilon_{\mathbf{k}}^2$ because the average of $\cos \mathbf{k}_a \cos \mathbf{k}_b$ is a $1/d$ correction). The net effect is an extra factor of $1/2$. The summation over $\mathbf{k}$ can be written as an integral over energy with a weighting factor of $\rho(\epsilon)a^2t^{*2}/2d$. This yields for the optical conductivity

$$\begin{aligned}
\sigma_{ab}(\nu) = \frac{e^2a^2t^{*2}}{4\pi d}\delta_{ab}\int_{-\infty}^{\infty} d\omega \int_{-\infty}^{\infty} d\epsilon\rho(\epsilon)\frac{f(\omega) - f(\omega+\nu)}{\nu} \\
\times \mathrm{Re}\Big\{G_{\mathbf{k}}(\omega)G_{\mathbf{k}}(\omega + \nu) - G_{\mathbf{k}}^*(\omega)G_{\mathbf{k}}(\omega + \nu)\Big\}.
\end{aligned} \tag{2.146}$$

The integrand actually involves just the product of the imaginary parts of the two Green's functions. If we recall the definition of the spectral function, and we define the constant $\sigma_0 = e^2\pi^2/2hda^{d-2}$ (which has all of the dimensionful constants restored), then the conductivity becomes

$$\sigma_{ab}(\nu) = \frac{\sigma_0}{2}\delta_{ab}\int_{-\infty}^{\infty} d\omega \int_{-\infty}^{\infty} d\epsilon\rho(\epsilon)\frac{f(\omega)-f(\omega+\nu)}{\nu}A_{\mathbf{k}}(\omega)A_{\mathbf{k}}(\omega+\nu). \tag{2.147}$$

Using the fact that $\mathrm{Im}z = (z-z^*)/2i$, and expanding the integrand by partial fractions allows the integral to be performed over ϵ

$$\sigma_{ab}(\nu) = \frac{\sigma_0}{4\pi^2}\delta_{ab}\int_{-\infty}^{\infty} d\omega\frac{f(\omega)-f(\omega+\nu)}{\nu} \tag{2.148}$$
$$\times\ \mathrm{Re}\left[-\frac{G(\omega)-G(\omega+\nu)}{\nu+\Sigma(\omega)-\Sigma(\omega+\nu)}+\frac{G^*(\omega)-G(\omega+\nu)}{\nu+\Sigma^*(\omega)-\Sigma(\omega+\nu)}\right].$$

The final step for the *dc*-conductivity is to take the limit of $\nu \to 0$. Using the facts that

$$\lim_{\nu\to 0}\frac{f(\omega)-f(\omega+\nu)}{\nu} = -\frac{df(\omega)}{d\omega}, \tag{2.149}$$

and

$$\lim_{\nu\to 0}\frac{G(\omega)-G(\omega+\nu)}{\nu+\Sigma(\omega)-\Sigma(\omega+\nu)} = -2+2[\omega+\mu-\Sigma(\omega)]G(\omega), \tag{2.150}$$

produces our final result

$$\sigma(0) = \sigma_0\int_{-\infty}^{\infty} d\omega\left(-\frac{df(\omega)}{d\omega}\right)\tau(\omega), \tag{2.151}$$

with the exact many-body relaxation time $\tau(\omega)$ defined by

$$\tau(\omega) = \frac{1}{4\pi^2}\left(\frac{\mathrm{Im}G(\omega)}{\mathrm{Im}\Sigma(\omega)}+2-2\mathrm{Re}\{[\omega+\mu-\Sigma(\omega)]G(\omega)\}\right). \tag{2.152}$$

Our calculations have been performed for the hypercubic lattice. All of the steps are the same for other lattices up to Eq. (2.145) except there is an additional $\mathbf{v}_{\mathbf{k}}^2$ factor. In cases where we need to take into account the cross terms $\cos\mathbf{k}_a\cos\mathbf{k}_b$, the analysis is even more complex. Details are worked out explicitly for the simple cubic lattice in Problem A.16.

We examine the relaxation time for a Mott insulator phase for $U = 3/\sqrt{2}$ on the hypercubic lattice and for $U = 3$ on the Bethe lattice (which correspond to similar insulating phases) for a range of w_1 values and $T = 0$ in Fig 2.10. Note the unphysical behavior for the hypercubic lattice. The relaxation time behaves like a power law in the "gap region" because the DOS is exponentially small, but the lifetime of the states is exponentially

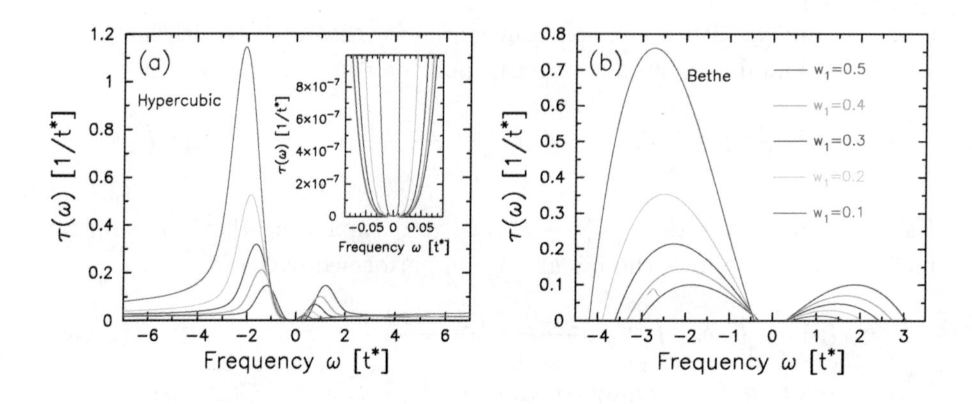

Fig. 2.10 Exact many-body relaxation time for similar Mott insulators with (a) $U = 3/\sqrt{2}$ on a hypercubic lattice and (b) $U = 3$ on a Bethe lattice [the legend in panel (b) also applies to panel (a)]. The hypercubic lattice has a number of unphysical behaviors due to the fact that it has an infinite bandwidth, and many of the "rare" states have long lifetimes, which can contribute significantly to the transport. This can be seen in the inset to panel (a) where the ω^4 behavior of $\tau(\omega)$ is clearly seen in the "pseudogap region".

large, so they can carry current. Also, the relaxation time does not go to zero at large frequency for the same reason. The Bethe lattice behaves more as is expected: the gap region is well defined with $\tau(\omega) = 0$ within the gap, and the relaxation time vanishes outside the band.

In addition to the *dc*-conductivity, we are also interested in thermal transport. Since electrons carry energy with them as they move through the lattice, they transport heat as well as charge. Since the weighting factor that determines the energy carried by an electron is different from the weighting factor that counts the number of electrons that move through the lattice, one can transport heat in the absence of a particle current and *vice versa*. The two thermal transport quantities we want to calculate are the thermopower S which is the thermal analog of the Hall effect: since the weighting factors for the particle and heat currents differ, as described above, we can have a situation where there is no particle current flow, but there is heat current flow. So we can apply a thermal gradient over an isolated piece of material (that carries no charge current) and measure a voltage. The ratio of the voltage difference to the temperature difference is given by the thermopower. In addition, we can examine the electronic contribution to the thermal conductivity κ_e. Like electrical current, which flows when there is a voltage difference, heat current flows when there is a temperature difference. Real materials have two carriers for heat current —

the electrons and the phonons. In this book we will not consider the phonon contribution, but it often can contribute significantly at low temperatures.

The standard approach to charge and thermal transport is to relate these experimental quantities to relevant particle-current–particle-current (L_{11}), particle-current–heat-current ($L_{12} = L_{21}$), and heat-current–heat-current (L_{22}) correlation functions. We have already determined L_{11} which satisfies $\sigma_{dc} = e^2 L_{11}$. The thermopower and thermal conductivity satisfy

$$S = -\frac{k_B}{|e|T}\frac{L_{12}}{L_{11}}, \quad \kappa_e = \frac{k_B^2}{T}\left[L_{22} - \frac{L_{12}L_{21}}{L_{11}}\right]. \tag{2.153}$$

A theorem by Jonson and Mahan [Jonson and Mahan (1980); Jonson and Mahan (1990)], states that if one can determine the exact many-body relaxation time, then the thermal coefficients satisfy the so-called Mott non-interacting form [Chester and Thellung (1961)]

$$L_{ij} = \frac{\sigma_0}{e^2}\int_{-\infty}^{\infty} d\omega \left(-\frac{df(\omega)}{d\omega}\right)\tau(\omega)\omega^{i+j-2}. \tag{2.154}$$

This result implies that we don't even need to determine the heat-current operator in order to calculate the thermal transport; we only need to know the relaxation time for the charge transport multiplied by the appropriate power of frequency. What is remarkable is that one can perform a brute-force proof of the Jonson-Mahan theorem for the Falicov-Kimball model in DMFT, by exactly calculating all relevant correlation functions and showing they add up to reproduce Eq. (2.154) [Freericks and Zlatić (2001)]. The details are quite technical and will not be reproduced here. Instead, a sketch of the original Jonson-Mahan proof, as formulated in [Mahan (1998)] is presented in Prob. A.17.

We end this section with a brief discussion about optical sum rules. The optical conductivity in Eq. (2.147) satisfies a sum rule. The integral of $\sigma(\nu)$ from 0 to ∞ is proportional to the average kinetic energy [Maldague (1977)]. This is important because it serves as a useful test of the numerics. If we independently calculate the average kinetic energy and the optical conductivity, then we can verify that the sum rule holds. This is a very useful step in ensuring that there are no errors in one's computational codes. Unfortunately it is a step that is often neglected by many researchers. We urge all readers of this book to perform this check when they are calculating transport properties.

2.9 Metal-Insulator Transitions within DMFT

The strongly correlated metal-insulator transition is one of the most thoroughly studied problems in condensed matter physics. Even so, its solution has remained elusive in three dimensions. DMFT has shed light onto this problem by illustrating a number of ways in which the transition occurs, and by describing a number of different scenarios for the transition itself.

Fermi liquid theory [Landau (1956)] is one of the hallmarks of condensed matter physics. It says that if the interactions are weak, then a fraction of the Fermi-like quasiparticles of the noninteracting system remain in the interacting system, with minor changes: their lifetime is still infinite at the Fermi energy (and $T = 0$), but their mass can be modified, and their spectral weight can be reduced from 1, with the remainder of the weight being pushed to higher-energy incoherent excitations. This is the basis of much of band theory, which neglects the electron-electron interactions, and only includes the electron-ion interactions. In DMFT there are two classes of metals: (i) Fermi-liquid metals, which have the same Fermi surface as the noninteracting system has (from the Luttinger theorem [Luttinger (1962)] and the fact that the self-energy has no momentum dependence) and the DOS at the Fermi energy is unchanged from the noninteracting value at $T = 0$ [Müller-Hartmann (1989a); Müller-Hartmann (1989b); Müller-Hartmann (1989c)] and (ii) a non-Fermi-liquid metal, which has a finite lifetime at the Fermi surface (and $T = 0$), and whose DOS at the Fermi energy may be modified from the noninteracting value (when $T = 0$) because the Luttinger theorem no longer applies. The Hubbard model and the periodic Anderson model fit into the former category, while the Falicov-Kimball model fits into the latter.

As the interactions increase, if we have, on average, one electron per lattice site, then the electron-electron interaction can localize the electrons by freezing out the double-occupancies on each lattice site due to a large Coulomb repulsion [Gebhard (1997)]. This is seen in the DOS by it being suppressed to zero at the chemical potential and $T = 0$. On lattices that have a finite bandwidth, like the Bethe lattice, a true gap can open as the interactions are increased further. While on lattices that have an infinite bandwidth, like the hypercubic lattice, the DOS can only vanish right at the chemical potential (it is exponentially suppressed elsewhere within a "gap region") and the system only has a pseudogap. Since the DOS vanishes at the pseudogap, and since the noninteracting DOS has an infinite bandwidth, it is clear from the Hilbert transform, that the only way to get a vanishing DOS is for the Green's function itself to vanish, which implies the self-energy is infinitely large. This occurs from the formation of a pole in the self-energy at a critical value of the interaction strength $U_{\mathrm{c}}^{\mathrm{pole}}$ (see Problem A.13). At half filling, it turns out that the MIT occurs when a pole forms in the self-energy on the Bethe lattice as well, but away from half filling, in a particle-hole asymmetric MIT, the critical interaction strength for gap formation $U_{\mathrm{c}}^{\mathrm{gap}}$, is always less than the critical interaction strength for pole formation [Demchenko, Joura and Freericks (2004)] (see Problem A.19).

The Hubbard model is probably the most studied problem for a strongly correlated MIT. The $U = 0$ state is always metallic, while the $U = \infty$ state is insulating at half filling, since there is one electron per site and no double occupancy allowed. The fundamental questions are: (i) what is the critical value of U for the MIT and (ii) is the transition continuous or discontinuous? There are only two limits where this problem has been solved exactly. In one-dimension, there is a Bethe-ansatz solution [Lieb and Wu (1968); Gebhard (1997)], which shows that the critical value of U is $U \to 0^+$, so it is difficult to study the behavior near the critical interaction strength. In infinite-dimensions, there is extensive numerical work, which culminated in the NRG analysis at $T = 0$ [Bulla (1999)], which explicitly showed the evolution of the MIT, confirming the qualitative features of an earlier perturbative analysis [Zhang, *et al.* (1993); Georges, *et al.* (1996)], which has two critical values of U, one where the insulating solution $U_{\mathrm{c}}^{\mathrm{ins}}$ becomes unstable, and one where the metallic solution becomes unstable $U_{\mathrm{c}}^{\mathrm{met}}$. Since $U_{\mathrm{c}}^{\mathrm{ins}} < U_{\mathrm{c}}^{\mathrm{met}}$, there is a coexistence region, where both metallic and insulating phases can exist, but the metallic phase is the global minimum of the free energy. Hence the MIT occurs at $U_{\mathrm{c}}^{\mathrm{met}}$. The "gap region" appears to have a discontinuous jump at the transition, but the numerical results are unable to resolve exponentially small DOS at finite frequency ω, which are known to exist on the hypercubic lattice and may be required by Fermi-liquid theory just before the transition [Kehrein (1998)].

We summarize the results of [Jarrell (1992); Zhang, *et al.* (1993); Georges, *et al.* (1996); Bulla (1999); Bulla, Costi and Vollhardt (2001)] and some previously unpublished results [Bulla (unpublished)] in a series of figures for the hypercubic lattice where $U_{\mathrm{c}} \approx 4.1$. In Fig. 2.11, we plot the DOS on the hypercubic lattice near the MIT. The DOS displays both upper and lower Hubbard bands, and in the metallic phase it has a narrow quasiparticle resonance, with the DOS at $\omega = 0$ pinned to the noninteracting value. In the lower set of panels, we plot the imaginary part of the

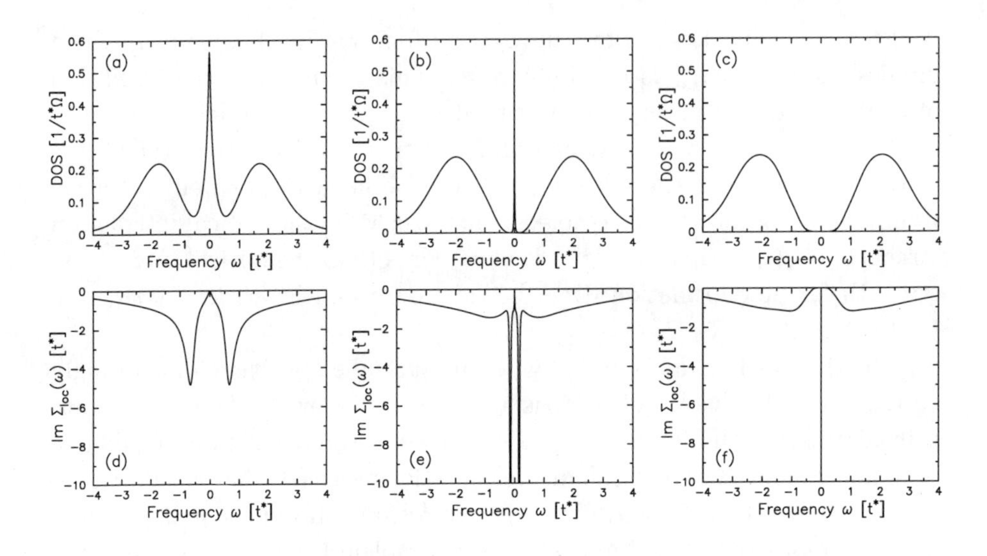

Fig. 2.11 $T = 0$ DOS at half filling for the Hubbard model on a hypercubic lattice with (a) $U = 0.8U_c$, (b) $0.99U_c$, and (c) $1.1U_c$. Note how the DOS remains at the noninteracting value in the metallic phase, and how there are well-defined upper and lower Hubbard bands before the transition. The transition occurs by the width of the quasiparticle peak shrinking to zero and disappearing. The insulator appears to have a well-defined gap at the transition. In the lower panels [(d–f)], we plot the imaginary part of the self-energy, which shows a narrow region with Fermi-liquid like behavior (quadratic in ω and vanishing at $\omega = 0$ [in panels d and e]) which gives way to a pole at $\omega = 0$ characterized by the appearance of a delta function [in panel f]. *Adapted with permission from* [Bulla (1999)] (original figure © 1999 the American Physical Society).

self-energy for the same values of interaction strength, which show the evolution from a Fermi liquid to a Mott insulator. Recent work [Karski, Raas and Uhrig (2005)], employing the density-matrix renormalization group to solve for the dynamics of the impurity problem, indicates that there is some additional sharp structure in the DOS near the band-gap edges in the correlated metal close to the critical U for the MIT; this structure is tied to a collective effect between the Fermi-liquid excitations and the incoherent excitations across the gap. This additional structure cannot be resolved with NRG.

We show the finite-temperature dependence of the DOS in Fig. 2.12 just below the critical value of U. Note how the behavior is quite anomalous for a metal. The DOS initially decreases as the temperature is lowered, until about $T = 0.02$, where it discontinuously jumps and increases to ultimately form a quasiparticle peak, which saturates at the noninteracting value as $T \to 0$. This behavior will lead to significant anomalies in the transport

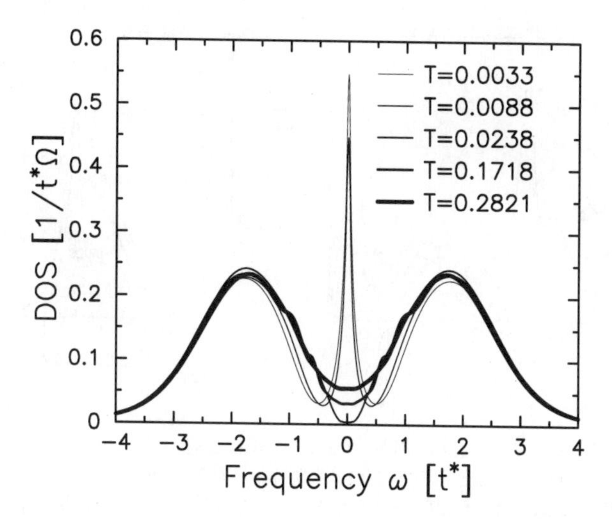

Fig. 2.12 DOS at half filling for the Hubbard model on a hypercubic lattice at $U = 3.54 = 0.86U_c$ and various temperatures. Note how the DOS has a dip at the chemical potential for high temperature, which initially deepens as T is lowered, and then has a sudden, discontinuous increase as the system is cooled below the Fermi-liquid coherence temperature and the quasiparticle gap is formed [Bulla (unpublished); *adapted with permission from data used in* Freericks, Devereaux and Bulla (2001)].

properties for temperatures above the coherence temperature, where the Fermi peak starts to form.

In the left panel of Fig. 2.13, we plot the DOS as the system is doped away from half filling for U slightly above the Mott transition value. As the system is doped, the chemical potential rapidly moves into the lower Hubbard band, and a quasiparticle peak is superimposed on the lower Hubbard band structure. In the right panel, we show the temperature evolution for $\rho_e = 0.915$. It shows how the peak sharpens and develops as T is lowered.

In summary, the Hubbard model displays rich physics near the Mott transition, that depends crucially on the temperature, the doping, and the interaction strength. In general, the ground state is a Fermi liquid on the metallic side, and the system is always metallic when doped away from half filling. But the Fermi temperature, where the quasiparticle peak develops gets pushed down toward $T = 0$ as the MIT is approaches, which means that the metallic phase can display quite anomalous behavior. As the quasiparticle peak disappears, by having its width reduced to zero, the insulating phase appears to have a well-formed "gap region" already in place. This

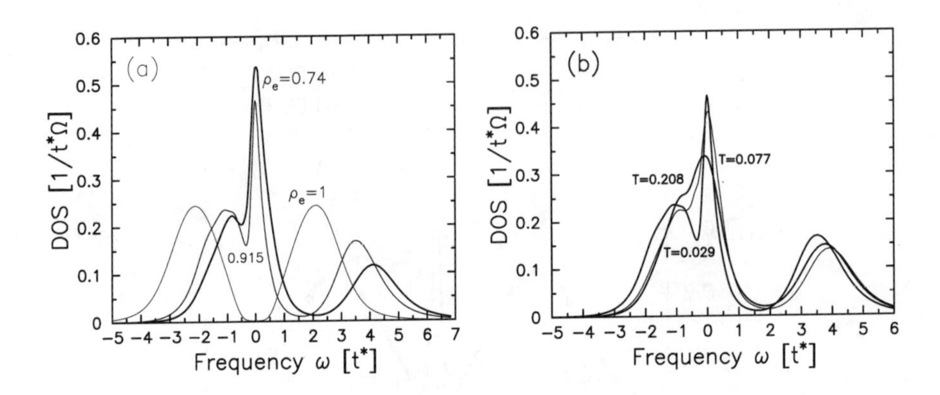

Fig. 2.13 DOS away from half filling for the Hubbard model on a hypercubic lattice with $U = 4.24 = 1.03U_c$ (half-filling). The left panel (a) shows the evolution with doping away from half filling, with a fixed temperature $T = 0.029$, and the right panel (b) shows the evolution with temperature at a fixed filling $\rho_e = 0.92$. *Adapted with permission from* [Freericks, *et al.* (2003a)] (original figure © 2003 the American Physical Society).

is a remnant of the coexistence of the insulating and metallic phases for $U_c^{\text{ins}} < U < U_c^{\text{met}}$.

It is also important to examine situations where the MIT does not occur at the particle-hole symmetric point, since nearly all real materials are not particle-hole symmetric [Hirsch (1993)]. There are two ways to break the particle-hole symmetry: (i) add next-nearest neighbor hopping, which breaks the bipartite symmetry of the lattice, or add extra bands that will break the particle-hole symmetry; or (ii) modify the model to make it explicitly particle-hole asymmetric. For illustrative purposes, we choose to examine the second option here. We study the Falicov-Kimball model with $\rho_e + w_1 = 1$, but $w_1 \neq 1/2$. Since there is on average one particle per site, and since the mobile electrons have a Coulomb interaction with the localized electrons, then if the interaction is strong enough, the electrons will be frozen on the lattice, and unable to conduct. Does the breaking of particle-hole symmetry change the character of the MIT? It does not do so on the hypercubic lattice, because on the hypercubic lattice, the only way to form a pseudogap is to have the self-energy develop a pole, so the MIT and the pole formation in the self-energy always occur at the same critical value of U. But on any lattice with a finite bandwidth, like the Bethe lattice, we find that the opening of a true gap occurs *before* a pole forms in the self-energy.

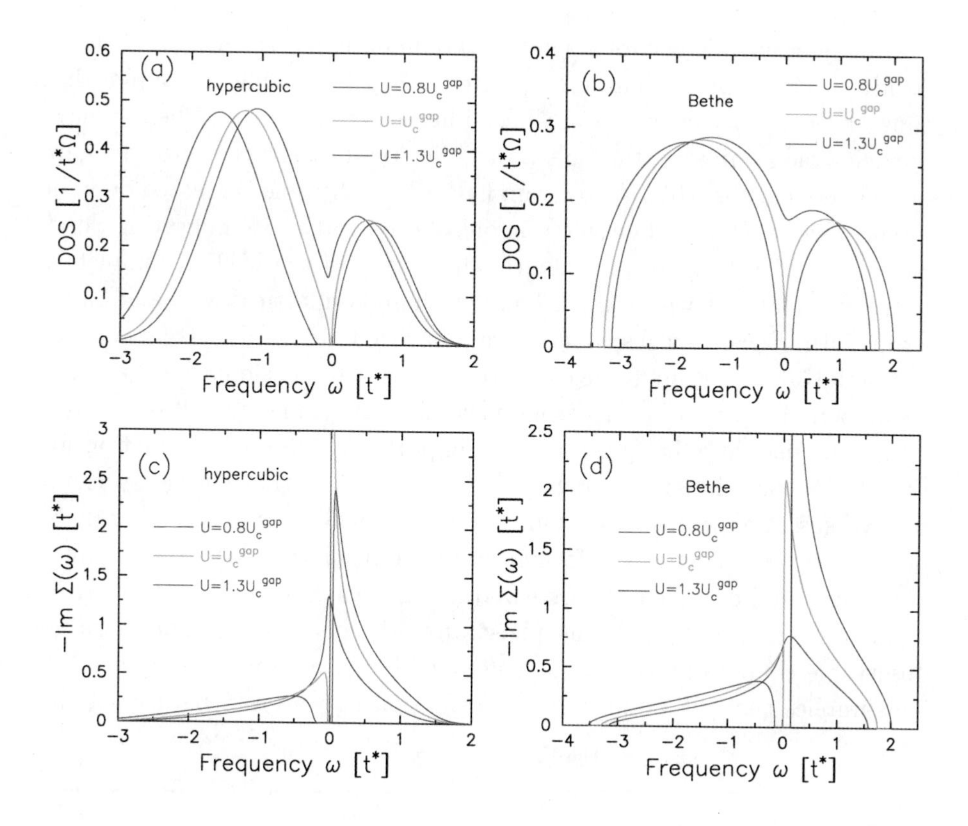

Fig. 2.14 MIT scenarios for the Falicov-Kimball model away from half filling on the hypercubic (left panels) and the Bethe lattice (right panels). Panels (a) and (b) show the $T = 0$ DOS at $w_1 = 0.25$ and $\rho_e = 0.75$ for three values of U: ($U = 0.8U_c^{\text{gap}}$ [red], $U = U_c^{\text{gap}}$ [green] and $U = 1.3U_c^{\text{gap}}$ [blue]); in the last case $U > U_c^{\text{pole}}$ for the Bethe lattice. The corresponding imaginary parts of the self-energies are shown in panels (c) and (d). Note that the chemical potential lies at $\omega = 0$ on these curves. The pole for $U = 1.3U_c^{\text{gap}}$ on the Bethe lattice [the blue curve in panel (d)] is so close to the right band edge, that it cannot be separately distinguished in the plot. *Panels (c) and (d) adapted with permission from* [Demchenko, Joura and Freericks (2004)] (original figure © 2004 the American Physical Society).

In Fig. 2.14, we summarize the results for the MIT at $w_1 = 0.25$ on the hypercubic and Bethe lattices, where $U_c^{\text{gap}} = U_c^{\text{pole}} \approx 1.633$ on the hypercubic lattice and where $U_c^{\text{gap}} \approx 1.908$ and $U_c^{\text{pole}} \approx 2.309$ on the Bethe lattice. It is interesting to note that on the hypercubic lattice, the MIT occurs at larger U values as w_1 is reduced from $1/2$, while on the Bethe lattice, the MIT occurs at smaller U values as w_1 is reduced from $1/2$, but the pole formation occurs at higher values of U. So there is a region in U where the system is a Mott insulator, but the pole has not yet developed in the self-energy; this occurs whenever the noninteracting DOS has a finite bandwidth. The phase diagram is plotted in Fig. 2.15 along with a plot that shows how the pole evolves with U for different values of w_1. When we have particle-hole symmetry, the pole always lies right at the center of the gap, and appears at the MIT. Away from half filling, the pole is created only for larger values of U, and it enters from one of the band edges, migrating closer to the center as U increases. Hence, the nature of the MIT is completely different on lattices with a finite bandwidth and on lattices with an infinite bandwidth. Because real systems always have a finite bandwidth, we expect the Bethe lattice results to be closer to real three-dimensional systems than the hypercubic lattice results when particle-hole symmetry is broken.

So one may ask the question, how important is the pole formation for the MIT? Does the system behave differently when the self-energy has a pole versus when it does not? On the hypercubic lattice it definitely does, because the pole signals the formation of the insulating (or semi-metallic) phase, but on the Bethe lattice, we find no significant change in the behavior of the system in the insulating phase after the pole forms. There are no observable changes in any of the common charge or transport properties. This implies that although the pole formation appeared to be driving the

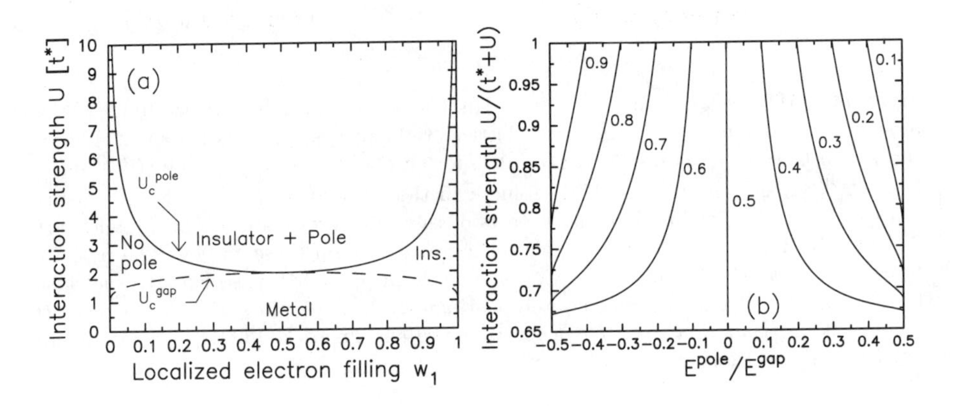

Fig. 2.15 Phase diagram showing the three regimes on the Bethe lattice as a function of U and w_1 in panel (a): (i) the metallic region (below the dashed line); (ii) the insulating region with no pole in the self-energy (between the solid and dashed lines); and (iii) the insulating region with a pole in the self-energy (above the solid line). In panel (b), we plot the relative location of the pole within the gap as a function of U. The curves are for different values of w_1 and they asymptotically approach $1/2 - w_1$ as $U \to \infty$. Note how the $w_1 = 0.5$ case has qualitative differences with all other cases. *Reprinted with permission from* [Demchenko, Joura and Freericks (2004)] (© 2004 the American Physical Society).

Mott physics at half filling, the behavior changes away from half filling, and it is likely that the pole formation is not a defining property of the Mott phase, it only happens to appear at some critical value of U and it has limited effects on the character of the insulating phase.

2.10 Bulk Charge and Thermal Transport

We begin our discussion with bulk transport in the Falicov-Kimball model [Freericks, *et al.* (2003b); Joura, Demchenko and Freericks (2004)]. As discussed above, the transport properties are likely to be different on the hypercubic and the Bethe lattices because the relaxation time $\tau(\omega)$ has anomalous behavior on the hypercubic lattice—it goes to zero like ω^4 in the "insulating" phase and it goes to a constant for large $|\omega|$. We expect more normal behavior on the Bethe lattice which has a physically correct relaxation time (vanishing inside the gap, and outside the band edges). In our model, the conduction electrons scatter off of a fixed concentration of local particles with a given interaction strength U^{FK}. But the scattering is treated in an annealed statistical ensemble, which cannot be thought of as an independent scattering model, where the thermopower would be independent of the concentration of scatterers (because the L_{11} and L_{12} coefficients would each be proportional to the concentration of scatterers, and hence their ratio would be independent of the scatterer concentration). Instead, we will see strong variations in the thermopower as a function of the concentration of scatterers.

In Fig. 2.16, we plot the conductivity versus temperature for the case of half-filling $\rho_e = w_1 = 0.5$ on the hypercubic lattice for a variety of different U values. Note how the conductivity at $T = 0$ continuously goes to zero as we pass through the Mott transition at $U = \sqrt{2}$. The behavior for weak scattering is a constant conductivity versus temperature at low T, which is expected for scattering off of static defects. As we approach the MIT, the conductivity starts to rise as T increases, similar to that of an insulator, even though it remains metallic down to $T = 0$. This occurs, in part, because the system has a strong dip in the DOS at the Fermi level on the metallic side of the MIT; hence heating the sample provides more phase space for particles to transport current.

In Fig. 2.17, we plot the logarithm of the conductivity versus inverse temperature on the hypercubic and Bethe lattices for similar values of U. The corresponding relaxation time was already plotted in Fig. 2.10. Since

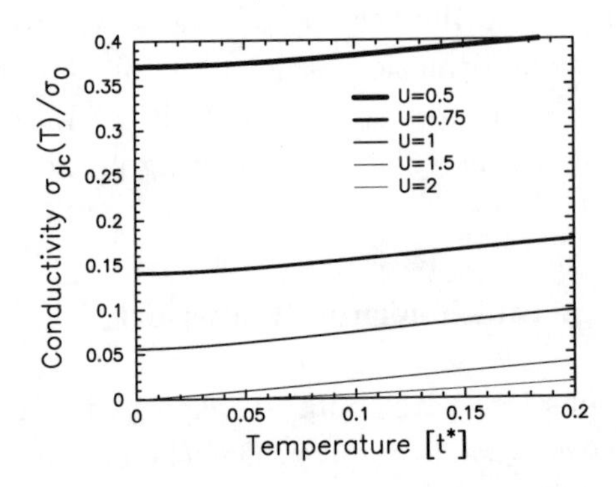

Fig. 2.16 DC conductivity versus T for the Falicov-Kimball model at half-filling ($\rho_e = w_1 = 0.5$) on a hypercubic lattice. We show results for $U = 0.5$, 0.75, 1, 1.5, and 2. The Mott transition occurs at $U = \sqrt{2}$. Note how the conductivity goes to zero continuously as we approach the MIT and how the metallic phase has anomalous behavior (of increasing the conductivity as T increases) for a wide range of U values because the scattering is quite strong, even at $U = 0.5$.

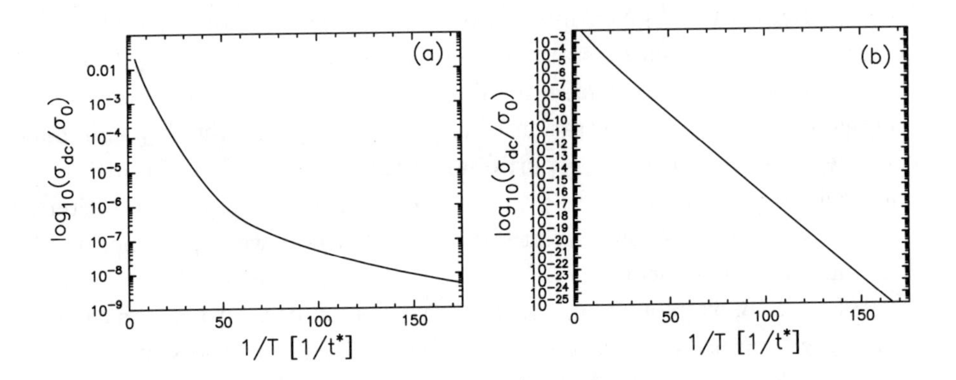

Fig. 2.17 Arrhenius plot of the logarithm of the DC conductivity versus $1/T$ for the Falicov-Kimball model at half-filling ($\rho_e = w_1 = 0.5$). Panel (a) is for the hypercubic lattice with $U = 3/\sqrt{2}$ and panel (b) is for the Bethe lattice with $U = 3$. The relaxation time appears in Fig. 2.10. Note how the conductivity does not have an Arrhenius form for the hypercubic lattice, but it does for the Bethe lattice (with an activation energy of 0.315 which is approximately equal to the barrier height of 0.306).

the hypercubic lattice has long-lived states close to the Fermi level and at large energies, the relaxation time has anomalous behavior, which translates into anomalies in the transport. In particular, we do not see exponentially activated behavior in the conductivity on the hypercubic lattice because of the quartic behavior of the relaxation time at low frequency; instead, the conductivity behaves like T^4 for low temperature in the insulating phase. On the Bethe lattice, where the relaxation time is identically equal to zero within the gap, we do see nice activated behavior for temperatures well below the gap with an activation energy approximately equal to half of the gap in the DOS (since the chemical potential lies at the center of the gap).

We encounter even more interesting behavior when we examine the thermal transport. In Fig. 2.18, we plot the thermopower for the cases $\rho_e = 1 - w_1$ on the hypercubic lattice. In the metallic case [$U = 1$, panel (a)], the behavior is as one would expect for a metal with scattering. The thermopower is larger the more particle-hole asymmetric the system is, and it vanishes linearly with T as $T \to 0$, due to the shrinking of the Fermi window. When we go to a correlated insulator [$U = 2$, panel (b)], the behavior looks similar at high temperatures, but is quite different at low temperature. It develops a sharp low-temperature peak for some values of w_1. The origin of this peak is most likely arising from a sharp T-dependence of the chemical potential at temperatures smaller than the "gap". As the chemical potential changes sharply with T, the thermopower also changes sharply, because the slope of the relaxation time grows as the chemical potential moves away from the center of the gap. This low-temperature peak in the thermopower can be important for applications if it gives rise to a large figure of merit (see below). In both cases, the thermopower curves

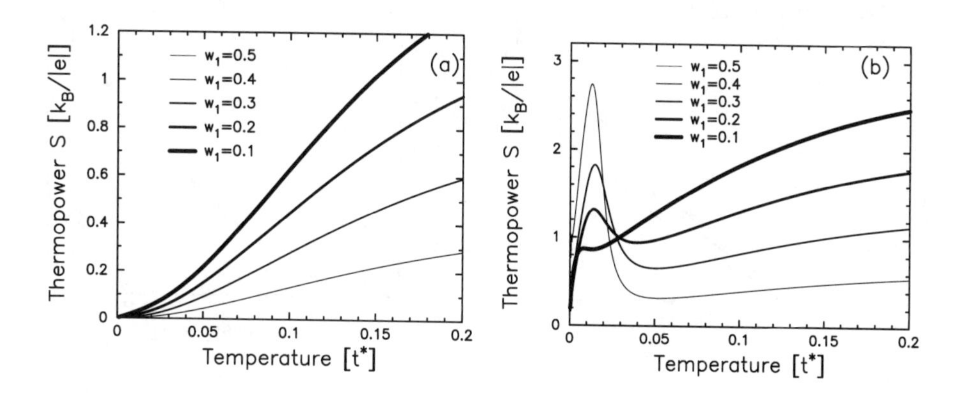

Fig. 2.18 Thermopower versus T for the Falicov-Kimball model with $\rho_e = 1 - w_1$ and various values of w_1 (0.1, 0.2, 0.3, 0.4, and 0.5) on the hypercubic lattice. Panel (a) is for a metallic phase with $U = 1$, while panel (b) is for an insulating phase with $U = 2$. The thermopower vanishes when $w_1 = 0.5$ due to particle-hole symmetry; it is largest at high T for the most asymmetric cases ($w_1 = 0.1$ here). Note the low-temperature peak in the insulating case [panel (b)].

have a broad peak at slightly higher temperatures (between 0.25 and 0.4; not shown here). The maximal thermopower can be quite large in both cases (up to about $3k_B/|e|$), and it tends to increase as the correlations increase.

In Fig. 2.19, we plot the electronic contribution to the thermal conductivity at the critical interaction strength for the metal-insulator transition (at $w_1 = 0.5$) on the Bethe lattice [$U = 2$, panel (a)] and on the hypercubic lattice [$U = \sqrt{2}$, panel (b)]. The behavior of the thermal conductivity is similar on both lattices at low temperature and does not have strong dependence on w_1, even though most of the Bethe lattice curves are insulators, while most of the hypercubic lattice curves are metals. At higher temperature (not shown), the behavior differs. On the Bethe lattice, the thermal conductivity has a peak around 0.3 and then slowly decreases. On the hypercubic lattice, the thermal conductivity continues to increase because of the anomalous behavior of the relaxation time for large frequencies.

The relative efficiency of a thermoelectric cooler is measured with its figure of merit ZT, which is defined to be

$$ZT = \frac{T\sigma_{dc}S^2}{\kappa_e + \kappa_l}, \tag{2.155}$$

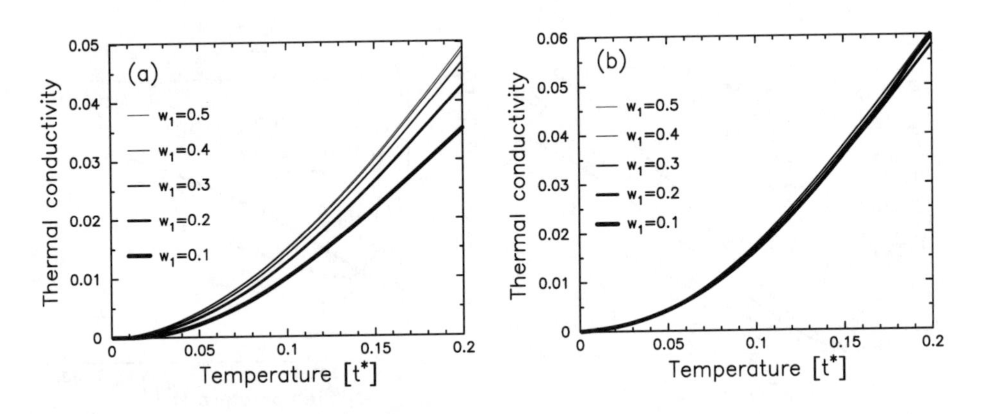

Fig. 2.19 Electronic contribution to the thermal conductivity versus T for the Falicov-Kimball model with $\rho_e = 1 - w_1$ and various values of w_1 (0.1, 0.2, 0.3, 0.4, and 0.5) at the critical interaction strength for the metal-insulator transition at $w_1 = 0.5$. Panel (a) is for the Bethe lattice with $U = 2$, where most of the fillings are insulators, while panel (b) is for the hypercubic lattice with $U = \sqrt{2}$, where all of the fillings except $w_1 = 0.5$ are metals. Note how there is little dependence of the thermal conductivity on the concentration of scatterers here.

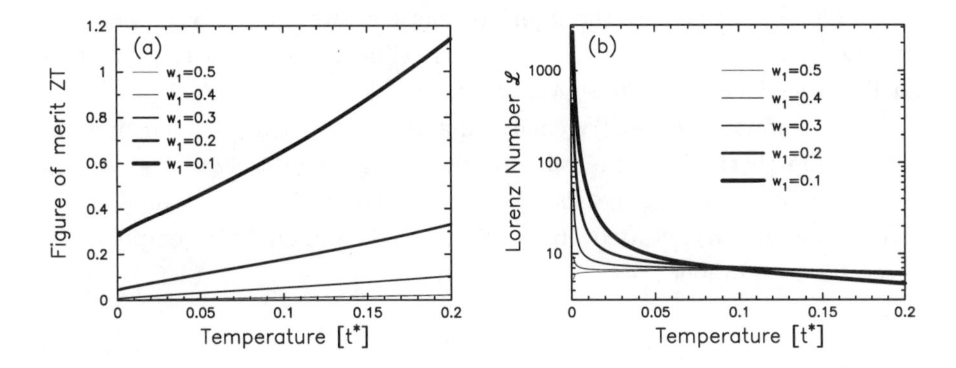

Fig. 2.20 (a) Figure of merit versus T and (b) Lorenz number versus T (semi-logarithmic plot) for the Falicov-Kimball model with $\rho_e = 1 - w_1$ and various values of w_1 (0.1, 0.2, 0.3, 0.4, and 0.5) at the critical interaction strength for the metal-insulator transition at $w_1 = 0.5$ on the Bethe lattice ($U = 2$). The figure of merit vanishes for $w_1 = 0.5$ because $S = 0$ due to particle-hole symmetry.

which depends on the sum of the electronic (κ_e) and lattice (κ_l) thermal conductivities. In our calculations, we will examine the electronic version of the figure of merit, which sets $\kappa_l = 0$; this approach always overestimates the figure of merit, but should be reasonable whenever the electronic thermal conductivity dominates over the lattice thermal conductivity (ordinarily this will be at high temperature). The figure of merit arises from the competition between the transfer of heat that is "dragged" along with an electrical current versus the transfer of heat via standard thermal conduction. For a thermoelectric cooler application, we need to be able to drive the heat to flow from cold to hot, while the thermal conductivity process always goes from hot to cold. If the thermopower is large enough, we can achieve cooling. It is often stated that one needs $ZT > 1$ for a viable device. While it is true that most commercial devices do operate in this regime, one can use devices that operate over narrower temperature ranges effectively with $ZT < 1$. Determining how large ZT must be for any given application is not an easy task. But one good rule of thumb is that the maximal temperature difference between a hot reservoir T_h and a cold reservoir T_c satisfies $T_h - T_c = ZT_c^2/2$, so the size of ZT and the temperature of the cold reservoir both determine how hot the hot reservoir can be, and thereby yield the degree of cooling possible. For example, the figure of merit needs to be pushed up to $ZT > 4$ to be competitive with conventional coolant based refrigerators at room temperature. In the results shown in Fig. 2.20 (a), one can see that as the system is made more particle hole asymmetric, the figure of merit grows. It also grows as T is increased, because the Lorenz number is reduced. In any case, one does not find very large figures-of-merit here.

Wiedemann and Franz [Wiedemann and Franz (1853)] noticed that the ratio of the thermal conductivity to the charge conductivity was a constant for many different metals when compared at the same temperature. Lorenz [Lorenz (1872)] discovered that the ratio was linear in temperature, so the Lorenz number

$$\mathcal{L} = \frac{\kappa_e}{\sigma_{dc} T}, \tag{2.156}$$

is a constant for many different metals. When $\mathcal{L}$ is low, the figure of merit is high, and *vice versa*. Of course, whenever the thermopower has a sign change, the figure of merit is small in that vicinity.

In Fig. 2.20, we plot the figure of merit and the Lorenz number on the Bethe lattice at $U = 2$. The figure of merit does not need to vanish as $T \to 0$ in an insulator. Note how the Lorenz number becomes huge at low temperature in these insulators (the Wiedemann-Franz relation need not hold in an insulator). A large Lorenz number will suppress the figure of merit unless the thermopower grows fast enough to compensate for the reduction.

We now go on to study a different system, where many-body effects are even more important. The Kondo effect [Kondo (1964)] is one of the oldest and most studied many-body physics problems. As early as the 1930s, it was discovered that small concentrations of magnetic impurities in metals often led to a minimum of the resistivity as a function of temperature. Since the resistivity of a conventional Fermi-liquid metal increases monotonically with temperature, the appearance of a minimum was puzzling. In the early 1960s, Kondo showed, via perturbation theory, that a simple model, where the conduction electrons interact with the magnetic impurity spin (also called a local moment), leads to the resistivity minimum (when conventional phonon scattering is included). Anderson [Anderson (1961)] developed a simple model, called the single-impurity Anderson model, to describe the physical behavior for the general case—the Kondo model was a limiting case of the Anderson model when the Coulomb interaction between f-electrons was large, and there was, on average, one f-electron at the impurity site. For years physicists struggled with finding an exact solution to these problems. They finally yielded to a scaling approach of Anderson and Yuval [Anderson and Yuval (1969);

Yuval and Anderson (1970); Anderson, Yuval and Hamann (1970)] followed by a numerical solution by Ken Wilson [Wilson (1975)]. These results were then verified by an explicit Bethe ansatz solution [Andrei (1980); Wiegmann and Tsvelick (1983)].

The physical explanation for the Kondo effect turned out to be rather simple: each impurity system has a characteristic temperature called the Kondo temperature (T_K), which can vary over orders of magnitude in different systems. As the temperature is lowered from high temperature to the Kondo temperature, the resistivity decreases as in a normal metal (when one includes phonon scattering, which is reduced as the phase space for phonons decreases at low T). Once the Kondo temperature is approached, the impurity acts like a strong scatterer, which eventually binds an electron from the conduction sea to form a bound spin singlet at $T = 0$. This strong binding leads to the increase in the resistivity at low temperature. Furthermore, the results of different systems are determined solely by the Kondo temperature, and there is a universal form for the resistivity as a function of T/T_K. (If the phonon scattering was ignored, the high-temperature limit would correspond to disordered spin scattering, and there would be no clear resistance minimum.)

Soon thereafter, a class of materials called heavy Fermions was discovered. These materials include $4f$ or $5f$ electrons in partially filled shells (usually Ce, Yb, or U materials), and they possess strongly renormalized properties, like an effective electron mass that is 1000 times larger than the bare mass. Their transport properties are also unusual. The resistivity often shows a low temperature increase (as T increases) to a broad maximum that slowly decreases (*i. e.*, a resistance maximum). The thermopower often has a sharp peak at low temperatures. These systems are believed to be dense Kondo or Anderson model systems, with an "impurity" at every site. The model that describes these systems is called the periodic Anderson model. Less is known about the solutions of this model, but much progress has been achieved with DMFT using quantum Monte Carlo [Jarrell (1995); Tahvildar-Zadeh, Jarrell and Freericks (1997); Tahvildar-Zadeh, Jarrell and Freericks (1998); Tahvildar-Zadeh, *et al.* (1999)] and NRG [Pruschke, Bulla and Jarrell (2000); Grenzebach, *et al.* (2006)]. The behavior of the dense Kondo systems is quite different from the dilute impurity case. When one is well away from half-filling, but with an f-electron density near 1, there is a high temperature scale, similar to the Kondo temperature, where the localized electron starts to have its spin moment screened. Then there is a second, often much lower, temperature called the coherence temperature, below which the system becomes a Fermi liquid, and the screening of the local moments is complete. Because there aren't enough conduction electrons near the Fermi energy to screen all of the local moments, the screening process must be a many-body collective effect; it is this difficulty that the conduction electrons have in screening that in part causes the coherence temperature to be so low. This physical process is called the principle of exhaustion and was originally proposed by Nozieres [Nozieres (1985); Nozieres (1998)]. In a Fermi liquid, the resistivity grows like T^2 at low temperature, and vanishes at $T = 0$ (in real materials, residual disorder scattering always creates a finite resistivity). As the coherence disappears, the resistivity often decreases again, in the high-temperature limit. So the heavy Fermion compounds often have a resistivity maximum, rather than the minimum seen in the dilute limit. On the other hand, when one is at half-filling, and the f-electron filling is near 1, the system becomes a Kondo insulator characterized by a single energy scale, which is often enhanced relative to the single-impurity Kondo scale. Finally, if the f-electron filling deviates significantly from one, the system is in the intermediate-valence regime, where the spin screening leading to the Kondo effect competes against charge excitations, and there is no complete consensus on how these systems behave.

After this brief introduction to the physical behavior of Kondo-Anderson models and heavy Fermions, we show some recent numerical results for

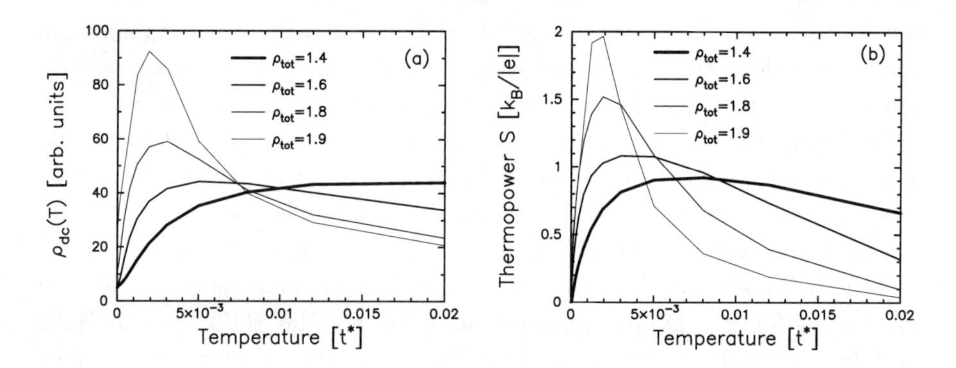

Fig. 2.21 (a) DC resistivity versus T and (b) thermopower versus T for the periodic Anderson model with $V = 0.056$, $E_f = -0.5$, $U = 1$ and various values of the total filling ρ_{tot} (1.4, 1.6, 1.8, and 1.9) on the hypercubic lattice. The DMFT is solved by the NRG with a small imaginary part added to the frequency 0.0001 to smooth out the raw data for the DOS. *Adapted from* [Grenzebach, *et al.* (2006)].

transport [Grenzebach, *et al.* (2006)] that were obtained for the spin-one-half periodic Anderson model on a hypercubic lattice using NRG as the impurity solver for the DMFT. This problem has numerous numerical challenges that we will not discuss in detail here, but we will summarize one of the most challenging ones. In order to properly calculate transport, one needs the relaxation time in a frequency window on the order of the temperature (the so-called Fermi window). The NRG approach determines the temperature scale by the size of the finite chain on which a given set of calculations is performed. Since the NRG produces discrete spectra composed of delta functions, the DOS and self-energy can have oscillatory structure within the Fermi window (because there isn't a dense enough set of delta functions due to the finite chain length), which can cause problems for properly calculating the transport. The solution taken for the numerical results shown here is to evaluate the Green's functions and self-energies along a line parallel to, but slightly above the real axis. The small imaginary part of the frequency helps smooth out these oscillations, and allow calculations to be performed in a consistent fashion.

The data corresponds to a number of different regimes, each based on the filling of the electrons. The f-electron filling does not depend too strongly on temperature for the temperatures considered here. We will be considering four different total fillings of the electrons (conduction plus f), ranging from 1.4 to 1.9; the f-filling satisfies $\rho_f \approx 0.77$ for $\rho_{tot} = 1.4$, $\rho_f \approx 0.88$ for $\rho_{tot} = 1.6$, $\rho_f \approx 0.96$ for $\rho_{tot} = 1.8$, and $\rho_f \approx 0.98$ for $\rho_{tot} = 1.9$. When $\rho_f \approx 1$, we are in the Kondo regime, where local moments are well-formed at the f-electron sites; when ρ_f differs from 1, we are in the intermediate-valence regime, where the Kondo effect is less well developed, and instead, charge fluctuations of the f-electron become more important. For the numerical data presented here, the $\rho_{tot} = 1.8$ and 1.9 data are probably in the Kondo regime, but the other fillings are not. The f-electron level lies at $E_f = -0.5$ (measured relative to the conduction band center) and the f-f Coulomb repulsion is $U = 1$; the chemical potential at low T lies above E_f for all of these cases, but is quite close to E_f for the $\rho_{tot} = 1.4$ and 1.6 cases. As the filling approaches 2, the system becomes an insulator. Hence, the effective correlations are stronger for larger fillings and the system also becomes more Kondo-like. Examining the resistivity data in Fig. 2.21(a), shows that the resistivity develops a maximum at low T as the effective correlations are increased. The low-T region is consistent with Fermi-liquid behavior, since the resistivity increases as T increases, but it doesn't give the expected T^2 growth (the resistivity appears to be finite as $T \to 0$; this is a numerical artifact of the finite imaginary part used in the calculation). At higher T, the resistivity becomes anomalous due to the unbinding of the spin singlets, which cause significant scattering throughout the lattice. As the effective correlations increase (ρ_{tot} gets larger), this Fermi-liquid regime shrinks, and is pushed to lower T because the coherence temperature vanishes as the correlations become large enough. Note that the much stronger temperature dependence of the resistivity for the periodic Anderson model versus the Falicov-Kimball model arises because the PAM has Fermi-liquid behavior at low temperature, which governs the low-T behavior, while the Falicov-Kimball model does not have such low-T behavior.

The thermopower in Fig. 2.21(b), has a broad maximum that develops into a sharper peak at low T as the effective correlations increase. The magnitude of the thermopower is similar to that seen in the Falicov-Kimball model, but in the PAM case, the results are rather generic for a wide range of metallic systems and are pushed to much lower temperatures, while in the Falicov-Kimball model they arise only from a "fine-tuning" of the parameters for the particle-hole asymmetric insulator. Because the thermopower is large for a metal, the electronic figure-of-merit might not change much if the lattice thermal conductivity is included, since the Wiedemann-Franz law indicates that the phonon contribution to the thermal conductivity is

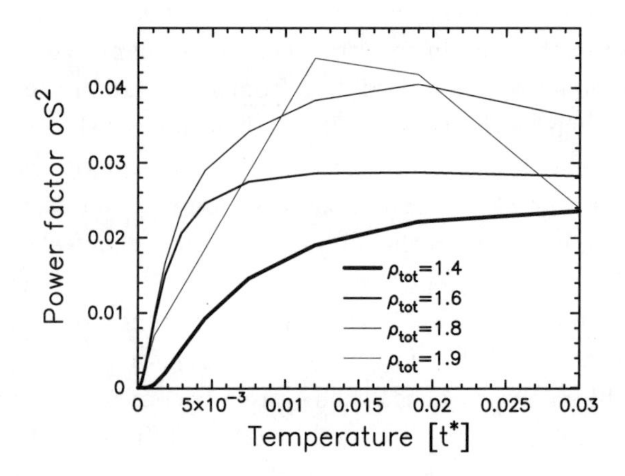

Fig. 2.22 Power factor $\sigma_{dc}S^2$ in arbitrary units. The parameters are identical to those in Fig. 2.21. Note how the maximal power factor occurs above the maximum of the resistivity. It is highest for the $\rho_{tot} = 1.9$ data set. *Adapted from* [Grenzebach, *et al.* (2006)].

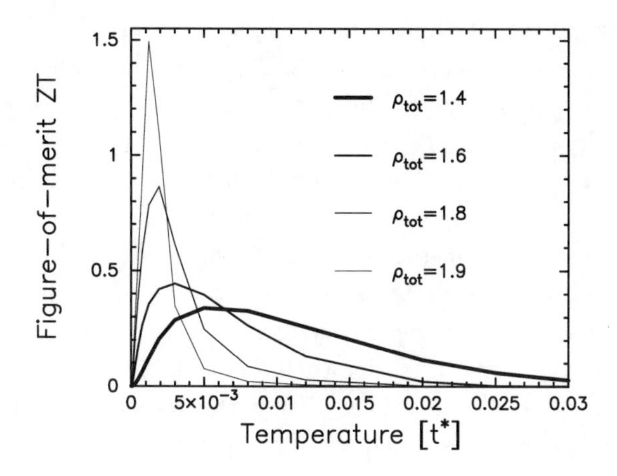

Fig. 2.23 Figure-of-merit ZT. The parameters are identical to those in Fig. 2.21. Note how the maximal figure-of-merit occurs for the most strongly correlated data set, and how it is pushed downward in T as the effective correlations increase. *Adapted from* [Grenzebach, *et al.* (2006)].

generically small. Note that these results for the PAM are independent of the Mahan-Sofo conjecture [Mahan and Sofo (1996)] that a sharp peak in the f-electron DOS just off of the chemical potential can give rise to a large thermopower. This is because the f-electrons do not directly participate in the transport if the hybridization is independent of momentum; instead, it is the conduction electron DOS that solely determines the transport, and that DOS actually has a minimum near the chemical potential. In this sense, the thermopower behaves in many respects like that seen in the Falicov-Kimball model, except now, the peak develops as T is reduced, so the T dependence can be stronger, and we have not tuned the chemical potential to lie at the bottom of the minimum, just close to it. Finally, we comment on possible applications. For a thermoelectric, we want the charge conductivity and the thermopower high and the thermal conductivity to be low. In these systems, as we increase the effective correlations, the minimum of the conductivity is in close proximity to the maximum of the thermopower, so these two effects work against one another. In the most correlated system, the conductivity is rising where the thermopower has a peak, so in this regime the system may have better prospects for thermal cooling or power generation applications. This issue is emphasized with a plot of the so-called power factor $\sigma_{dc}S^2$ (in arbitrary units) in Fig. 2.22, which has a maximum at low temperatures and is highest for $\rho_{tot} = 1.9$.

In addition, we also plot the electronic figure-of-merit for the thermoelectric for the same parameters in Fig. 2.23. Note how the peak develops and is pushed lower in temperature as the system becomes more strongly correlated.

In any case, if the Wiedemann-Franz law holds (as it does for most metals), then the thermopower needs to be larger than about 2 in these units to become useful (in other words, to have $ZT > 1$). These results show that it may be possible to achieve this goal, but it will be a challenge. Often in real materials it is necessary to reduce the lattice contribution to the thermal conductivity in order to have a viable device. Multilayered nanostructures might achieve this goal if the interfaces cause significant phonon scattering, but are tuned so as to not affect the electron transport significantly. We evaluate transport in such devices later (but based on the Falicov-Kimball model).

Finally, we show the DOS for the conduction and the f-electrons in Fig. 2.24. The plot is for $T = 0.00003$, and fillings $\rho_{tot} = 1.4$, 1.6, 1.8, and 1.9. We plot both the conduction electron DOS (panel a) and the f-electron DOS (panel b). Note how the conduction electron DOS develops a dip near the chemical potential, while the f-electron DOS is strongly enhanced. The dip at the chemical potential brings in anomalous features to the resistivity. In particular, as the Fermi window widens at higher T, there

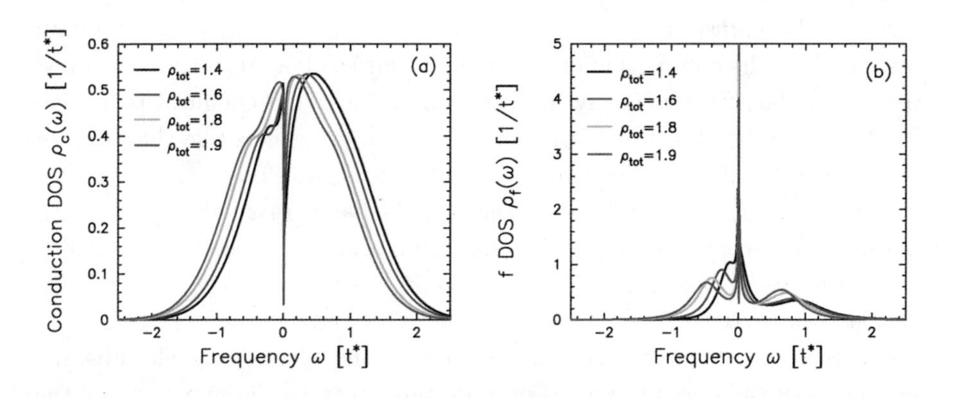

Fig. 2.24 (a) Conduction electron DOS versus frequency at $T = 0.00003$ for $\rho_{tot} = 1.4$, 1.6, 1.8, and 1.9 (b) f-electron DOS for the same parameters. Note how the conduction DOS develops a dip near the chemical potential, while the f-electron DOS develops a peak. The transport is determined by the conduction electrons, because they carry the current. Since the PAM is a Fermi liquid at low temperatures when it is not an insulator, the conduction DOS at $\omega = 0$ is equal to the noninteracting result, even though it displays a dip in the interacting system. *Adapted from* [Grenzebach, *et al.* (2006)].

are more states available for transporting electrons (in conventional metals the number of states doesn't change much with T). This enhancement primarily causes a reduction in the resistivity, independent of the enhancement that arises from scattering effects. This helps explain the anomalous temperature dependence of the resistivity. Similarly, because the DOS is asymmetric near the chemical potential, with a large derivative, the thermopower becomes large at low temperature (temperatures smaller than the width of the dip). This helps explain the peaks seen in the thermopower above. Furthermore, since both the anomalies in the resistivity and the peaks in the thermopower are arising from the dip in the DOS, it shouldn't be a surprise that the two features are correlated together in some way.

Chapter 3

Dynamical Mean-Field Theory of a Multilayered Nanostructure

3.1 Potthoff-Nolting Approach to Multilayered Nanostructures

In 1999, Potthoff and Nolting introduced the formal developments required to solve inhomogeneous dynamical mean-field theory [Potthoff and Nolting (1999a)]. The original problem they focused on was the Hubbard model on a two-dimensional surface of a three-dimensional finite crystal. Their interest was in the stability of the Mott transition as the surface is approached. After developing the formalism, they studied the surface Mott transition in detail [Potthoff and Nolting (1999b); Potthoff and Nolting (1999c); Potthoff (2002)] and the problem of the Mott transition directly in a thin film [Potthoff and Nolting (1999d)].

We will apply the Potthoff-Nolting approach to multilayered nanostructures, which involve translationally invariant x–y planes stacked in the longitudinal z-direction (see Fig. 3.1). Hence we have perfect periodicity in the x and y directions, but we allow inhomogeneity in the z-direction. All interactions must also be translationally invariant within each plane, but can change from one plane to the next. Potthoff and Nolting introduced the idea of a mixed basis for inhomogeneous DMFT: Fourier transform the x and y coordinates to wavevectors $\mathbf{k}_x$ and $\mathbf{k}_y$ but keep the z-component in real space; we describe the z-coordinate with a Greek letter (α, β, γ and so on). Our requirements are then that each interaction (say a Coulomb interaction U_α, for example), the hopping within a plane $t_\alpha^{\parallel}$, and the hopping between planes $t_{\alpha\alpha+1}$ and $t_{\alpha-1\alpha}$, are all fixed for a given plane, but can change as a function of the plane index α; we do require Hermiticity though, so $t_{\alpha\alpha+1} = t^*_{\alpha+1\alpha}$ (in the normal state the hopping matrix elements will always be real; they can be complex in a superconductor as will be dis-

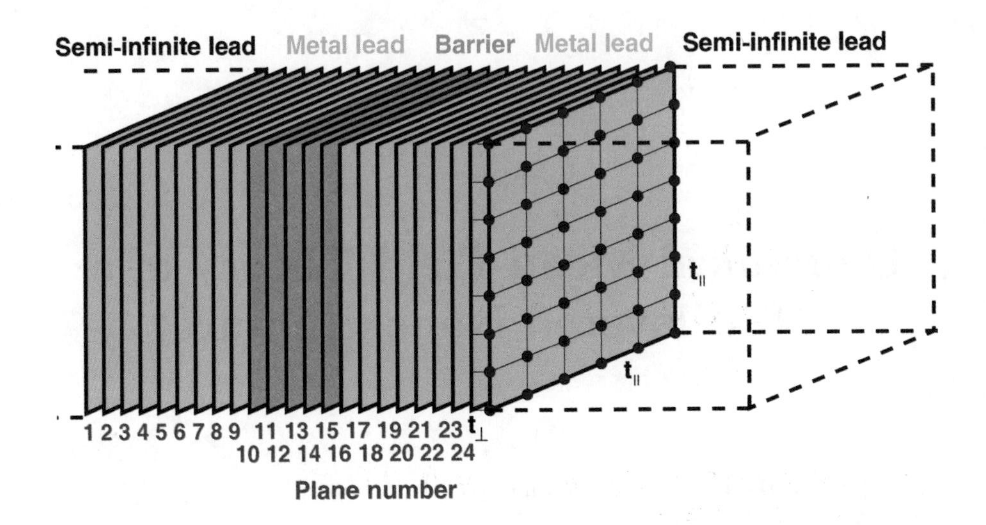

Fig. 3.1 Schematic representation of the longitudinally stacked planes of the multilayered nanostructure. The lattice sites are located at the positions of a simple cubic lattice (indicated by dots on the far right plane), but the materials (or models) that describe the different planes need not be the same. Since we chose the simple cubic lattice for the locus of lattice sites, each plane is a square-lattice plane. The kinetic energy is described by a nearest-neighbor hopping term, which is the same for every x and y direction on each plane (but can vary from plane to plane) and is denoted $t_{\|}$ in the figure, and which can vary from plane to plane for hopping to the right or to the left in the longitudinal direction (denoted by $t_{\perp}$). We use Greek labels α, β, γ ... to denote the planes (which are numbered in blue here), and Fourier transform the representation in the x and y directions. The figure shows a situation of a metal–barrier–metal junction, with metal leads in pink and the barrier in green. Those colored planes are included in the self-consistent calculations (although in most of the data presented here, the pink region extends 30 planes on each side, rather than the 10 depicted in the figure). The dashed lines indicate the semi-infinite parts of the leads, which are taken to be identical to the bulk metal, and are not self-consistently determined; in this sense, they form the left and right boundary conditions for the calculation.

cussed when we develop the theory for Josephson junctions). Because of the translational invariance in each two-dimensional plane, we can describe the intraplane hopping via a two-dimensional bandstructure

$$\epsilon^{\|}_{\alpha(\mathbf{k}_x,\mathbf{k}_y)} = -2t^{\|}_{\alpha}[\cos\mathbf{k}_x + \cos\mathbf{k}_y], \tag{3.1}$$

where we have chosen the planes to be square lattices.

The starting point for all of our calculations is to determine the equations satisfied by the local Green's functions. In the case of inhomogeneous DMFT, the formalism is somewhat more complicated than for the bulk. First we note that the noninteracting real-space Green's function

$G_{ij}^{\text{nonint}}(z)$, with z lying anywhere in the complex plane, satisfies

$$\sum_k [(z+\mu)\delta_{ik} + t_{ik}]\, G_{kj}^{\text{nonint}}(z) = \delta_{ij}, \tag{3.2}$$

with $-t_{ij}$ the hopping matrix on the simple cubic lattice (organized into hopping within a plane and hopping between planes, as described above and in Fig. 3.1). Using the mixed basis of Potthoff and Nolting, we Fourier transform the x- and y-directions to $\mathbf{k}_x$ and $\mathbf{k}_y$, respectively, and keep the z-direction in real space. We denote the two-dimensional wavevector by $\mathbf{k}^{\|} = (\mathbf{k}_x, \mathbf{k}_y, 0)$ and find

$$\sum_\gamma \left[(z+\mu-\epsilon^{\|}_{\alpha\mathbf{k}^{\|}})\delta_{\alpha\gamma} + t_{\alpha\alpha+1}\delta_{\alpha+1\gamma} + t_{\alpha\alpha-1}\delta_{\alpha-1\gamma}\right] G_{\gamma\beta}^{\text{nonint}}(\mathbf{k}^{\|}, z) = \delta_{\alpha\beta}. \tag{3.3}$$

This has the form we expect; the inverse Green's function (in a matrix sense) $G_{\alpha\gamma}^{\text{nonint}}(\mathbf{k}^{\|}, z)^{-1}$ is determined by the quantity in the square brackets of Eq. (3.3).

The next step we need to take is to examine the Dyson equation in the remaining real space

$$G_{\alpha\beta}(\mathbf{k}^{\|}, z) = G_{\alpha\beta}^{\text{nonint}}(\mathbf{k}^{\|}, z) + \sum_\gamma G_{\alpha\gamma}^{\text{nonint}}(\mathbf{k}^{\|}, z)\Sigma_{\gamma\gamma}(z) G_{\gamma\beta}(\mathbf{k}^{\|}, z), \tag{3.4}$$

where we make the DMFT assumption that the self-energy is local, but it can vary from one plane to another (*i. e.*, $\Sigma_{\alpha\beta} = 0$ if $\alpha \neq \beta$, but $\Sigma_{\alpha\alpha}$ can vary with α). Multiplying on the left by the inverse of the noninteracting Green's function, substituting in its explicit form from Eq. (3.3), and performing the sum over γ explicitly, gives our final equation

$$\begin{aligned}[z+\mu-\Sigma_{\alpha\alpha}(z) - \epsilon^{\|}_{\alpha\mathbf{k}^{\|}}]G_{\alpha\beta}(\mathbf{k}^{\|}, z) &+ t_{\alpha\alpha+1}G_{\alpha+1\beta}(\mathbf{k}^{\|}, z) \\ &+ t_{\alpha\alpha-1}G_{\alpha-1\beta}(\mathbf{k}^{\|}, z) = \delta_{\alpha\beta}. \end{aligned}\tag{3.5}$$

When viewed as a matrix in α and β, this equation has a tridiagonal structure to it. In other words, for each fixed value of parallel momentum $\mathbf{k}^{\|}$, the problem decouples into a one-dimensional chain in the z-direction, given by the tridiagonal structure of the equations. It is precisely this decoupling that allows us to efficiently solve the inhomogeneous DMFT.

The algorithm for solving inhomogeneous DMFT is similar to that of the bulk case, with a number of complications. Given a set of local self-energies for each plane in the simulation, we use the solution of Eq. (3.5), to determine the local Green's function on each plane (this is analogous to the Hilbert transform, and is solved with the quantum "zipper" algorithm in the next section). Once we have the local self-energy and the local Green's function, we use the conventional algorithm of extracting the effective medium, solving the time-dependent impurity problem, and then determining the impurity self-energy, just as in the bulk; but it must now be done for each plane in the simulation to get the set of all self-energies. This loop is then iterated until it converges.

It seems like this problem becomes infinite in extent, since we have an infinite number of sites for an inhomogeneous system that is attached to extensive leads. But we can make a simple approximation that restores tractability to the procedure—we simulate only a finite number of planes and then attach the leftmost piece of the nanostructure to a semi-infinite bulk lead which has a self-energy that does not change with plane number as we move further to the left, and do a similar thing on the right. We simply need to choose the width of the simulated region to be large enough that the system has effectively healed itself to the bulk limit to the right and to the left of the simulated region. When performed in this fashion, the calculations are always in the thermodynamic limit, so all local DOS are continuous functions of frequency, and do not require any artificial broadening. This is the "top-down" approach that we take to multilayered nanostructures—we build in the inhomogeneity on top of the thermodynamic limit, rather than a "bottom-up" approach that starts from atoms and increases the number of atoms to get to the nano regime.

3.2 Quantum Zipper Algorithm (Renormalized Perturbation Expansion)

Tridiagonal systems can have any matrix element of the inverse constructed directly via continued fractions and recursion relations. The technique, as originally applied to Bethe lattice systems, was called the renormalized perturbation expansion, and it is described in detail, in that context, in the classic text [Economou (1983)]. This name is an odd choice, because the algorithm is exact for calculating the inverse matrix elements of any tridiagonal system (no periodicity is required). Here, we prefer to call it the quantum zipper algorithm, because we use two sets of recursion relations, one from the right and one from the left, that move through the planes, in turn, like a zipper is moved.

The starting point is the EOM derived above in Eq. (3.5). The equation with $\beta = \alpha$ is different from the equations with $\beta \neq \alpha$. The former is solved directly via

$$G_{\alpha\alpha}(\mathbf{k}^{\parallel}, z) = \frac{1}{z + \mu - \Sigma_{\alpha}(z) - \epsilon^{\parallel}_{\alpha\mathbf{k}^{\parallel}} + \frac{G_{\alpha\alpha-1}(\mathbf{k}^{\parallel},z)}{G_{\alpha\alpha}(\mathbf{k}^{\parallel},z)} t_{\alpha-1\alpha} + \frac{G_{\alpha\alpha+1}(\mathbf{k}^{\parallel},z)}{G_{\alpha\alpha}(\mathbf{k}^{\parallel},z)} t_{\alpha+1\alpha}}, \tag{3.6}$$

and the latter equations can all be put into the form

$$-\frac{G_{\alpha\alpha-n+1}(\mathbf{k}^{\parallel}, z) t_{\alpha-n+1\alpha-n}}{G_{\alpha\alpha-n}(\mathbf{k}^{\parallel}, z)} = z + \mu - \Sigma_{\alpha-n}(z) - \epsilon^{\parallel}_{\alpha-n\mathbf{k}^{\parallel}} + \frac{G_{\alpha\alpha-n-1}(\mathbf{k}^{\parallel}, z) t_{\alpha-n-1\alpha-n}}{G_{\alpha\alpha-n}(\mathbf{k}^{\parallel}, z)}, \tag{3.7}$$

for $n > 0$, with a similar result for the recurrence to the right. In these equations, we have used Σ_{α} to denote the local self-energy $\Sigma_{\alpha\alpha}$ on plane α. We define the left function

$$L_{\alpha-n}(\mathbf{k}^{\parallel}, z) = -\frac{G_{\alpha\alpha-n+1}(\mathbf{k}^{\parallel}, z) t_{\alpha-n+1\alpha-n}}{G_{\alpha\alpha-n}(\mathbf{k}^{\parallel}, z)} \tag{3.8}$$

and then determine the recurrence relation from Eq. (3.7)

$$L_{\alpha-n}(\mathbf{k}^{\parallel}, z) = z + \mu - \Sigma_{\alpha-n}(z) - \epsilon^{\parallel}_{\alpha-n\mathbf{k}^{\parallel}} - \frac{t_{\alpha-n\alpha-n-1} t_{\alpha-n-1\alpha-n}}{L_{\alpha-n-1}(\mathbf{k}^{\parallel}, z)}. \tag{3.9}$$

We solve the recurrence relation by starting with the result for $L_{-\infty}$, and then iterating Eq. (3.9) up to $n = 0$. Of course we do not actually go out infinitely far in our calculations. We assume we have semi-infinite metallic leads, hence we can determine $L_{-\infty}$ by substituting $L_{-\infty}$ into both the left and right hand sides of Eq. (3.9), which produces a quadratic equation for $L_{-\infty}$ that is solved by

$$L_{-\infty}(\mathbf{k}^{\parallel}, z) = \frac{z + \mu - \Sigma_{-\infty}(z) - \epsilon^{\parallel}_{-\infty\mathbf{k}^{\parallel}}}{2} \pm \frac{1}{2}\sqrt{[z + \mu - \Sigma_{-\infty}(z) - \epsilon^{\parallel}_{-\infty\mathbf{k}^{\parallel}}]^2 - 4t^2_{-\infty}}. \tag{3.10}$$

The sign in Eq. (3.10) is chosen to yield an imaginary part less than zero for z lying in the upper half plane, and *vice versa* for z lying in the lower half plane. If $L_{-\infty}$ is real, then we choose the root whose magnitude is larger than $t_{-\infty}$ (the product of the roots equals $t^2_{-\infty}$). In our calculations, we usually assume that the left function is equal to the value $L_{-\infty}$ found in the bulk, until we are within thirty planes of the first interface. Then we allow those thirty planes to be self-consistently determined with L_α possibly changing, and we include a similar thirty planes on the right hand side of the last interface, terminating with the bulk result to the right as well. This approach is accurate, when the system heals to its bulk values within those thirty planes on either side of the interface. If this healing has not occurred, then one needs to include more planes before one terminates the problem with the semi-infinite bulk solution.

In a similar fashion, we define a right function and a recurrence relation to the right, with the right function

$$R_{\alpha+n}(\mathbf{k}^{\|}, z) = -\frac{G_{\alpha\alpha+n-1}(\mathbf{k}^{\|}, z)t_{\alpha+n-1\alpha+n}}{G_{\alpha\alpha+n}(\mathbf{k}^{\|}, z)} \tag{3.11}$$

and the recurrence relation

$$R_{\alpha+n}(\mathbf{k}^{\|}, z) = z + \mu - \Sigma_{\alpha+n}(z) - \epsilon^{\|}_{\alpha+n\mathbf{k}^{\|}} - \frac{t_{\alpha+n\alpha+n+1}t_{\alpha+n+1\alpha+n}}{R_{\alpha+n+1}(\mathbf{k}^{\|}, z)}. \tag{3.12}$$

We solve the right recurrence relation by starting with the result for R_∞, and then iterating Eq. (3.12) up to $n = 0$. As before, we determine R_∞ by substituting R_∞ into both the left and right hand sides of Eq. (3.12), which produces a quadratic equation for R_∞ that is solved by

$$\begin{aligned} R_\infty(\mathbf{k}^{\|}, z) = & \frac{z + \mu - \Sigma_\infty(z) - \epsilon^{\|}_{\infty\mathbf{k}^{\|}}}{2} \\ & \pm \frac{1}{2}\sqrt{[z + \mu - \Sigma_\infty(z) - \epsilon^{\|}_{\infty\mathbf{k}^{\|}}]^2 - 4t_\infty^2}. \end{aligned} \tag{3.13}$$

The sign in Eq. (3.13) is chosen the same way as for Eq. (3.10). In our calculations, we also usually assume that the right function is equal to the value R_∞ found in the bulk, until we are within thirty planes of the first interface. Then we allow those thirty planes to be self-consistently determined with R_α possibly changing, and we include a similar thirty planes on the left hand side of the last interface, terminating with the bulk result to the left as well.

Using the left and right functions, we finally obtain the Green's function

$$G_{\alpha\alpha}(\mathbf{k}^{\|}, z) = \frac{1}{L_\alpha(\mathbf{k}^{\|}, z) + R_\alpha(\mathbf{k}^{\|}, z) - [z + \mu - \Sigma_\alpha(z) - \epsilon^{\|}_{\alpha\mathbf{k}^{\|}}]} \tag{3.14}$$

where we used Eqs. (3.9) and (3.12) in Eq. (3.6). The local Green's function on each plane is then found by summing over the two-dimensional momenta, which can be replaced by an integral over the two-dimensional density of states (DOS):

$$G_{\alpha\alpha}(z) = \int d\epsilon_{\alpha}^{\parallel} \rho^{2d}(\epsilon_{\alpha}^{\parallel}) G_{\alpha\alpha}(\epsilon_{\alpha}^{\parallel}, z), \tag{3.15}$$

with

$$\rho^{2d}(\epsilon_{\alpha}^{\parallel}) = \frac{1}{2\pi^2 t_{\alpha}^{\parallel} a^2} \mathbb{K}\left(\sqrt{1 - \frac{(\epsilon_{\alpha}^{\parallel})^2}{(4t_{\alpha}^{\parallel})^2}}\right), \tag{3.16}$$

and $\mathbb{K}(x)$ the complete elliptic integral of the first kind. If $t_{\alpha}^{\parallel}$ varies in the nanostructure, then changing variables to $\epsilon = \epsilon_{\alpha}^{\parallel}/t_{\alpha}^{\parallel}$ in Eq. (3.15) produces

$$G_{\alpha\alpha}(z) = \int_{-4}^{4} d\epsilon \frac{1}{2\pi^2 a^2} \mathbb{K}\left(\sqrt{1 - \frac{\epsilon^2}{16}}\right) G_{\alpha\alpha}(t_{\alpha}^{\parallel}\epsilon, z), \tag{3.17}$$

so that we can take the ϵ variable to run from -4 to 4 for the integration on every plane, and we just need to introduce the corresponding $t_{\alpha}^{\parallel}\epsilon$ substitution (for $\epsilon_{\alpha}^{\parallel}$) into the left and right recurrence relations. In the bulk limit (simple-cubic-lattice), where we use $t_{\alpha} = t$, we find that the local Green's function found from Eqs. (3.15) and (3.14) reduce to the well-known expressions for the three-dimensional Green's functions on a simple cubic lattice [Economou (1983)], with a hopping parameter t.

In this section, we showed how to calculate the local Green's function on each plane. The quantum zipper algorithm can also determine all off-diagonal Green's functions as well. This is done by taking the local solutions, and building up to the off-diagonal ones by employing Eq. (3.5) and the definitions in Eqs. (3.8) and (3.11). For details, see Prob. A.23.

3.3 Computational Methods

The computational issues for a nanostructure are more complex than for calculations in the bulk. We describe numerical implementation of the Potthoff-Nolting algorithm here.

Since we have to work with a finite set of computer resources, the nanostructure calculations must be finite in extent. In all of the examples described in this work, we take a bulk system to the left and attach it to a self-consistently determined nanostructure in the center which has 30 planes from the lead of the device on the left before the first nontrivial interface.

Then we have a barrier region, which can be as complicated as desired. Finally we cap of the right end with another 30 planes and then a bulk lead to the right. This means we must simulate 60 planes plus the number of barrier planes. Typically, the barrier region is not larger than 200 planes. Nevertheless, since we must perform the DMFT algorithm for each plane, in order to find the self-energy on each plane, the effort scales with the number of planes chosen for the self-consistent region.

The DMFT algorithm starts with a self-energy on each plane (see Fig. 3.2). Next, we use the quantum zipper algorithm to find the local Green's function on each plane. This step is the inhomogeneous nanostructure equivalent to the Hilbert transform. Once the local Green's function is known on each plane, we extract the local effective medium via the scalar relation

$$G_{0\alpha}(z)^{-1} = G_{\alpha}(z)^{-1} + \Sigma_{\alpha}(z), \tag{3.18}$$

on each plane. Next, we need to solve the local impurity problem for the given Hamiltonian on the αth plane with the given effective medium. This will produce a new local Green's function for each plane, and a new self-

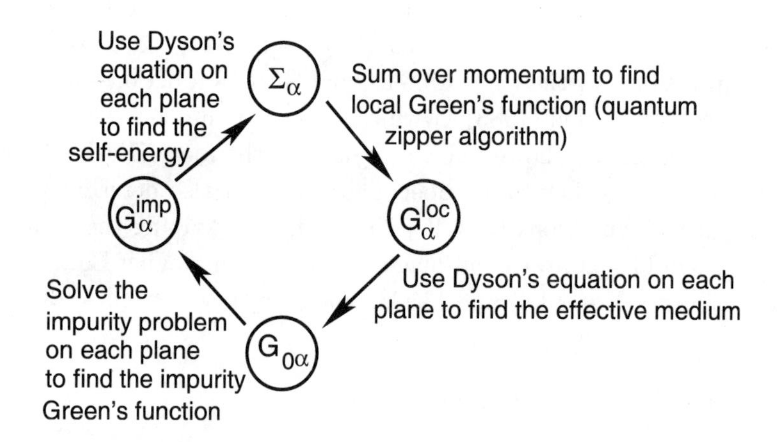

Fig. 3.2 Flow diagram for the DMFT algorithm in a multilayered nanostructure. Starting with an initial self-energy on each plane, we use the quantum zipper algorithm to calculate the local Green's function on each plane. Next, the effective medium for the impurity is extracted from the current self-energy and the local Green's function via Dyson's equation for each plane. The impurity problem is solved with the impurity solver chosen for the given model on each plane. Since the chemical potential is set by the filling of the bulk leads, it is input and not updated. Finally, Dyson's equation is used to find the new self-energy from the impurity Green's function and the effective medium for each plane.

energy for each plane. We check to see whether the self-energy has changed, and then iterate all of the steps to convergence.

The biggest complication that enters into the nanostructure calculations is the numerical treatment of the determination of the local Green's function from the local self-energy. When we evaluate Green's functions at the Matsubara frequencies, the only singularity in the integral comes from the logarithmic divergence of the two-dimensional DOS at zero energy. This logarithmic singularity is easy to remove by making a power law substitution into the quadrature routine near zero (see Prob. A.25). Hence there are not many subtle numerical issues on the Matsubara frequency axis.

When the Green's function is evaluated on the real axis, there is a possibility that the integrand has additional singularities that behave like the inverse of the square-root of $(a-\epsilon)^2 - b^2$. This behavior occurs when the imaginary part of the self-energy vanishes, as in the bulk simple-cubic lattice. In cases where the imaginary part of the self-energy is nonzero, the singularity may be avoided by the nonzero imaginary part; nevertheless, it is safest to use a quadrature routine that will automatically take care of these kinds of singularities, so they will not enter into calculations for some ranges of parameters. An outline for how to accomplish this is sketched in Prob. A.25. Since the Green's functions for each plane require the L and R functions in order to calculate them, we need to perform the recursion for all planes to get the relevant functions. Hence, it makes sense to calculate the integrations for each plane in parallel, in order to save the labor needed in determining the L and R functions for each plane. Such an algorithm will be vastly more efficient than one that calculates for each plane separately. It is also easy to implement, because we simply set up an integration grid over the range where ϵ runs from -4 to 4. The same integration grid is used for every plane, so the L and R functions that are generated for a given value of ϵ can be used in the calculations for every plane. Note that the formalism simplifies somewhat if we assume all $t^{\|}_{\alpha\alpha+1}$ are independent of α. Then at each plane, the parallel component of the bandstructure is the same, and we can use ϵ instead of $\mathbf{k}^{\|}$ to label the relevant Green's functions.

This approach creates an algorithm that is embarrassingly parallel. When we perform the integration over the two-dimensional DOS, each ϵ value is independent of other values, so we can use a master-slave format, where the master sends an ϵ value to a slave node, which determines the contribution of $G_\alpha(\epsilon, z)$ for every α, and sends them back to the master node, which accumulates the results and then sends a new ϵ value to the slave node.

Numerical accuracy of the integration routine is also important. As the self-energy develops a sharp structure in frequency on the real axis, the integration routine requires more and more points to properly converge. In calculations that we have performed, we usually use 5,000 to 10,000 points in the integration grid, but have increased to 1,000,000 points when needed for frequencies where the self-energy has sharp structures. One always has to be careful and check that the integrations have properly converged (which can only be achieved by redoing the calculation on a finer mesh and comparing it with previous results) in order to ensure that the results are correct. Otherwise, it is easy for unsuspecting errors to creep into results when one is close to the metal-insulator transition.

An alternative scheme for the numerical quadrature would be to employ integrators that are adaptive and estimate the error of the integration. It is not easy to use these integrators on every plane "at once" as can be done when the integration grid for each plane is the same, but there do exist integrators that can handle singularities at endpoints, and which minimize the number of evaluations of the functions in the integrand. The drawback is that one has to repeat this effort for each plane in the simulated part of the nanostructure, and that can become quite time consuming even on a parallel machine. We prefer using the fixed grid for each plane whenever possible, although there are circumstances, especially when a system has electronic charge reconstruction at an interface, where the grid must change for each plane.

3.4 Density of States for a Nanostructure

In this section, we give some examples of solutions of the inhomogeneous DMFT algorithm for the case of strongly correlated multilayered nanostructures that have no electronic charge reconstruction at the interfaces. The systems are tuned so that their chemical potentials match for all T. This is only possible in general if there is a symmetry of the Hamiltonian that pins the chemical potential to the same constant value for all parts of the nanostructure and for all T. We choose to work at the particle-hole symmetric point of half filling; our model consists of ballistic-metal leads to the left and to the right and of a FK model with half filling for both the delocalized and localized particles. In this case, we fix the chemical potential at $\mu = 0$ for the leads, which yields half filling. The FK model, will be half-filled when the chemical potential equals $U/2$, which would provide a mismatch. We shift the bands of the FK-model barrier downward by $U/2$ to compensate for this. Hence, we write the FK model potential energy as

$$\hat{V}'_{\mathrm{FK}} = \sum_{\alpha} \sum_{i \in \mathrm{plane}} U_\alpha \left(c^\dagger_{\alpha i} c_{\alpha i} - \frac{1}{2} \right) \left(w_{\alpha i} - \frac{1}{2} \right). \tag{3.19}$$

In Eq. (3.19), we use a notation where α denotes the plane number and i denotes a site on the plane. In our calculations we always include 30 self-consistent metal lead planes to the left and 30 to the right. We set $U^{\mathrm{FK}}_\alpha = 0$ for the metal lead planes.

In Fig. 3.3, we plot the DOS for four different planes in the metallic leads as a function of frequency [Freericks (2004b)]. Note how there are large amplitude oscillations in the "flat part" of the DOS that are created by the change in the quantum-mechanical character of the wavefunctions as we move from the metal to the barrier. The number of peaks in the oscillations increases by one for each plane as we move away from the interface, and the amplitude decays. By the time we have reached the thirtieth plane, the DOS looks quite similar to a simple cubic DOS, as expected. Note that there are oscillations generated in the other parts of the DOS, with similar behavior as well.

These oscillations are Friedel-like oscillations that arise from the sharp change in character of the device at the interfaces. In Fig. 3.4, we show these oscillations at $\omega = 0$ as a function of the plane position for $U^{\mathrm{FK}} = 2$ and 6 nanostructures with a barrier of 20 planes (left panels) [Tahvildar-Zadeh, Freericks and Nikolić (2006)]. The circles are the data points, and the line is a fit of the oscillations to the Friedel-like form

$$A_\alpha(\omega = 0) = A + B \sin(ka\alpha - \delta) \frac{1}{(a\alpha + b)^\gamma}. \tag{3.20}$$

In the ballistic-metal leads, the decay of the amplitude depends inversely on the distance from the interface; in the barrier, the amplitude of the oscillations decays much faster, being power law when the barrier is metallic, and becoming exponential when it is insulating. In the right panels, we show the DOS inside the barrier up to the center of the barrier at plane number 40 (the results for planes 41 to 80 are a symmetric mirror image). For $U^{\mathrm{FK}} = 6$, we use an exponential plot to show the decay of the DOS as one moves into the barrier. The decay length of the exponential is independent of the width of the barrier [Freericks (2004b)]. For $U^{\mathrm{FK}} = 2$, we use a linear scale. Note how there are clear oscillations when the barrier is

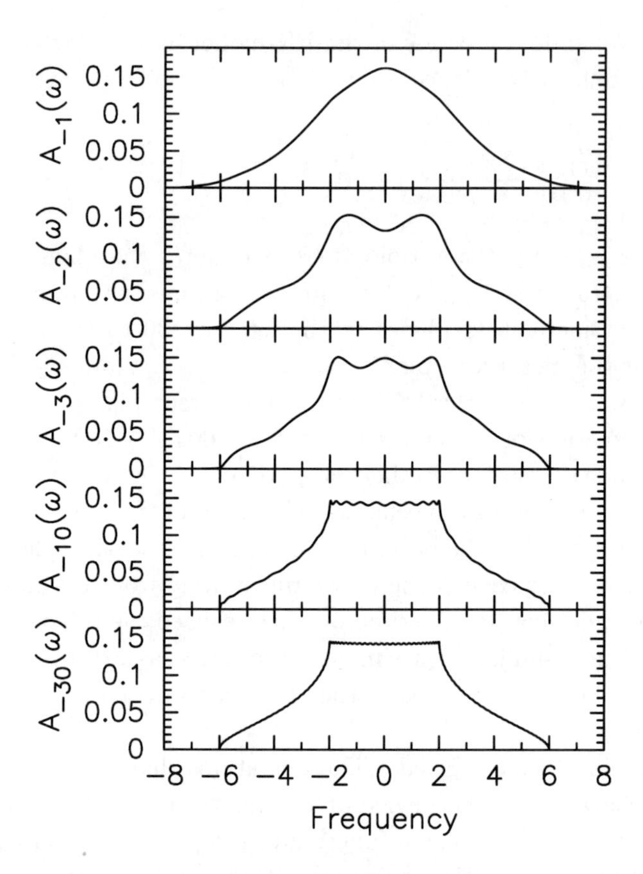

Fig. 3.3 Lead DOS for an $N = 5$ barrier device with $U = 6$. The different panels show the DOS in the first metal plane to the left of the barrier, in the second, the third, the tenth and the thirtieth. The DOS has oscillations with a large amplitude that decreases as we move away from the interface, so that the system approaches the bulk cubic DOS. A careful examination of the panels shows that the "flat" region with $|\omega| < 2$ shows a half-period oscillation for each unit of distance from the current plane to the interface, but the amplitude shrinks dramatically as we move further from the interface. Note that there are also oscillations (with the same kind of increase in the number of half periods with the distance from the interface) in the region with $|\omega| > 2$. *Reprinted with permission from* [Freericks (2004b)] (© 2004 the American Physical Society).

metallic, but the oscillations disappear for insulating barriers (the MIT for the FK model on a simple cubic lattice occurs at $U^{\mathrm{FK}} \approx 4.92$).

In Fig. 3.5 (top), we plot a false color contour plot of the DOS for a nanostructure with 20 planes in the barrier and $U^{\mathrm{FK}} = 2$. The ripples from the Friedel-like oscillations are most apparent in the central red region of the metallic leads. One can also see how the "bands" widen in the barrier, but

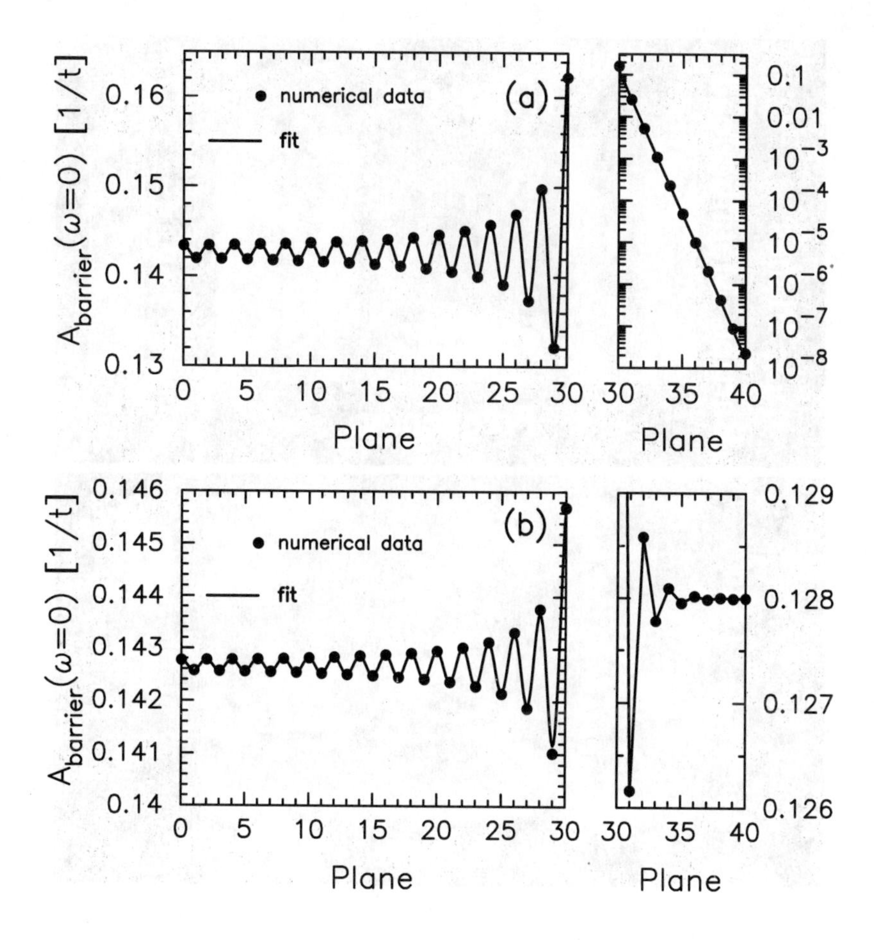

Fig. 3.4 DOS at $\omega = 0$ for multilayered nanostructures with 30 self-consistent planes in each metallic lead and for 20 planes in the barrier, which is described by the Falicov-Kimball model at half filling with $U^{\mathrm{FK}} = 6$ (upper panels) and $U^{\mathrm{FK}} = 2$ (lower panels). Note how there are Friedel-like oscillations in the metallic leads that decay inversely with the distance from the interface; in the barrier the decay of the amplitude is more rapid and becomes an exponential decay when the barrier is insulating. *Adapted with permission from* [Tahvildar-Zadeh, Freericks and Nikolić (2006)].

are depressed near the chemical potential. In the bottom panel, we enhance the oscillations present in the nanostructure DOS by taking the difference with the corresponding bulk DOS and assigning colors to amplitudes close to the origin. Now, one can see oscillations develop in all regions of the DOS, including within the barrier.

In Fig. 3.6, we plot a similar false-color plot for the correlated insulator case of $U^{\mathrm{FK}} = 6$. The oscillations in the metallic lead are similar, but the

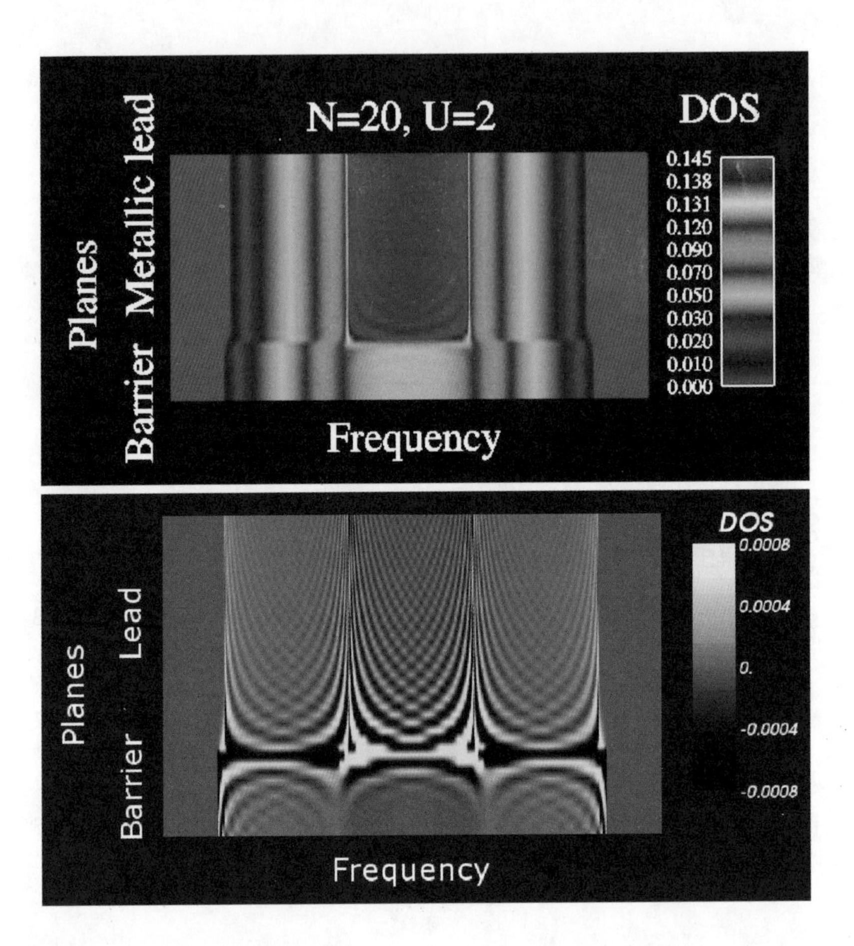

Fig. 3.5 (Top) False-color contour plot of the DOS for a $N = 20$ barrier plane device with $U^{\mathrm{FK}} = 2$. The barrier planes are the lowest ten planes at the bottom of the figure, while the thirty metallic planes lie on top. Note how the ripples of the Friedel oscillations are most visible in the central region, where the DOS has a plateau (red region). (Bottom) False-color plot of the difference between the nanostructure DOS and the corresponding bulk DOS for the lead and the barrier. By taking this difference, oscillations (on top of the bulk DOS) are enhanced and more easily imaged. We assign colors to amplitudes close to the origin to bring out the oscillatory behavior more clearly. *Bottom figure prepared with the assistance of S. Boocock.*

barrier region looks quite different — it has exponential decay near the chemical potential and oscillations near the metallic-lead band edge. In the lower panel, we plot the difference between the nanostructure DOS and the bulk DOS for the lead or for the barrier (depending on the plane location). The oscillations are further enhanced by focusing on the low amplitudes near the origin. Note how this visualization technique now brings out the

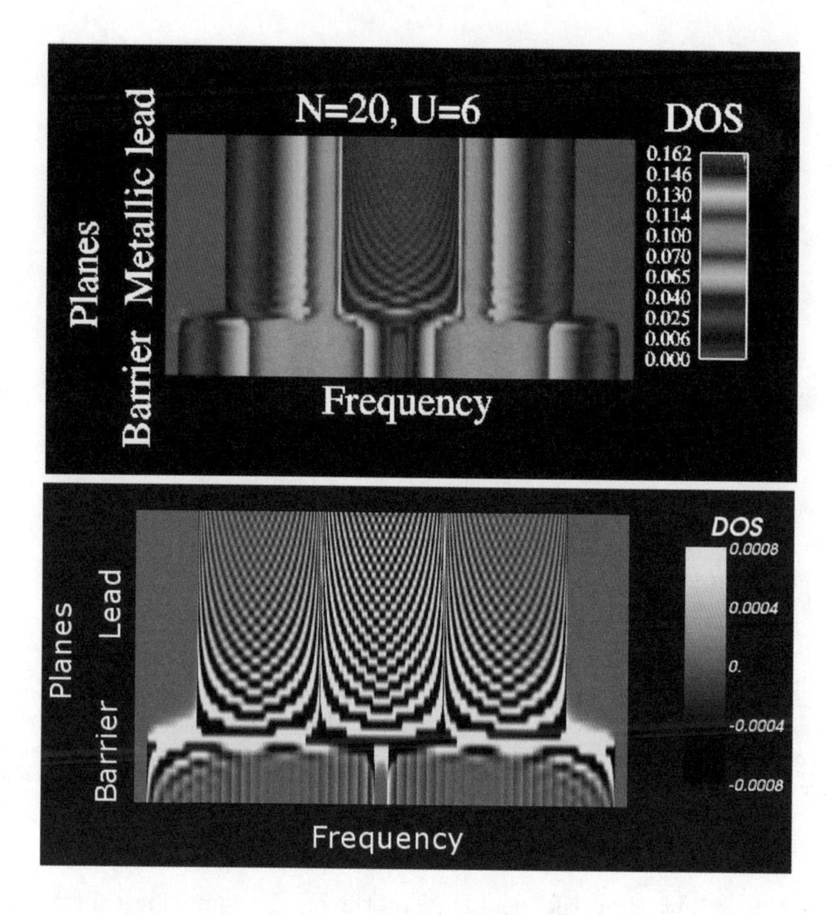

Fig. 3.6 (Top) False-color contour plot of the DOS for a $N = 20$ barrier plane device with $U^{\mathrm{FK}} = 6$. The barrier planes are the lowest ten planes at the bottom of the figure, while the thirty metallic planes lie on top. Note how the ripples of the Friedel oscillations are most visible in the central region, where the DOS has a plateau (yellow region). Note how the oscillations do not change too much in the metallic lead (from Fig. 3.5), but the barrier DOS changes dramatically because of the exponential suppression near the chemical potential (purple region). *Reprinted with permission from* [Freericks (2004b)] (© 2004 the American Physical Society). (Bottom) False-color plot of the difference of the nanostructure DOS from its bulk values. Note how in this picture the oscillations on top of the bulk DOS are strongly enhanced and can be seen to occur in all regions of energy, and even some can be seen within the barrier near the band edges. *Bottom figure prepared with the assistance of S. Boocock.*

oscillations present in the upper and lower regions of the DOS in addition to those already identified in the central region (in the top panel). Also, the oscillations within the barrier can now be clearly seen near the interface and near the band edges.

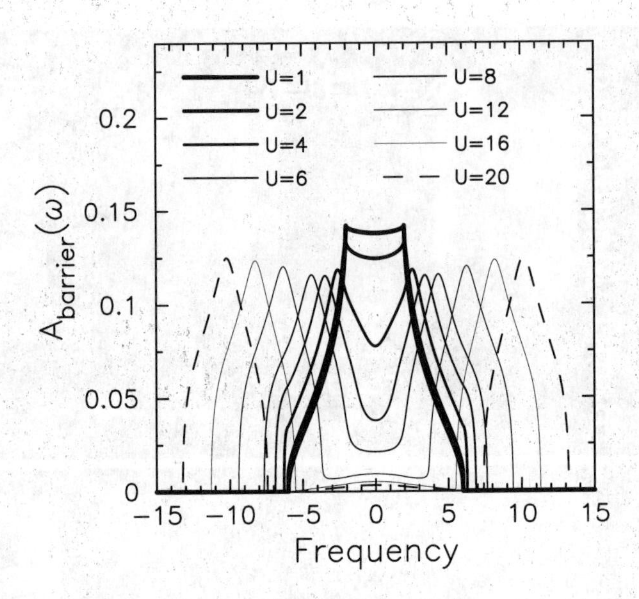

Fig. 3.7 Local DOS (vertical axis) as a function of frequency (horizontal axis) in the barrier for a single barrier plane multilayered nanostructure. The barrier is described by the Falicov-Kimball model at half filling with various U^{FK} values. Note how the DOS at the chemical potential initially decreases, but then becomes a local maximum (while still decreasing with increasing interaction strength). The DOS always remains metal-like as demonstrated in Prob. A.34. *Reprinted with permission from* [Freericks (2004b)] (© 2004 the American Physical Society).

In Fig. 3.7, we show the local DOS on the barrier plane for a single-plane nanostructure, whose barrier is described by the Falicov-Kimball model with different values of the interaction strength. Note how the double-peak structure of the Mott insulator develops as U^{FK} increases, but the central region of the DOS does not go to zero. Instead, it becomes a local maximum and appears metal-like for large interaction strength. This behavior arises from the delocalizing effect of the metallic leads, and can be thought of as a normal-state proximity effect of the leads — they do not allow an insulating state to form because there are interface localized states present for all interactions strengths. Hence, as one tries to confine the Mott insulator into a thin layer, it "spreads out" slightly and presents a conducting channel at low energies. These junctions never display insulating behavior in their resistance. For a more quantitative analysis of this system, look at Prob. A.34.

We have a few comments to make about these results. First off, the Falicov-Kimball model is not a Fermi liquid, so one might ask what will happen to these nanostructures if the system is described by a model (like the Hubbard model) that has a Fermi-liquid ground state. In the bulk, we will see the development of the Fermi-liquid coherence peak at low temperatures (below the renormalized coherence temperature). In general, we expect such a peak, and such a coherence effect to survive in these nanostructures as well. At low enough temperatures, these systems will become good conductors.

Next, we discuss what to expect for the transport when the DOS starts to exponentially decay in the barrier. When the barrier is composed of a bulk insulator, then the DOS will decrease with the thickness of the barrier, and the transport should be dominated by quantum-mechanical tunneling at low temperature. This tunneling gives way to incoherent thermally activated transport (described by Ohmic scaling) when the temperature is high enough [Freericks (2004a)].

Finally, one might wonder how this many-body approach of inhomogeneous DMFT compares to other approaches such as the Landauer approach, which is based on a single-particle picture with particles moving through fixed potential barriers, often determined in a phenomenological way. From the fact that the DOS of the leads does not change too much with the character of the insulating phase, we can conjecture that the treatment of the lead as a semi-infinite metal in the Landauer approach is probably quite reasonable. Further, in the insulating phase, we expect there to be an effective potential barrier to the electrons, which might be able to be modeled (at low T) by an energy-independent barrier as in the Landauer approach. So if one chooses the barrier height and shape properly, the results of a Landauer approach are likely to be able to reproduce some of the main features of the many-body calculation. But this approach always requires some kind of fitting to determine the barrier height and cannot be viewed as a fully microscopic model for the transport. Nevertheless, it would be interesting to try to quantitatively compare these two approaches.

3.5 Longitudinal Charge Transport Through a Nanostructure

The derivations that led up to Eqs. (2.151) and (2.152) were performed for bulk systems that have the full translational symmetry of the lattice. In addition, the conductivity was calculated for a uniform ($\mathbf{q} = 0$) electric field. In a more general case, where the field has momentum dependence, the *dc* conductivity gains some momentum dependence $\sigma(\mathbf{q})$. Maxwell's equations relate the charge current density with the electric field and the conductivity. In particular, for a translationally invariant system, the relation is that $\mathbf{j}^c(\mathbf{q}) = \sigma(\mathbf{q})\mathbf{E}(\mathbf{q})$. Hence the current is proportional to the product of the conductivity and the electric fields in Fourier space. If we invert the Fourier transform and express the result in real space, then we find that the product of two Fourier transforms is transformed into the convolution of the two functions in real space, so we have $\mathbf{j}^c(\mathbf{R}) = \int d^dR'\sigma(\mathbf{R}-\mathbf{R}')\mathbf{E}(\mathbf{R}')$. The conductivity depends on the difference of the two spatial arguments since we have translational invariance on the lattice. The most general form on a lattice, when there is inhomogeneity, is then

$$\mathbf{j}^c_\alpha = a\sum_\beta \sigma_{\alpha\beta}\mathbf{E}_\beta. \tag{3.21}$$

Equation (3.21) shows that the relationship between the current and the electric field is an inherently nonlocal relationship. This result is often presented in elementary texts on electromagnetism, but it can be easily forgotten, since it may seem reasonable to assume that the current at a given location in space depends solely on the electric field at that location. That assumption is false, because fields at other locations can affect the current at a given position.

Another important observation we need to make is that when the current is flowing in a steady state, we have that the charge density is fixed as a function of time (like water flowing through a hose—the amount of water in the hose at any given time is the same, even though it is moving, because the faucet continually supplies more water to make up for the water that exits the end of the hose). If the charge density is a constant, then the equation of continuity $\nabla\cdot\mathbf{j} = \partial_t\rho = 0$ implies that the number current is conserved throughout the sample. This means that the charge current that is incident on any plane must be the same as the charge current incident on any other plane. This fact also will play a crucial role in our analysis of transport in inhomogeneous systems.

We also assume, at the moment, that we have no temperature change through the device (*i.e.*, we consider the isothermal conductivity), and we assume that there is no electronic charge reconstruction, so the charge density is uniform in each of the materials that make up the device (in practical situations, this is only guaranteed at all T for the case of half filling, where the chemical potential is independent of T due to particle-hole symmetry).

So we must formulate the transport problem in real space. The first step is to determine what the number current operator is for longitudinal flow in the multilayered nanostructures. Taking the steps that led to the bulk current operator in Eq. (2.119), we see that there is no change for the real-space expression in a nanostructure, we simply need to restrict the direction of the current to lie in the longitudinal direction (for current flow perpendicular to the planes), so we have

$$\mathbf{j}^{\text{long}} = \sum_{\alpha} \mathbf{j}_{\alpha}^{\text{long}}, \quad \mathbf{j}_{\alpha}^{\text{long}} = -iat_{\alpha\alpha+1} \sum_{i\in\text{plane}} \left(c_{\alpha i}^{\dagger} c_{\alpha+1i} - c_{\alpha+1i}^{\dagger} c_{\alpha i} \right), \quad (3.22)$$

where the summation is over all the sites i of the two-dimensional planes, and the notation αi denotes a lattice site on the αth plane at the ith location on the plane. Note that the subscript α on the current operator denotes the total current operator flowing through the αth plane, and does not indicate a Cartesian coordinate of the current operator; the current operator is always taken in the z-direction for the longitudinal flow. The current at plane α is thus defined to be the total number of electrons flowing to the left minus the total flowing to the right. For a derivation of Eq. (3.22), see Prob. A.27.

A comment is in order about the choice given in Eq. (3.22) for the current associated with the αth plane. First note that the form chosen is not the same as the choice that would arise from taking the commutator of the polarization operator (at the αth plane) with the Hamiltonian. The direct result from the commutator $\hat{j}_{\alpha} = i[\mathcal{H}, \sum_{i\in\text{plane}} z_{\alpha} c_{\alpha i}^{\dagger} c_{\alpha i}]$

$$\hat{j}_{\alpha} = -i \sum_{i\in\text{plane}} [t_{\alpha\alpha+1}(c_{\alpha+1i}^{\dagger} c_{\alpha i} - c_{\alpha i}^{\dagger} c_{\alpha+1i}) + t_{\alpha-1\alpha}(c_{\alpha-1i}^{\dagger} c_{\alpha i} - c_{\alpha i}^{\dagger} c_{\alpha-1i})] z_{\alpha}, \quad (3.23)$$

does not seem reasonable, because it is weighted by the z-coordinate of the αth plane, rather than involving the difference of currents moving in opposite directions (at the αth plane). When we have full translational symmetry, we derive the conventional form for the current operator by shifting the spatial index of one of the terms, to explicitly carry out the cancellation of the spatial coordinates (just take the summation of the above result over α, and shift $\alpha \to \alpha + 1$ in the last two terms). More reflection on this issue, shows that the explicit form of the local current operator that will enter the Kubo formula actually originates from the $-\mathbf{j} \cdot \mathbf{A}$ term that corresponds to the perturbation of the Hamiltonian due to the electric field; this is because we evaluate the expectation value of the total current with the perturbation of the Hamiltonian due to the external field and that field enters via the vector potential value at a specific plane [see Eq. (2.128)]. Hence the conductivity matrix is defined from the piece of the total current operator that couples to the field at plane α and, since the total current will be the sum of the currents at each plane, the current-current correlation function for the conductivity matrix involves the local current operators that couple to the vector potential. Hence, we choose the perturbation of the Hamiltonian to be

$$\mathcal{H}'(t) = -i|e|a \sum_{\alpha i} t_{\alpha\alpha+1}(c^\dagger_{\alpha+1i}c_{\alpha i} - c^\dagger_{\alpha i}c_{\alpha+1i})A_\alpha(t), \tag{3.24}$$

where we have taken the vector potential along the z-direction, and independent of the intraplane coordinates, because the field is uniform for each plane. We feel this choice makes good physical sense because we couple the vector potential to the physical current between the αth and $\alpha+1$st planes. Alternatively, one can view this as a coupling of the current between the αth and $\alpha+1$st plane to the electric vector potential located halfway between those two planes (in this interpretation, we would use $[A_\alpha + A_{\alpha+1}]/2$ as the coupling field). Finally, in the spirit of how we break up the energy polarization into pieces associated with each lattice site, one can take the local current operator to be $j_\alpha^{\text{long,sym}} = -iat_{\alpha-1\alpha}(c^\dagger_\alpha c_{\alpha-1} - c^\dagger_{\alpha-1}c_\alpha)/2 - iat_{\alpha\alpha+1}(c^\dagger_{\alpha+1}c_\alpha - c^\dagger_\alpha c_{\alpha+1})/2$, corresponding to the average of the currents located just to the left and to the right of plane α. This choice sounds like the most physical choice, but the calculations for it are somewhat more complicated, and it is not likely the end results are too different from our first choice. The difference between the two approaches is actually quite simple. In the first approach, one should envision the spatial indices α and β to correspond to $z_\alpha + a/2$ and $z_\beta + a/2$; that is, they are shifted to the right by half the distance between the planes. In the second, symmetrized approach, the α and β indices denote the planar indices. For this reason, we don't expect the final results to be too different for either approach. Due to the simplicity, we choose to take the current operator to be the current between the αth and $\alpha+1$st planes for our derivations below, and we show how to get the corresponding symmetrized results at the end.

As before, we will use the Kubo formula to find the *dc* conductivity matrix. We express it in terms of a polarizability matrix, which we first evaluate on the imaginary time axis, then we Fourier transform to the Matsubara frequencies, and finally we take the analytic continuation and the limit of the frequency going to zero to find the conductivity matrix. Hence

$$\sigma_{\alpha\beta}(0) = \lim_{\nu \to 0} \mathrm{Re} \frac{ie^2 \Pi_{\alpha\beta}(\nu)}{\nu}, \quad \Pi_{\alpha\beta}(i\nu_l) = \int_0^\beta d\tau e^{i\nu_l \tau} \langle \mathcal{T}_\tau \mathbf{j}_\alpha^{\mathrm{long}\dagger}(\tau) \mathbf{j}_\beta^{\mathrm{long}}(0) \rangle. \tag{3.25}$$

Since the vertex corrections vanish in the bulk DMFT when we are in the infinite-dimensional limit, we are going to make the approximation here that we neglect vertex corrections for the nanostructure as well. This step isn't even valid in the infinite-dimensional limit, because it relied on the velocity operators being odd in parity and the Green's functions being even, but parity is not a good quantum number when there is lattice inhomogeneity except for special planes, if the overall structure has a mirror-plane symmetry. Hence, the arguments used before to guarantee the vertex corrections would vanish will not hold any longer. Furthermore, we are looking at a three-dimensional system, so there is no guarantee that the vertex corrections vanish anymore, in any case. We continue to neglect them, nevertheless, because we expect they will be small, since they vanish in certain limits, and for a pragmatic reason, we don't have any straightforward way to evaluate them for all models (although they could be systematically calculated for the Falicov-Kimball model). A longer discussion of this issue can be found in the last section of this chapter.

Since we are neglecting the vertex corrections, the polarizability matrix will correspond to the bare function, which has the same functional form when represented in terms of Green's functions, as the noninteracting case. Hence, we can evaluate the polarizability matrix by employing Wick's theorem, which pairs each of the electron creation operators with a respective annihilation operator in the four-operator expectation value, and writes the four-operator expectation value as the sum of products of all different two-operator expectation values (*i.e.* Green's functions). We introduce this technique here, because it is an alternative way to derive our equations, and we have already seen Wick's theorem when we discussed the jellium problem.

Substituting the expressions for the current operators into the Kubo formula gives

$$\Pi_{\alpha\beta}(i\nu_l) = a^2 t_{\alpha\alpha+1} t_{\beta\beta+1} \int_0^\beta d\tau e^{i\nu_l\tau} \times \sum_{ij\in\text{plane}} \left\langle \mathcal{T}_\tau \left[c^\dagger_{\alpha+1i}(\tau)c_{\alpha i}(\tau) - c^\dagger_{\alpha i}(\tau)c_{\alpha+1i}(\tau)\right] \times \left[c^\dagger_{\beta j}(0)c_{\beta+1j}(0) - c^\dagger_{\beta+1j}(0)c_{\beta j}(0)\right]\right\rangle. \quad (3.26)$$

Now we perform the pairings as directed by Wick's theorem (this process is called evaluating the contractions) and we need only worry about the pairings between operators at time τ and those at time 0, because the other pairings can be shown to cancel themselves out. We also assume that we pair only creation and annihilation operators together, because we are in the normal state. Employing the notation

$$G_{\alpha\beta ij}(\tau) = -\langle \mathcal{T}_\tau c_{\alpha i}(\tau) c^\dagger_{\beta j}(0)\rangle, \quad (3.27)$$

we find that the polarizability matrix becomes

$$\Pi_{\alpha\beta}(i\nu_l) = a^2 t_{\alpha\alpha+1} t_{\beta\beta+1} \int_0^\beta d\tau e^{i\nu_l\tau} \sum_{ij\in\text{plane}} \Big[- G_{\beta+1\alpha+1ji}(-\tau)G_{\alpha\beta ij}(\tau) + G_{\beta\alpha+1ji}(-\tau)G_{\alpha\beta+1ij}(\tau) + G_{\beta+1\alpha ji}(-\tau)G_{\alpha+1\beta ij}(\tau) - G_{\beta\alpha ji}(-\tau)G_{\alpha+1\beta+1ij}(\tau)\Big]. \quad (3.28)$$

Expanding the Green's functions in the Fourier series in terms of the Matsubara frequency Green's functions, and performing the integral over τ yields

$$\Pi_{\alpha\beta}(i\nu_l) = a^2 t_{\alpha\alpha+1} t_{\beta\beta+1} T \sum_m \sum_{ij\in\text{plane}} \Big[- G_{\beta+1\alpha+1ji}(i\omega_m)G_{\alpha\beta ij}(i\omega_m + i\nu_l) + G_{\beta\alpha+1ji}(i\omega_m)G_{\alpha\beta+1ij}(i\omega_m + i\nu_l) + G_{\beta+1\alpha ji}(i\omega_m)G_{\alpha+1\beta ij}(i\omega_m + i\nu_l) - G_{\beta\alpha ji}(i\omega_m)G_{\alpha+1\beta+1ij}(i\omega_m + i\nu_l)\Big]. \quad (3.29)$$

The sum over the planar indices i and j can now be performed. Since the Hamiltonian is translationally invariant in the planar directions, the summations involve functions of $\mathbf{R}_i - \mathbf{R}_j$ and of $\mathbf{R}_j - \mathbf{R}_i$. The product of these two functions has a simple Fourier transform, so we get

$$\begin{aligned}\Pi_{\alpha\beta}(i\nu_l) = a^4 t_{\alpha\alpha+1} t_{\beta\beta+1} T \sum_m \sum_{\mathbf{k}^{\parallel}} \\ \times \Big[- G_{\beta+1\alpha+1}(\mathbf{k}^{\parallel}, i\omega_m) G_{\alpha\beta}(\mathbf{k}^{\parallel}, i\omega_m + i\nu_l) \\ + G_{\beta\alpha+1}(\mathbf{k}^{\parallel}, i\omega_m) G_{\alpha\beta+1}(\mathbf{k}^{\parallel}, i\omega_m + i\nu_l) \\ + G_{\beta+1\alpha}(\mathbf{k}^{\parallel}, i\omega_m) G_{\alpha+1\beta}(\mathbf{k}^{\parallel}, i\omega_m + i\nu_l) \\ - G_{\beta\alpha}(\mathbf{k}^{\parallel}, i\omega_m) G_{\alpha+1\beta+1}(\mathbf{k}^{\parallel}, i\omega_m + i\nu_l) \Big]. \end{aligned} \tag{3.30}$$

Now we need to perform the analytic continuation. We rewrite the Matsubara summations using contour integrations as in Fig. 2.9, substitute $f(\omega - i\nu_l) \to f(\omega)$, then let $i\nu_l \to \nu + i\delta$. After performing the substitution $\omega \to \omega + \nu$ in the relevant integrals, we are left with the final formula for the polarizability

$$\begin{aligned}\Pi_{\alpha\beta}(\nu) = \frac{a^4 t_{\alpha\alpha+1} t_{\beta\beta+1}}{\pi} \sum_{\mathbf{k}^{\parallel}} \int d\omega \\ \Big[f(\omega) \Big\{ - \mathrm{Im} G_{\beta+1\alpha+1}(\mathbf{k}^{\parallel}, \omega) G_{\alpha\beta}(\mathbf{k}^{\parallel}, \omega + \nu) \\ + \mathrm{Im} G_{\beta\alpha+1}(\mathbf{k}^{\parallel}, \omega) G_{\alpha\beta+1}(\mathbf{k}^{\parallel}, \omega + \nu) \\ + \mathrm{Im} G_{\beta\alpha}(\mathbf{k}^{\parallel}, \omega) G_{\alpha+1\beta}(\mathbf{k}^{\parallel}, \omega + \nu) \\ - \mathrm{Im} G_{\beta\alpha}(\mathbf{k}^{\parallel}, \omega) G_{\alpha+1\beta+1}(\mathbf{k}^{\parallel}, \omega + \nu) \Big\} \\ + f(\omega + \nu) \Big\{ - G^*_{\beta+1\alpha+1}(\mathbf{k}^{\parallel}, \omega) \mathrm{Im} G_{\alpha\beta}(\mathbf{k}^{\parallel}, \omega + \nu) \\ + G^*_{\beta\alpha+1}(\mathbf{k}^{\parallel}, \omega) \mathrm{Im} G_{\alpha\beta+1}(\mathbf{k}^{\parallel}, \omega + \nu) \\ + G^*_{\beta+1\alpha}(\mathbf{k}^{\parallel}, \omega) \mathrm{Im} G_{\alpha+1\beta}(\mathbf{k}^{\parallel}, \omega + \nu) \\ - G^*_{\beta\alpha}(\mathbf{k}^{\parallel}, \omega) \mathrm{Im} G_{\alpha+1\beta+1}(\mathbf{k}^{\parallel}, \omega + \nu) \Big\} \Big]. \end{aligned} \tag{3.31}$$

Using the definition of $\sigma_{\alpha\beta}(0)$ in Eq. (3.25), produces our last formula for the conductivity matrix, after noting that $-\partial_\omega f(\omega) = 1/4T\cosh^2(\beta\omega/2)$,

$$\sigma_{\alpha\beta}(0) = \frac{e^2}{ha^2}\frac{1}{2T}\int d\epsilon^{\|}\rho^{2d}(\epsilon^{\|})\int d\omega\frac{1}{\cosh^2\frac{\beta\omega}{2}}t_{\alpha\alpha+1}t_{\beta\beta+1}$$
$$\Big\{ -\mathrm{Im}G_{\alpha\beta+1}(\epsilon^{\|},\omega)\mathrm{Im}G_{\beta\alpha+1}(\epsilon^{\|},\omega)$$
$$-\mathrm{Im}G_{\alpha+1\beta}(\epsilon^{\|},\omega)\mathrm{Im}G_{\beta+1\alpha}(\epsilon^{\|},\omega)$$
$$+\mathrm{Im}G_{\alpha\beta}(\epsilon^{\|},\omega)\mathrm{Im}G_{\beta+1\alpha+1}(\epsilon^{\|},\omega)$$
$$+\mathrm{Im}G_{\beta\alpha}(\epsilon^{\|},\omega)\mathrm{Im}G_{\alpha+1\beta+1}(\epsilon^{\|},\omega)\Big\}, \tag{3.32}$$

where we extracted the relevant dimensionful constants (the integrand is taken to be dimensionless, all of the dimensions are included in the prefactor). Note that the indices α and β (for the conductivity matrix, not the Green's functions) correspond to the spatial locations between the α and $\alpha+1$ planes and the β and $\beta+1$ planes, respectively. If we want the symmetrized version of the conductivity matrix, it satisfies $\sigma^{\mathrm{sym}}_{\alpha\beta}(0) = [\sigma_{\alpha-1\beta-1}(0) + \sigma_{\alpha-1\beta}(0) + \sigma_{\alpha\beta-1}(0) + \sigma_{\alpha\beta}(0)]/4$. In this case, the spatial indices on the left-hand side are always the planar indices.

In order to determine the conductivity matrix, we need to compute the off-diagonal Green's functions that correlate different planes together. This is described in Prob. A.23. Note that one needs to specify the actual matrix dimensions of the conductivity matrix before calculating it. Since the barrier region is often the region with the largest contribution to the resistance (especially in situations where the leads are made from ballistic metals), we start with the planes at the center of the device and work outwards. In the case of ballistic leads, which provide only a contact resistance, since they have vanishing bulk resistivity, we need to take the matrix out to include only the first plane of the interface within the lead. In other cases, one takes the matrix size out to the point where the probes used to measure the voltage drop in the experiment lie.

Given the conductivity matrix in Eq. (3.32), we now show how to extract the resistance of the multilayered nanostructure. We begin from the relation between the electric field and the expectation value of the planar current density (per unit cell) in linear response

$$\langle \dot{\mathbf{j}}^{\mathrm{long,c}}_\alpha\rangle = a\sum_\beta \sigma_{\alpha\beta}(0)\mathbf{E}_\beta. \tag{3.33}$$

Assuming that the conductivity matrix is invertible, we multiply by its matrix inverse to find the electric field at each plane in the steady-state

$$\mathbf{E}_\alpha = \frac{1}{a}\sum_\beta [\sigma(0)]^{-1}_{\alpha\beta}\langle \mathbf{j}^{\mathrm{long,c}}_\beta\rangle, \tag{3.34}$$

where we can drop the index β from the current operator because it is conserved, and hence identical on each plane. Now we integrate the electric field over all of the planes to find the voltage drop across the device

$$V = a\sum_\alpha \mathbf{E}_\alpha = \sum_{\alpha\beta}[\sigma(0)]^{-1}_{\alpha\beta}\langle \mathbf{j}^{\mathrm{long,c}}\rangle \tag{3.35}$$

(the voltage drop is the difference of the voltage from the left and right leads, which accounts for the positive sign for the first equation above). Ohm's law says that $V = R_n I$, so we can immediately extract the resistance-area per unit cell product of the nanostructure

$$R_n a^2 = \sum_{\alpha\beta}[\sigma(0)]^{-1}_{\alpha\beta}. \tag{3.36}$$

The validity of this analysis, by demonstrating that the conductivity matrix is indeed invertible, will be presented in detail below.

Note that it is a resistance R_n that is extracted from a nanostructure calculation, and not a resistivity. This occurs because the nanostructure has a specific geometrical arrangement, and the measurement technique involves the measurement of the total current through the nanostructure and the voltage across it, leading to the extraction of the resistance. If this was a homogeneous system, we could multiply the resistance by appropriate geometrical factors and extract a resistivity, but that procedure makes no sense for an inhomogeneous system.

It is interesting, nevertheless, to take the bulk limit of the derivation that led up to Eq. (3.36). Doing so will allow us to extract a resistance and should yield Ohm's law [Ohm (1827)]. We first examine the case of a ballistic metal that has no scattering. Even though the resistivity vanishes in that case, a ballistic lead can have a finite voltage drop, and hence a finite resistance over it when we examine the voltage drop over the semi-infinite lead. The corresponding resistance is called a contact resistance, and was first discussed by Sharvin [Sharvin (1965)].

To get quantitative, we first note that in the bulk, the Green's function for a ballistic metal satisfies

$$\begin{aligned} G_{\alpha\alpha+n}(\epsilon^{\|},\omega) &= -\frac{i}{\sqrt{4-(\omega+\mu-\epsilon^{\|})^2}} \\ &\times \left[-\frac{\omega+\mu-\epsilon^{\|}}{2}+i\frac{\sqrt{4-(\omega+\mu-\epsilon^{\|})^2}}{2}\right]^{|n|} \\ &= -\frac{i}{\sqrt{4-(\omega+\mu-\epsilon^{\|})^2}}e^{i\theta|n|}, \end{aligned} \tag{3.37}$$

when $|\omega+\mu-\epsilon^{\|}| \leq 2$, which is when there is a nonzero imaginary part. The phase factor θ is determined by the term in the square brackets, which lies on the unit circle. If we examine the $\alpha\alpha+n$ component to the *dc* conductivity in the bulk at low temperatures (where we can treat the derivative of the Fermi factor as a delta function and integrate over ω and simply set $\omega = 0$), we find

$$\begin{aligned} \sigma_{\alpha\alpha+n}(0) &= \frac{e^2}{ha^2}\int d\epsilon^{\|}\rho^{2d}(\epsilon^{\|})\theta(|\mu-\epsilon^{\|}| \leq 2) \\ &\times \frac{4}{4-(\mu-\epsilon^{\|})^2}\left[-\cos(n+1)\theta\cos(n-1)\theta+\cos n\theta\cos n\theta\right]. \end{aligned} \tag{3.38}$$

The term in the brackets becomes $\sin^2\theta = 1-(\mu-\epsilon^{\|})^2/4$ which cancels the term in the denominator, so the integral becomes just an integral of the two-dimensional DOS from $\mu-2$ to $\mu+2$. Evaluating the integral at half-filling ($\mu = 0$) produces $\sigma_{\alpha\beta}(0) = 0.63e^2/ha^2$. This agrees with the Sharvin result that can be calculated directly from the Fermi surface area [the Sharvin conductance is $(k_F^2/4\pi)$ (e^2/h) per spin with k_F^2 a suitable average of the square of the Fermi wavevector for the noninteracting Fermi surface] [Freericks, Nikolić and Miller (2001)].

It is possible to show that in the general bulk case, where there is scattering in the system and a nonvanishing self-energy, the conductivity matrix is a function of $\alpha-\beta$ (because we now have a homogeneous problem). Now we want to work in Fourier space rather than real space to perform calculations. We write the conductivity matrix as

$$\sigma_{\alpha\beta}(0) = \sum_q \sigma_q(0)e^{iq(z_\alpha-z_\beta)} \tag{3.39}$$

with $\sigma_q(0) = \sum_{\alpha\beta}\sigma_{\alpha\beta}(0)\exp[-iq(z_\alpha-z_\beta)]/N^2$, and N the number of planes used in the system. Similar expressions can be written for $\sigma^{-1}_{\alpha\beta}(0)$,

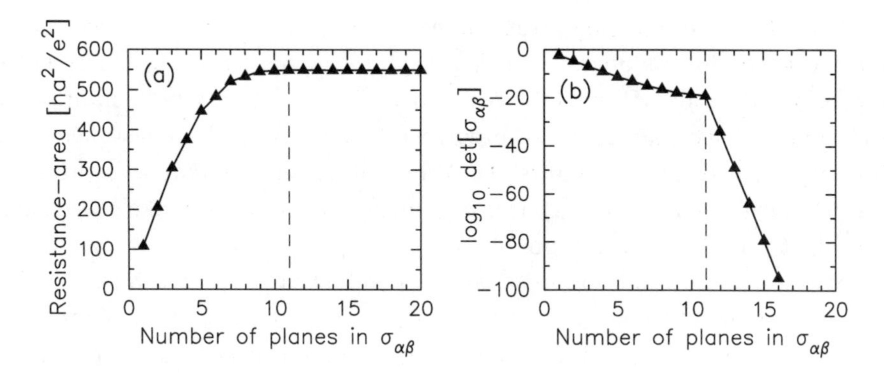

Fig. 3.8 Numerical properties of the conductivity matrix for a nanostructure with thirty planes in each lead and ten planes in the barrier. The leads are ballistic metals and the barrier is described by the Falicov-Kimball model at half filling with $U = 5$. The dashed line shows when the conductivity matrix includes the first plane of the lead. In panel (a), we plot the resistance-area-per-unit-cell product as calculated from the truncated conductivity matrix. Note how we approach the final limiting value once we reach the barrier-lead interface. In panel (b), we plot the logarithm of the determinant of the truncated conductivity matrix. Note how the determinant changes sharply when we enter the lead, which is expected because there is no scattering in the lead.

and a direct calculation shows that $\sigma_q^{-1}(0) = 1/(\sigma_q(0)N^2)$. Then we can directly compute the resistance, and find it satisfies the Ohm's law scaling $R_n = \rho_{\text{bulk}} Na/a^2$, with $\rho_{\text{bulk}}^{-1} = a\sum_{\alpha\beta}\sigma_{\alpha\beta}(0)/N$ (recall the length of the system is $L = Na$). Hence, our formalism has the correct bulk limit.

One issue of these calculations is does the truncated conductivity matrix have a well-defined matrix inverse so that the resistance calculation is well defined? We illustrate that this is so in Fig. 3.8. This is a plot of properties of the conductivity matrix for a ten-plane barrier described by the Falicov-Kimball model with $U = 5$. Note how the conductivity rapidly converges to its final value as the number of planes in the truncated matrix becomes larger than the size of the barrier. In panel (b), we plot the logarithm of the determinant. Note how it starts to sharply decrease once the truncated conductivity matrix includes the barrier and the first plane in the lead. This is expected, since the conductivity matrix in the lead assumes one of two values depending on whether α is larger or smaller than β and quickly becomes nearly singular. Nevertheless, the computation of the resistance-area-per-unit-cell product is completely robust and well-defined. As a general rule, one should truncate the matrix at the size of the barrier plus one plane on each side to calculate the resistance of a junction.

Note that we can now argue *a posteriori* that the choice we made for the local current operator in Eq. (3.22) is a good choice. Our results reproduce the Sharvin contact resistance, have a proper bulk limit, and the conductivity matrix has the correct behavior for the resistance of a junction with ballistic metal leads. While other choices may also produce these results, we feel the physical arguments given above make the choice we take to be the most reasonable one.

We make one final comment about conductivity calculations. In the case where the chemical potential is either fixed at $\mu = 0$ for symmetry reasons, or if we are working at low enough temperatures that the chemical potential does not change much with T, then there is a huge simplification in the computation of the resistance as a function of T. Since the DOS and self-energies of the nanostructure do not depend on temperature for the Falicov-Kimball model, we can perform the calculation of the resistance for a range of temperatures in parallel, by simply modifying the value of the derivative of the Fermi factor that goes into the integral for the truncated conductivity matrix. This saves much computation time, since we do not need to recalculate all of the off-diagonal Green's functions for each temperature. Unfortunately, this speed up will work only when both the DOS and the chemical potential are (nearly) independent of T.

3.6 Charge Reconstruction (Schottky Barriers)

In this section, we describe how to modify the DMFT procedure to include the charge reconstruction that generically takes place at each interface in a multilayered nanostructure. The reason why is simple—the nanostructure has its chemical potential set by the leads, which extend out to infinity in each direction. If the chemical potential of the barrier region is different from that of the leads, then the barrier is in an unstable electronic state, which requires charge to rearrange itself, forming screened dipole layers at each interface, until the system reaches a steady state with a static redistribution of charge. This effect is called the Schottky effect [Schottky (1940)] when the interface is between a conventional semiconductor and an *sp* metal. Here we show how to determine this effect for strongly correlated systems.

The approach will be a semiclassical one. Since long-range interactions can be treated in a mean-field sense in DMFT, such an approach is consistent with the local approximation for the self-energy that we have been taking throughout the text. We will first calculate the electronic charge on each plane via the Green's function approach of Potthoff and Nolting [Potthoff and Nolting (1999a)]. This calculation can be performed on the imaginary (Matsubara) frequency axis because the charge can be found by summing the Green's functions over all frequencies (with an appropriate regularization scheme). For some models, like the Hubbard model, or the PAM, one cannot perform the calculations solely on the imaginary axis with the NRG, so they will need to use a real-axis code to find the charge densities on each plane (this complication will be described in detail below). Next, we find the charge deviation on each plane; namely, we determine whether extra charge has entered or left the plane. Since the positive background charge of the ions remains the same, the charge deviation will give rise to an electric field. There are two different ways to treat this field. The simplest is to assume the electric charge is uniformly spread over the plane [Nikolić, Freericks and Miller (2002a)]. Then the electric field is constant, perpendicular to the plane, and pointing away from it in both directions if the net charge density is positive, while pointing toward the plane if the net charge density is negative. The second method uses the actual distribution of the ions, and the spatial profile of the electrons, if available, to calculate the charge [Okamoto and Millis (2004a); Okamoto and Millis (2004b)]. This approach is closer to an Ewald-like summation [Ewald (1921)] of the charge densities. It is expected that the two treatments should yield similar results.

The magnitude of the field when treated as a plane of uniform charge density is just

$$|\mathbf{E}| = \frac{|e||\rho_\alpha - \rho_\alpha^{\mathrm{bulk}}|a}{2\epsilon_0\epsilon_{\mathrm{r}\alpha}}, \tag{3.40}$$

where ρ_α is the quantum-mechanically calculated electron number density at plane α, $\rho_\alpha^{\mathrm{bulk}}$ is the bulk electron number density for the material that plane α is composed of, ϵ_0 is the permittivity of free space, and $\epsilon_{\mathrm{r}\alpha}$ is the relative permittivity of plane α. The contribution to the electric potential $V^{\mathrm{c}}(z)$ from this field satisfies

$$E = -\frac{d}{dz}V^{\mathrm{c}}(z). \tag{3.41}$$

Since the electric field is constant in magnitude, it is straightforward to compute the contribution to the Coulomb potential at plane β due to the change in the charge density at plane α:

$$V^{\mathrm{c}}_{\beta}(\alpha) = \frac{|e|(\rho_\alpha - \rho_\alpha^{\mathrm{bulk}} - \bar{\rho})a}{2\epsilon_0} \begin{cases} \sum_{\gamma=\alpha+1}^{\beta}[\frac{1}{2\epsilon_{\mathrm{r}\gamma}} + \frac{1}{2\epsilon_{\mathrm{r}\gamma-1}}], & \beta > \alpha \\ 0, & \beta = \alpha \\ \sum_{\gamma=\alpha-1}^{\beta}[\frac{1}{2\epsilon_{\mathrm{r}\gamma}} + \frac{1}{2\epsilon_{\mathrm{r}\gamma+1}}], & \beta < \alpha \end{cases} . \quad (3.42)$$

Note that if the relative permittivity is a constant, independent of the planes, then the potential energy is proportional to $-|\mathbf{z}_\alpha - \mathbf{z}_\beta|/\epsilon_{\mathrm{r}}$ as one might expect. The reason why we need to sum over two terms in Eq. (3.42) is because we envision the αth plane of charge to be infinitesimally thick, and go through the lattice sites of plane α, but the dielectric has a thickness of a and is centered around the plane of atoms. Hence, if the permittivity changes from one plane to another, a polarization charge develops half-way between the two planes where the dielectric is changing, and the electric field has a discontinuity at that point (see Fig. 3.9).

We have added a term $\bar{\rho}$ to Eq. (3.42). This is used to help the calculations converge more easily. The parameter $\bar{\rho} = \sum_\alpha(\rho_\alpha - \rho_\alpha^{\mathrm{bulk}})/N$ with N the total number of self-consistent planes in the simulation. The parameter $\bar{\rho}$ is the average excess electron density, spread uniformly through the system. When the algorithm has converged, the parameter $\bar{\rho}$ will be close to zero; this needs to be checked in any calculation, of course.

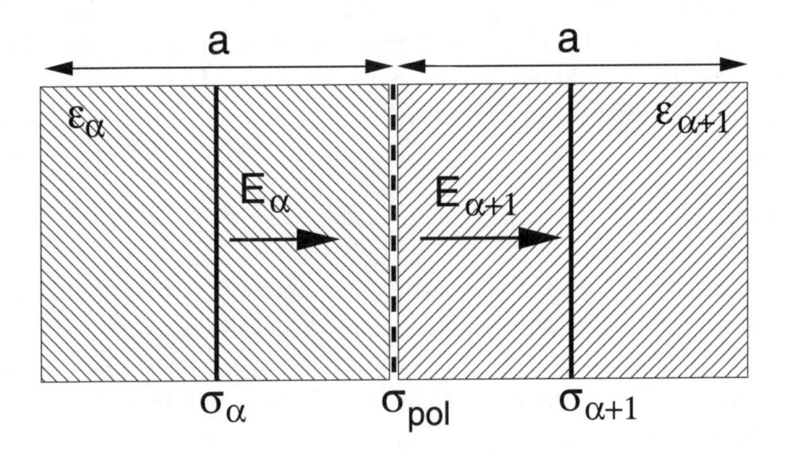

Fig. 3.9 Geometry taken for the classical electrostatics problem. We show the blow up of two planes, α and $\alpha+1$. Assuming the charge density on plane α is $(\rho_\alpha - \rho_\alpha^{\mathrm{bulk}})a = \sigma_\alpha$ and the relative permittivity is $\epsilon_{\mathrm{r}\alpha}$ (and similarly for the $\alpha+1$ plane), then the change in polarization at the interface between the two dielectric planes induces a polarization charge on the interface (denoted σ_{pol}) that leads to a discontinuous jump in the electric field halfway between the two lattice planes. Once the fields are known, we integrate to get the electric potentials. Note the discontinuity in the electric field occurs at the *midpoint* between the two lattice planes.

It is actually the potential energy $-|e|V^{\mathrm{c}} = V$ that shifts the chemical potential at each planar site. We define a parameter

$$e_{\mathrm{Schot}}(\alpha) = \frac{e^2 a}{2\epsilon_0 \epsilon_{\mathrm{r}\alpha}}, \tag{3.43}$$

which determines how the extra charge density decays away the interfaces. The parameter e_{Schot} has the units of an energy multiplied by an area; the product of e_{Schot} with the local DOS has units of the inverse of a length, and this is what determines the decay length of the charge profile. Using this parameter, we can immediately calculate the potential energy due to the Coulomb interaction

$$V_\beta = -\sum_\alpha (\rho_\alpha - \rho_\alpha^{\mathrm{bulk}} - \bar{\rho}) \begin{cases} \sum_{\gamma=\alpha+1}^{\beta} \frac{1}{2}[e_{\mathrm{Schot}}(\gamma) + e_{\mathrm{Schot}}(\gamma-1)], & \beta > \alpha \\ 0, & \beta = \alpha \\ \sum_{\gamma=\alpha-1}^{\beta} \frac{1}{2}[e_{\mathrm{Schot}}(\gamma) + e_{\mathrm{Schot}}(\gamma+1)], & \beta < \alpha \end{cases}. \tag{3.44}$$

Note that a similar analysis can be carried out if one uses the Ewald-like technique for determining the charge reconstruction. Note further that $\bar{\rho}$ is used to improve the convergence of the iterations and it vanishes for the (converged) final fixed point solution.

These potential energies modify the Hamiltonian by the long-range Coulomb interaction of the charge reconstruction. The additional piece of the Hamiltonian (due to the charge rearrangement) is

$$\mathcal{H}_{\mathrm{charge}} = \sum_\alpha V_\alpha \sum_{i \in \mathrm{plane}} c_{\alpha i}^\dagger c_{\alpha i}. \tag{3.45}$$

Hence, they can be treated by shifting the chemical potential $\mu \to \mu - V_\alpha$ on each plane depending on what the Coulomb potential energy is for the given plane. For consistency, we must have that the potentials go to zero as we move far enough into either of the leads. This requirement enforces overall charge conservation—any charge that moves out of the barrier remains in the leads, localized close to the interface, and *vice versa*. Of course, the potentials V_α that appear in the electronic charge reconstruction Hamiltonian in Eq. (3.45) must be determined self-consistently.

There will be no electronic charge reconstruction if the chemical potentials in the bulk of both the leads and the barrier match. In order to have freedom to adjust the mismatch of the chemical potentials, we need to be able to change the value of the band zero of the barrier relative to the band zero of the leads. This parameter is $\Delta E_{F\alpha}$, which vanishes in the

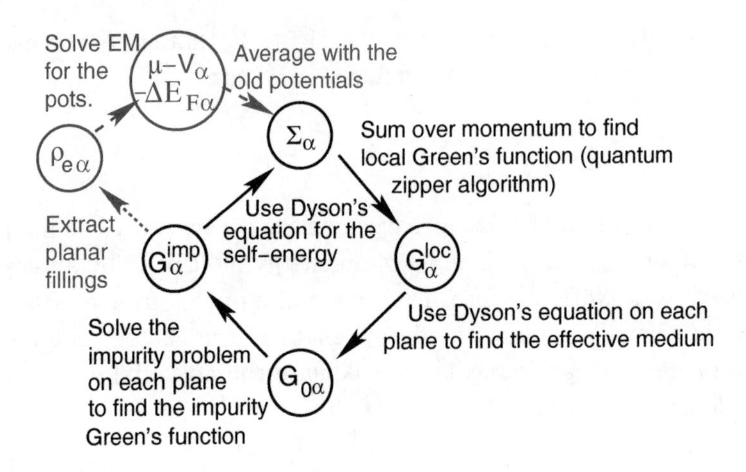

Fig. 3.10 Flow diagram for the DMFT algorithm in a multilayered nanostructure with electronic charge reconstruction. The main algorithm in black is the same as before, and is all that is needed on the real axis, if an imaginary-axis formulation is available. The red portion describes how to perform the self-consistent charge reconstruction. We determine the charges on each plane, determine how they differ from the bulk charge on the plane to find the excess or deficit charge. Then we use classical electromagnetism to find the electric potentials on each plane and then the contribution of the potential energy to the electrochemical potential on each plane. Then we average the potentials with a large damping factor so that the potentials are updated slowly (typically we take 99% or more of the old potential in the average). This then is all that is needed for the next loop of the algorithm.

leads, and is generically a nonzero constant in the barrier (independent of the temperature or the charge rearrangement). Hence we add an additional term $-\sum_\alpha \sum_{i\in\text{plane}} \Delta E_{F\alpha} c^\dagger_{\alpha i} c_{\alpha i}$ to the Hamiltonian.

The DMFT algorithm is then modified as shown in Fig. 3.10, with the new steps indicated in red. If there is a separate algorithm available for the Matsubara frequency Green's functions, then we use the red part on the imaginary axis, and we don't need it on the real axis. If such an algorithm does not exist (such as when we perform calculations with the NRG), then we would use the red part on the real axis. The new steps are to first find the electron density on each plane. Then we subtract the bulk charge density of each plane to find the excess or deficit charge on the given plane. Once the change in charge density is known, we can calculate the electrical potential, and then the contribution to the potential energy. This gets added to the chemical potential to determine the electrochemical potential at each plane. But we need to average that result with the old potentials in order to be able to slowly converge to the final answer. If the potentials are updated too much, then the system of equations migrates to an unphysical fixed point, or limit cycle. Typically, we use at least 99% of the old potential (or more) in each averaging. Then we have all the inputs for the next loop in the algorithm.

It is significantly more difficult to reach the correct physical fixed point when there is a charge reconstruction. If the iterations are performed without significant damping, to slow the updating of the potentials and the Green's functions, then they will not converge to the correct solution of developing screened dipole layers at each interface, but instead, will often converge to a screened dipole sitting in the middle of the barrier. While there is no way to fully guarantee a robust solution to this problem, we have found that the following strategies often yield good results: (i) the updating of the potentials must be significantly damped by the equation

$$V_\alpha^{next\ iteration} = \alpha_V V_\alpha^{old} + (1-\alpha_V)V_\alpha^{new}, \tag{3.46}$$

where we usually take $0.99 \le \alpha_V \le 0.999$; (ii) updating of the self-energy should continue to be damped as in Eq. (2.116); (iii) the equations should converge to a level of at least 5×10^{-8}; and (iv) the total charge deviation $\bar{\rho}$ (summed over all simulated planes) should be smaller than $2-3\times 10^{-5}$. We find that the above computational scheme seems to work in calculations where e_{Schot} does not change through the lattice, and is relatively small (≤ 0.5), and we choose $\alpha_V = 0.99$; the number of iterations needed is on the order of $2000-5000$. When e_{Schot} is larger on one set of planes, then we need to increase α_V (usually 0.995 is large enough), and we may need to anneal in the increase over a series of calculations. The annealing process would start by running the calculation for a uniform small value of e_{Schot}, then slowly increase it in the barrier (or in the leads), in small steps (like in steps of 0.5), using the potentials, Green's functions, and self-energies from the previous converged calculation as the starting point for the new calculation. When performed in this fashion, one can often stabilize what would otherwise yield an unstable iterative process.

The phenomenon of electronic charge reconstruction is illustrated in Fig. 3.11 for a barrier of 20 planes attached to leads. There is no scattering in any of the planes (all are ballistic metals), but the center of the band for the barrier is shifted by an amount ΔE_F with respect to the position it would need to have to yield the required filling [$\rho_e = 1.0$ for panels (a) and (b) and $\rho_e = 0.01$ for panels (c) and (d) and the chemical potential is set by the leads to be $\mu = 0$ in these examples] the exact shifts are written in the caption to the figure; we are including spin in these calculations. The top panels show the case for metals, while the bottom panels show the

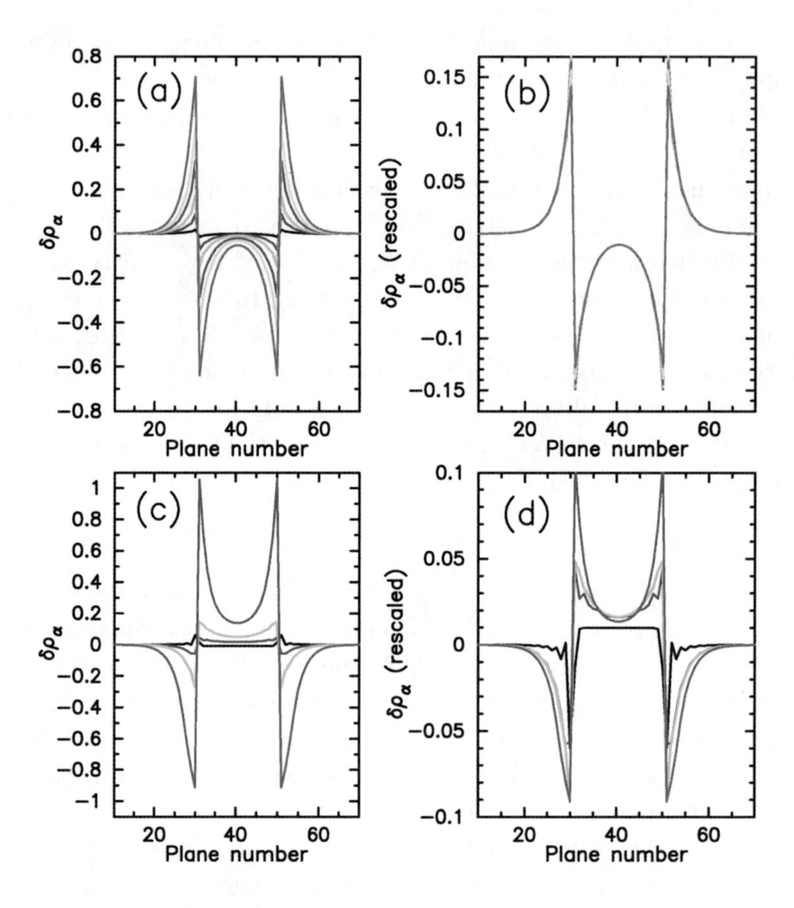

Fig. 3.11 Electronic charge reconstruction for a multilayered nanostructure consisting of a barrier region of 20 planes. The bulk charge density of the leads is equal to 1.0 (spin is included here) and of the barrier is equal to 1.0 in panels (a) and (b) and 0.01 in panels (c) and (d). The band for the barrier is shifted by an amount ΔE_F, which changes for the different curves [$\Delta E_F = 0.1$ (black), 0.5 (red), 1 (green), 2 (blue), 3 (cyan), and 5 (magenta) for panels (a) and (b) and $\Delta E_F = 1$ (black), -1 (red), -3 (green), and -10 (blue) for panels (c) and (d)]. Panels (a) and (c) show the electronic charge reconstruction, while panels (b) and (d) are rescaled by dividing the charge deviation by ΔE_F. The bands are uncorrelated here. *Figures adapted with permission from* [Freericks, Nikolić and Miller (2002)] (original figure ©World Scientific Publishing Co. Pte. Ltd., Singapore).

case of a metal-doped semiconductor-metal device. The screening length, as determined by the parameter $e_{\mathrm{Schot}}(\alpha)$, is about 2.2 lattice spacings. We choose $e_{\mathrm{Schot}}(\alpha) = 0.4$ throughout the device (in both the leads and the barrier).

The physics of the system is quite rich. To begin, note that the charge deviation grows as the mismatch of the chemical potentials increases. In the half-filled case [panels (a) and (b)] the results are symmetric with the sign of ΔE_F, as long as the sign of the charge deviation is changed as well. The charge deviations do not change much with respect to temperature, and the total charge deviation is smaller than $2-3\times10^{-5}$ in the self-consistent calculations. Since the various curves in panel (a) all seem to share the same general shape, we tried to rescale them by dividing by the shift of the band ΔE_F in panel (b). One can see that the rescaling works extraordinarily well, with deviations occurring only very close to the interfaces. Note as well, that as the chemical potential mismatch increases, the barrier is not thick enough to have the charge density heal to its bulk value. This kind of behavior is seen in the experimental results shown in the bottom panel of Fig. 1.16.

The behavior for the doped semiconductor case in panels (c) and (d) is similar, but has some notable differences. The results do not have any symmetry with respect to the sign of ΔE_F here. The curves appear to have a different shape as well, as we look at different chemical potential mismatches. Indeed, the scaling exercise, illustrated in panel (d) clearly shows that the system does not satisfy the same scaling as what was seen in the metallic case. This most likely is arising from the fact that the local DOS is nearly constant at half filling, but is quite asymmetric as we dope off of half filling, and sit close to the lower band edge. The asymmetry or nonconstant behavior is what probably leads to a breakdown of the scaling.

The effect of different screening lengths in the leads and in the barrier can be seen in Fig. 3.12. The left panel plots the potential, which develops a kink at the interface as the screening lengths are made to be different (see inset), and the right panel shows the charge density, which heals faster as the screening length decreases. In the inset one can see this illustrated by the charge at the center of the barrier. In these figures, the barrier is a small-gap insulator, with $w_1 = 0.5$ and $U^{\mathrm{FK}} = 6$, while the leads are ballistic. One can see that the screening and electronic charge reconstruction do not have a significant qualitative or quantitative change as interactions are turned on: compare the black curve on the right panel of Fig. 3.12 with the green curve in panel (a) of Fig. 3.11 (noting that there is spin included in the latter figure, but not in the former).

In Fig. 3.13, we show how the charge profile changes as a function of the thickness of the barrier. The leads are strongly correlated metals with $U^{FK} = 4$ and the barrier is a small-gap insulator with $U^{FK} = 6$.

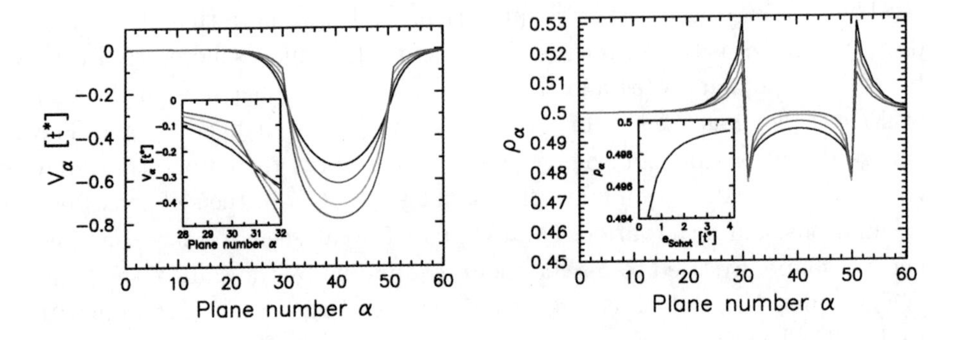

Fig. 3.12 Potential (left) and charge density (right) for the electronic charge reconstruction of a multilayered nanostructure consisting of a barrier region of 20 planes. The bulk charge density of the leads and the barrier is equal to 0.5 (no spin). The band for the barrier is shifted by an amount $\Delta E_F = -1.0$ and the barrier has a Falicov-Kimball interaction of $U^{\mathrm{FK}} = 6$ and $w_1 = 0.5$, which is just on the insulating side of the metal-insulator transition. The leads are ballistic, with $U^{\mathrm{FK}} = 0$. The screening parameter e_{Schot} is equal to 0.4 in the leads and varies from 0.4 (black) to 0.8 (red) to 2.0 (green) to 4.0 (blue) in the barrier. Note how the potential (left panel) goes to zero when deep in the leads, gets deeper as the screening length is reduced in the barrier, and develops a kink at the interface (plane number 30.5) as blown up in the inset. The kink determines the jump in the electric field at the interface. The right panel shows the charge density which heals much faster as the screening length is reduced. This is clarified in the inset, where the charge on the central plane of the barrier is plotted as a function of e_{Schot}.

The parameter e_{Schot} which determines the screening lengths is 0.4 in the leads and 2.0 in the barrier. Note how the charge deviation in the barrier cannot heal to zero for the thinner barriers even though the screening length is smaller than a lattice spacing in the barrier. This occurs because the total charge outside the barrier is too large to allow the charge within the barrier to be restored to its bulk value when the barrier is thin. Such a result is similar in qualitative nature to the experimental results shown in Fig. 1.16, except that the curvature of the charge profile in the barrier is opposite in the calculations and experiment. As the experimental barriers are made thicker, we expect that the curvature will change to be similar to the calculated results and indeed this is observed in experiment [Varela, *et al.* (2005)].

One other thing to note from these numerical examples is that the charge reconstruction at the interfaces of the nanostructure does not change much with the scattering strength within the different levels. The screened dipole layers that form at the interfaces are rather insensitive to changes in the correlations within the different planes. Hence, they can can be deter-

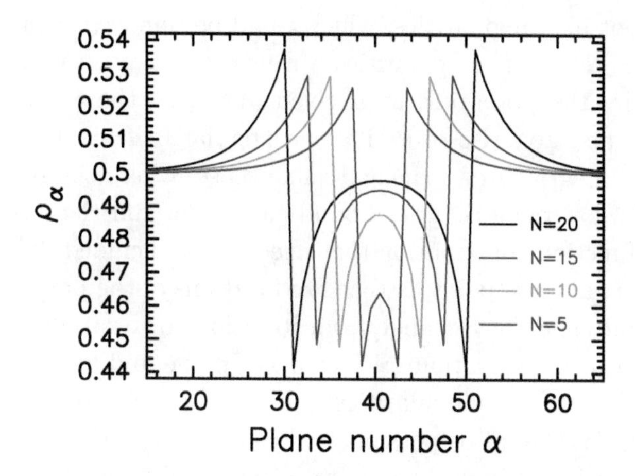

Fig. 3.13 Charge profile for a multilayered nanostructure with $U^{FK} = 6$ in the barrier and $U^{FK} = 4$ in the leads. The screening length is determined by e_{Schot} which is equal to 0.4 in the leads and 2.0 in the barrier. The mismatch of the chemical potentials is $\Delta E_F = 2.0$. The barrier thickness ranges from $N = 20$ (black) to $N = 15$ (red) to $N = 10$ (green) and $N = 5$ (blue). The different curves are offset so that the center of the barriers line up.

mined quite accurately from a noninteracting system. This observation can come in handy in trying to perform calculations for strongly correlated models that do not have a simple formulation on the imaginary-frequency axis. If we try to perform these kinds of calculations for a system where the impurity solver employs NRG, then we have to work directly on the real axis. This increases the numerical calculations for two reasons—first, the computational time for a real-axis solver is usually longer than for an imaginary-axis solver, and second, the inhomogeneous DMFT algorithm for the real axis when there is a charge reconstruction is significantly complicated because the numerical integrations now have numerous principal-value integrations, which are more difficult to handle than in the case when there is no charge reconstruction.

In order to determine the transport through the device we need to evaluate the real-axis results for the nanostructure with an electronic charge reconstruction. Unfortunately, the algorithms used when there is no electronic charge reconstruction cannot be simply employed for this case. The reason why is that the presence of the different potentials V_α on each plane causes the nature of the integrands over the two-dimensional DOS to have a different singular behavior than they had before. Previously, we focused on square-root-like singularities, which could be removed by a simple variable change. Now, the singularities are poles (because the denominators are shifted by the potentials at a given plane, so they vanish at different energies, and give rise to a different singular behavior), so we need to evaluate the integrals in a principal-value sense, where the real part is integrated with a symmetric grid around each pole, and the imaginary part has a delta-function contribution that needs to be included. The challenge, from a numerical standpoint, is that the locations of the poles change from plane to plane, and they change from iteration to iteration, so one needs to generate a new integration grid for every plane and for every iteration, to properly handle the singularities. No one has yet constructed such a code, although it should be possible to do so in principle. Instead, a "poor-man's approach" to the problem is to instead add a small imaginary part to the frequency (or equivalently to the self-energy), and evaluate the integrals slightly off the real axis. Then the imaginary part of the frequency smooths out the singularities, and they can be integrated with the same technique as used before. If the self-energy has an imaginary part larger in magnitude than the imaginary piece added to the frequency, the calculation is unchanged. It is only in regions where the self-energy has a small imaginary part that the calculation is modified. One needs to ensure that the quadrature grid spacing is small enough that one can properly capture enough points in the "near singularity" to properly integrate it with the quadrature rule. Then by performing the calculation for smaller imaginary parts, the results can be scaled down to the limit where the imaginary part vanishes. This is a more practical way to perform the calculations, but one should note that if there is a region where the DOS is becoming exponentially small, it is likely that adding the small imaginary part will change the value of the resistance, so one needs to check results for different sizes of the imaginary part to make sure they are robust, and represent the correct results. Details of solutions of the real-axis properties are given in Chapter 6 when we discuss thermal transport.

It is important to examine how the linear-response transport formalism is modified by the electronic charge reconstruction. We have taken the chemical potential as a constant throughout the multilayered nanostructure for thermodynamic equilibrium. One can directly show that the device carries no charge current even though there are nonzero electric fields arising from the electronic charge reconstruction (see Prob. A.26). No current flows because the putative current driven by the internal electric fields is canceled by an equal magnitude but oppositely directed current driven by the concentration gradient. The standard way to describe this result is via a phenomenological equation (for the case with no thermal gradients) [Onsager (1931a); Onsager (1931b)]

$$\langle j^{\mathrm{c,long}}\rangle = a\sum_{\beta}\sigma_{\alpha\beta}E_{\beta} - a|e|\sum_{\beta}\mathcal{D}_{\alpha\beta}\frac{\rho_{\beta+1}-\rho_{\beta}}{a} = -\frac{a}{|e|}\sum_{\beta}\sigma_{\alpha\beta}\frac{\tilde{\mu}_{\beta+1}-\tilde{\mu}_{\beta}}{a}, \tag{3.47}$$

where $\mathcal{D}_{\alpha\beta}$ is the diffusion constant for Fick's law of diffusion [Fick (1855)], and the second equality follows from the Einstein relation [Einstein (1905)] (or more correctly the Nernst-Einstein-Smoluchowski relation [Nernst (1889); Smoluchowski (1906)]) which relates the diffusion constant to the conductivity via

$$\sigma_{\alpha\beta} = e^2\mathcal{D}_{\alpha\beta}d\rho_{\beta}/d\mu. \tag{3.48}$$

The symbol $\tilde{\mu}_{\alpha} = \mu - V_{\alpha}$ is called the electrochemical potential. The Einstein relation can be derived by relating the gradient with respect to the chemical potential to the gradient with respect to the number concentration via the chain rule: $d\mu/dz = (d\mu/d\rho)d\rho/dz$, and the fact that the current vanishes in equilibrium [Luttinger (1964)].

Equation (3.47) implies that the condition for there to be no charge current is simply $d\tilde{\mu}/dz = 0$. The chemical potential is a constant, but it does vary with the filling, so if there is a change in electron concentration, then $d\tilde{\mu}/dz = (d\mu/d\rho)d\rho/dz - dV(z)/dz$, so the force from the electric field will be balanced by the force from the change in electron concentration. In addition, note that the current vanishes no matter how large the variation in the concentration is (*i. e.*, beyond the linear-response regime), so the conclusion is that the current generated by the internal electric field is always canceled by the current generated by the change in the electron concentration. Hence, for a linear-response treatment of transport, we can ignore the forces due to the internal electric fields and the concentration gradients, because they always cancel, and we can limit our focus to the effects of the external electric field only. This then implies that all of the analysis performed previously for the charge current continues to hold, and because the form of the charge current is unchanged when we have electronic charge reconstruction, the Kubo formula is identical as it was before (with the effects of the potentials V_{α} included, of course).

3.7 Longitudinal Heat Transport Through a Nanostructure

The basic idea behind a thermoelectric cooler or power generator is that there is a difference between the weighting factors that determine the bulk charge current and heat current. The charge current is weighted by the electron velocity, while the heat current is weighted by the velocity multiplied by the kinetic energy minus the chemical potential plus a term from the potential energy. Hence, one can create charge current without heat current, or *vice versa*; by carefully engineering the way electrons move through the device, one can control both the energy and charge flow.

The typical thermoelectric element consists of two metallic reservoirs connected by two legs of different materials—one leg is a n-type conductor, with the current carried by electrons, and the other leg is a p-type conduc-

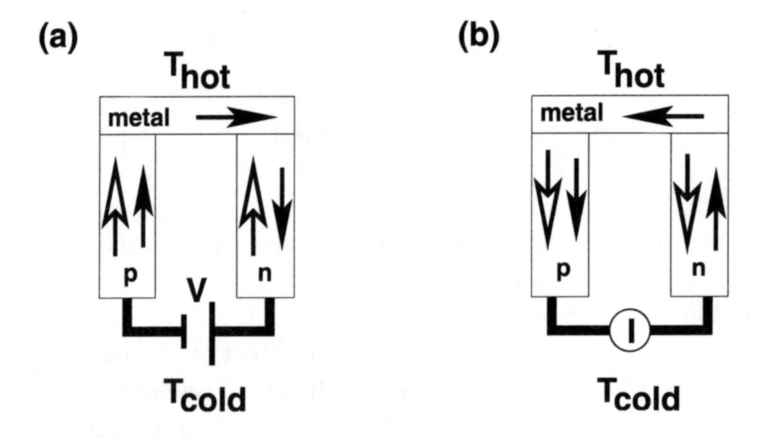

Fig. 3.14 Thermoelectric device schematics for (a) a cooler/refrigerator and (b) a power generator. The top is the hot reservoir, while the bottom is the cold reservoir. The right leg is an n-type material and the left leg is a p-type material. In the refrigerator in panel (a), a voltage source drives a charge current in a clockwise direction, which generates a heat current flow from the cold reservoir to the hot reservoir (because the p-type material carries heat flow in the same direction as the charge current flow, while the n-type material carries heat flow in the opposite direction as the charge current flow). This creates a net cooling effect if the thermo-electric driven heat current is larger than the conventional heat conducted from the hot to cold reservoir, and the internal Joule heating associated with the charge current flow. In the power generator in panel (b), a heat source maintains a temperature difference between the hot and cold reservoirs. The conventional thermal conductivity drives a heat current in both legs in the direction from the hot reservoir to the cold reservoir. This generates a counter-clockwise charge current flow, because of the way the charge current is "dragged" by the heat current in each leg. The current flow can be tapped similarly to the current flow from a battery for running electrical devices.

tor, with the current carried by holes. The basic set-up for a cooler (left) and a power generator (right) is given in Fig. 3.14. In the case of a cooler, a battery is attached between the legs, to drive current through the device in a clockwise direction. Heat is transported in the same direction as the charge current for the p-type conductor, while it is carried in the opposite direction for the n-type conductor. This produces a net heat flow from the bottom to the top, resulting in a lower T for the lower reservoir than for the upper. Reversing the direction of the current flow changes the direction of the heat flow, so the hot and cold reservoirs are switched. This implies that the thermoelectric device can be used as a cooler or as a heater just by changing the direction of the current flow, or similarly changing the polarity of the voltage source. Since conventional thermal conduction carries heat in the opposite direction from that driven by the charge current, one needs to compare the heat flow generated from the thermopower to that from conventional thermal conduction and to the bulk heat generated from Joule heating. The balance between these thermal heat flows determines the thermoelectric figure-of-merit ZT which optimizes the thermoelectric properties. Thermoelectric coolers are commercially available and are usually made from semiconductor alloys of Sb, Bi, and Te. They are much less efficient than compressor-based refrigerators, and the materials usually have a ZT value close to 1.

The power generator works with a different mode of operation. The upper reservoir is heated by a heat source, which causes heat to flow from the hot reservoir to the cold reservoir. Accompanying the heat flow is a charge current flow, which goes in a clockwise direction. The charge current flows until a voltage has developed between the legs that impedes further flow of charge. This device thus converts a temperature difference into a voltage which can be drawn on to provide current flow, just like a chemical battery. Once again, it is not as efficient as other battery sources, but because they have no moving parts, and heat sources (from radioactive materials) can have long lifetimes, they are the most reliable battery source for unmanned deep space probes. They also are being investigated as a means to recycle waste heat (from car exhaust systems, or industrial chimneys) and convert that heat into electrical power. This application is not yet commercially viable.

In actual devices, the basic thermoelectric element is joined together into a module, which cools in stages, with the total temperature difference coming from the difference of the hot reservoir in the first element and the cold reservoir in the last element. Cooling can be achieved down to about

150 K with such devices. For power generation, the module approach is also useful for generating high voltage sources, since the maximal voltage for a particular element will come from the temperature difference and the materials properties of the thermoelectric element (since thermopower is measured in μV/K, a single-stage thermoelectric power generator will usually generate less than 1 V).

In this text, we concentrate on nanostructures, which can be used to compose one of the legs of the thermoelectric device. This should be kept in mind as we describe different kinds of calculational techniques. We also concentrate solely on electronic transport mechanisms. In most thermoelectrics, the thermal conductivity from phonons can be large enough to significantly reduce the figure-of-merit. While we do not treat phonons in this book, it is expected that the phonon thermal conductivity will be further reduced in a nanostructure, because the interfaces in the nanostructures will cause significant phonon scattering if the masses of the ions in the different materials have a large mismatch [Hicks and Dresselhaus (1993)].

There is no simple way to derive the response of a strongly correlated system to both electrical fields and thermal gradients. The reason why is that the thermal gradient cannot be added as a field to the Hamiltonian like the electric field can, hence there is no way to follow the simple Kubo response theory developed for the charge current in an electric field (because the linear-response approach evaluates correlation functions at a fixed temperature, and a variation of the temperature with position is problematic to include within the formalism). Instead, we can couple a fictitious field to the heat-current operator, analogous to the vector potential that couples to the charge current, and determine the linear response with respect to this field. Then, we can compare the Kubo response to a phenomenological set of equations that relate the charge and heat currents to the electric field and the gradient of the temperature. We can then identify the relevant transport coefficients and how they are expressed in terms of correlation functions. This strategy was first adopted by [Luttinger (1964)].

In our nanostructures, we always have a charge reconstruction when we describe devices made from different materials, because the bulk chemical potentials will have different T dependence, and hence cannot always be equal (the only exception is for particle-hole symmetry at half-filling, but there the thermopower vanishes, so they are uninteresting for thermal transport). Hence the Hamiltonian must be modified to include the potential energy V_α on each plane, and the band offsets ΔE_F, as described in Sec. 3.6 [*i. e.*, we add $\sum_{\alpha i}(V_\alpha - \Delta E_{F\alpha})c^\dagger_{\alpha i}c_{\alpha i}$ to $\mathcal{H}$]. The band offsets are independent of T, and represent the difference in the band zeroes for the leads and the material placed at plane α. The potential energies V_α do depend on T, but they do not create any currents, because they correspond to the static potential associated with the electronic charge reconstruction (and the diffusion current generated by the change in electron concentration cancels the current from the electric field). But they do create internal electric fields that maintain the electronic charge redistribution amongst the planes.

The phenomenological study of currents caused by external electric fields or temperature gradients has been examined since the early 1800s. It was found that an electric field can drive a charge current (which is essentially Ohm's law [Ohm (1827)] with the conductivity as the phenomenological constant) and it can drive a heat current because the electrons carry heat with them as they move through the material (this phenomenon is called the Peltier effect [Peltier (1834)]). Similarly, a temperature gradient can drive heat conduction with the phenomenological thermal conductivity (called Fourier's law [Fourier (1822)]), and because the heat current generically carries charge, a temperature gradient can generate a charge current (called the Seebeck effect [Seebeck (1823)]). The phenomenological equations for the (linear response) longitudinal transport in a multilayered nanostructure are then

$$\langle \mathbf{j}_\alpha^{\mathrm{c,long}} \rangle = -|e|\langle \mathbf{j}_\alpha^{\mathrm{long}} \rangle = |e|a\sum_\beta L_{11\alpha\beta}\left[\frac{d\mu_\beta}{dT}\frac{T_{\beta+1}-T_\beta}{a} + |e|E_\beta\right] + |e|a\sum_\beta L_{12\alpha\beta}\frac{T_{\beta+1}-T_\beta}{aT_\beta}, \tag{3.49}$$

$$\langle \mathbf{j}_\alpha^{\mathrm{Q,long}} \rangle = -a\sum_\beta L_{21\alpha\beta}\left[\frac{d\mu_\beta}{dT}\frac{T_{\beta+1}-T_\beta}{a} + |e|E_\beta\right] - a\sum_\beta L_{22\alpha\beta}\frac{T_{\beta+1}-T_\beta}{aT_\beta}, \tag{3.50}$$

where the indices α and β denote the planar sites (or the midpoint between planar sites), the term $(T_{\beta+1} - T_\beta)/a$ is the discretized approximation of the temperature gradient and the L_{ij} coefficients can be thought of as the phenomenological parameters. We define the symbol $\mu_\beta = \mu - V_\beta - \Delta E_{F\beta}$, which may be thought of as the "local chemical potential" for plane β. Both μ and V_β depend on T, but the band offset $\Delta E_{F\beta}$ does not. The origin of the temperature derivative of μ_β entering into the phenomenological equations arises from the conventional $\nabla\mu$ term, which becomes $\nabla T d\mu/dT$ when the system is placed in a thermal gradient. The spatial derivative of the V_β terms does not drive any current, because it cancels with the current driven by the equilibrium concentration gradient (which we did not include in the above phenomenological equations), so the electric field E_β is the external field applied to the device (this is valid only in the linear-response regime of a small external electric field). Note, that there is a simple way to understand the signs that appear in Eqs. (3.49) and (3.50). First consider the external electric field, which can be written as the negative gradient of the electric potential. The current (whether of electrons or of holes), always runs down the potential hill. Since the conductivity is always positive, the first term in Eq. (3.49) must have a positive sign. The thermoelectric number current also runs downhill, so it is proportional to the negative temperature gradient. For electrons, the charge current is $-|e|$ times the number current, which gives rise to the positive sign for the last term in Eq. (3.49). Similarly, the thermal conductivity runs down the temperature "hill", so it has a negative sign in front of it. The Peltier effect term is the hardest to understand, but because the electrons are negatively charged, they actually move up the potential hill (the charge current runs down the hill because the electrons are negatively charged), so the heat is carried up the hill, and hence there is a minus sign in front of the term (recall the electric field is the negative gradient of the potential).

Our next step is to determine how to represent the thermal transport coefficients L_{ij} in terms of many-body correlation functions. We have already done this for the first coefficient, which is proportional to the conductivity matrix, and is represented by a current-current correlation function: $\sigma_{\alpha\beta} = e^2 L_{11\alpha\beta}$ (the modification of the Hamiltonian by the electronic charge reconstruction has no effect on the form of the charge current, or on the form of the correlations functions, but obviously creates additional scattering). Since this coefficient arises from an electric field, which can be added to the Hamiltonian, the derivation is rigorous. Similarly, if we follow all the steps that led up to the derivation of the conductivity matrix, but we examined the expectation value of the heat-current operator instead of the charge-current operator, we would find that the L_{21} correlation function was identical to the L_{11} correlation function except that it is a heat-current–charge-current correlation function instead of a charge-current–charge-current correlation function.

As we discussed above, there is no complete theory to determine the L_{12} and L_{22} coefficients for the phenomenological transport equations. But, classical nonequilibrium statistical mechanics has proved that there is a reciprocal relation between the "cross" terms in the transport equations [Onsager (1931a); Onsager (1931b)]. Written in the form we have them, this relation says that $L_{21} = L_{12}$. Knowing the form for L_{21}, we then conclude that L_{12} is the charge-current–heat-current correlation function. Keeping within this same vein, the natural conclusion is that the final transport coefficient L_{22} is a heat-current–heat-current correlation function (but we do not have a rigorous derivation of this result).

Evaluating these correlation functions would be a chore if we tried to calculate them directly (see [Freericks and Zlatić (2001)] for just such a calculation in the bulk). Instead, Jonson and Mahan [Jonson and Mahan (1980); Jonson and Mahan (1990)] showed that there is a simple relationship between all of the L_{ij} coefficients—if one is known exactly, then all can be found exactly. Similarly, if one is evaluated approximately, the Jonson-Mahan theorem provides the most reasonable and consistent way to calculate the others. The Jonson-Mahan theorem was originally proved for the L_{21} coefficient in the bulk. Here we extend this work to show a generalized Jonson-Mahan theorem that holds for inhomogeneous systems, and extends to the L_{22} coefficient as well.

In Sec. 3.5, we discussed the issue of how to define the "local" number current operator for a given plane, and settled on defining the number current operator between planes α and $\alpha+1$, as in Eq. (3.22). We also need the local heat current operator, which is derived in Prob. A.27 as

$$\begin{aligned}
\mathbf{j}_\alpha^{Q,\mathrm{long}} = iat_{\alpha\alpha+1}\Bigg\{ & -\sum_{ij}\frac{1}{2}(t^{\|}_{\alpha ij}+t^{\|}_{\alpha+1ij})(c^\dagger_{\alpha+1i}c_{\alpha j}-c^\dagger_{\alpha i}c_{\alpha+ij}) \\
& -\frac{1}{2}t_{\alpha+1\alpha+2}\sum_i(c^\dagger_{\alpha+2i}c_{\alpha i}-c^\dagger_{\alpha i}c_{\alpha+2i}) \\
& -\frac{1}{2}t_{\alpha-1\alpha}\sum_i(c^\dagger_{\alpha+1i}c_{\alpha i}-c^\dagger_{\alpha i}c_{\alpha+1i}) \\
& +\frac{1}{2}\sum_i(U_\alpha w_{\alpha i}+U_{\alpha+1}w_{\alpha+1i})(c^\dagger_{\alpha+1i}c_{\alpha i}-c^\dagger_{\alpha i}c_{\alpha+1i}) \\
& +\left[-\mu+\frac{1}{2}(V_\alpha+V_{\alpha+1})-\frac{1}{2}(\Delta E_{F\alpha}+\Delta E_{F\alpha+1})\right] \\
& \times(c^\dagger_{\alpha+1i}c_{\alpha i}-c^\dagger_{\alpha i}c_{\alpha+1i})\Bigg\},
\end{aligned} \tag{3.51}$$

for the Falicov-Kimball model. The heat current operator depends on the model being examined, because it involves commutators of the potential energy with the energy polarization, and we still subtract the chemical potential multiplied by the number current from the energy current to get the heat current. One might have thought we should subtract the "local chemical potential" multiplied by the local number current operator, but that would remove the extra terms in the heat current arising from the electronic charge reconstruction; one could have grouped those terms into either the Hamiltonian or the local chemical potential—we chose the former, so we subtract only $\mu j^{\rm long}$.

Now we need to determine the *dc* limit of the correlation functions L_{ij} on the real axis. The procedure is identical to the analytic continuation that was worked out for the bulk case. We start by defining a polarization operator on the imaginary axis, then we analytically continue to the real axis, we form the relevant transport coefficient, and then we take the limit of the frequency going to zero. We denote the four polarization operators by $\bar{L}_{ij\alpha\beta}(i\nu_l)$ according to

$$\begin{aligned}
\bar{L}_{11\alpha\beta}(i\nu_l) &= \int_0^\beta d\tau e^{i\nu_l\tau}\langle \mathcal{T}_\tau j_\alpha^{\rm c,long}(\tau) j_\beta^{\rm c,long}(0)\rangle, \\
\bar{L}_{12\alpha\beta}(i\nu_l) &= \int_0^\beta d\tau e^{i\nu_l\tau}\langle \mathcal{T}_\tau j_\alpha^{\rm c,long}(\tau) j_\beta^{\rm Q,long}(0)\rangle, \\
\bar{L}_{21\alpha\beta}(i\nu_l) &= \int_0^\beta d\tau e^{i\nu_l\tau}\langle \mathcal{T}_\tau j_\alpha^{\rm Q,long}(\tau) j_\beta^{\rm c,long}(0)\rangle, \\
\bar{L}_{22\alpha\beta}(i\nu_l) &= \int_0^\beta d\tau e^{i\nu_l\tau}\langle \mathcal{T}_\tau j_\alpha^{\rm Q,long}(\tau) j_\beta^{\rm Q,long}(0)\rangle, \qquad (3.52)
\end{aligned}$$

and the transport coefficients satisfy $L_{ij\alpha\beta} = \lim_{\nu\to 0}{\rm Re}[-i\bar{L}_{ij\alpha\beta}(\nu)/\nu]$. The Jonson-Mahan theorem [Jonson and Mahan (1980); Jonson and Mahan (1990)] can be straightforwardly generalized to treat this case. Begin by defining a generalized function

$$\begin{aligned}
F_{\alpha\beta}(\tau_1,\tau_2,\tau_3,\tau_4) = \Big\langle \mathcal{T}_\tau iat_{\alpha\alpha+1}\sum_{i\in{\rm plane}}\Big[c^\dagger_{\alpha+1i}(\tau_1)c_{\alpha i}(\tau_2) - c^\dagger_{\alpha i}(\tau_1)c_{\alpha+1i}(\tau_2)\Big] \\
\times\, iat_{\beta\beta+1}\sum_{j\in{\rm plane}}\Big[c^\dagger_{\beta+1j}(\tau_3)c_{\beta j}(\tau_4) - c^\dagger_{\beta j}(\tau_3)c_{\beta+1j}(\tau_4)\Big]\Big\rangle.
\end{aligned} \qquad (3.53)$$

Next, we determine the polarization operators by taking the appropriate limits and derivatives. Namely,

$$
\begin{aligned}
\bar{L}_{11\alpha\beta} &= \int_0^\beta d\tau_1 e^{i\nu_l\tau} F_{\alpha\beta}(\tau_1,\tau_1^-,0,0^-),\\
\bar{L}_{12\alpha\beta} &= \int_0^\beta d\tau_1 e^{i\nu_l\tau}\frac{1}{2}\left(\frac{\partial}{\partial\tau_3}-\frac{\partial}{\partial\tau_4}\right)F_{\alpha\beta}(\tau_1,\tau_1^-,\tau_3,\tau_4)\Big|_{\tau_3=0,\tau_4=0^-}\\
\bar{L}_{21\alpha\beta} &= \int_0^\beta d\tau_1 e^{i\nu_l\tau}\frac{1}{2}\left(\frac{\partial}{\partial\tau_1}-\frac{\partial}{\partial\tau_2}\right)F_{\alpha\beta}(\tau_1,\tau_2,0,0^-)\Big|_{\tau_2=\tau_1^-}\\
\bar{L}_{22\alpha\beta} &= \int_0^\beta d\tau_1 e^{i\nu_l\tau}\\
&\quad\times\frac{1}{4}\left(\frac{\partial}{\partial\tau_1}-\frac{\partial}{\partial\tau_2}\right)\left(\frac{\partial}{\partial\tau_3}-\frac{\partial}{\partial\tau_4}\right)F_{\alpha\beta}(\tau_1,\tau_2,\tau_3,\tau_4)\Big|_{\tau_2=\tau_1^-,\tau_3=0,\tau_4=0^-}.
\end{aligned}
\tag{3.54}
$$

This result holds because the $(\partial_\tau-\partial_{\tau'})/2$ operator converts the local charge current operator into the local heat current operator (see Prob. A.28). Using these identities, we can now explicitly evaluate the correlation functions in DMFT (neglecting vertex corrections), and perform the analytic continuation to determine the final result for the transport coefficients. Note that Eq. (3.54) is a general result, which holds for the exact correlation functions (when vertex corrections are included), because it is derived from an operator identity. Writing down the analytic continuation to the real axis says that we will get extra factors of ω or ω^2 in the integrals that calculate the transport coefficients versus the integrand for the L_{11} coefficient. While this statement is simple to understand and derive, we find it more useful to also derive the full expression for the transport coefficients; and to do that, we neglect the vertex corrections in our analysis, because they are expected to be small for these cases. One should always remember though, that the Jonson-Mahan theorem is an exact relation for the transport coefficients, when the vertex corrections are included.

The first step is to evaluate the expectation values of the Fermionic operators (in the definition of F) via contractions, because we neglect the vertex corrections. This yields

$$
\begin{aligned}
&F_{\alpha\beta}(\tau_1,\tau_2,\tau_3,\tau_4)\\
&\quad= a^2 t_{\alpha\alpha+1}t_{\beta\beta+1}\sum_{ij\in\text{plane}}\{G_{\beta\alpha+1ji}(\tau_4-\tau_1)G_{\alpha\beta+1ij}(\tau_2-\tau_3)\\
&\quad- G_{\beta+1\alpha+1ji}(\tau_4-\tau_1)G_{\alpha\beta ij}(\tau_2-\tau_3)\\
&\quad- G_{\beta\alpha ji}(\tau_4-\tau_1)G_{\alpha+1\beta+1ij}(\tau_2-\tau_3)\\
&\quad+ G_{\beta+1\alpha ji}(\tau_4-\tau_1)G_{\alpha+1\beta ij}(\tau_2-\tau_3)\}.
\end{aligned}
\tag{3.55}
$$

Next, we need to determine a spectral representation for the off-diagonal Green's function. Using the fact that

$$G_{\alpha\beta ij}(z) = -\frac{1}{\pi}\int d\omega \frac{\mathrm{Im}G_{\alpha\beta ij}(\omega)}{z-\omega}, \tag{3.56}$$

with z in the upper half plane (which can be shown by using the Lehmann representation), says that

$$G_{\alpha\beta ij}(\tau) = -\frac{1}{\pi}\int d\omega T \sum_n \frac{e^{-i\omega_n\tau}}{i\omega_n - \omega}\mathrm{Im}G_{\alpha\beta ij}(\omega). \tag{3.57}$$

Now we convert the sum over Matsubara frequencies into a contour integral (that surrounds each Matsubara frequency, but does not cross the real axis — the contour is then deformed into two contours, one running just above and the other just below the real axis), but we must be careful to ensure that the procedure is well-defined. If $\tau < 0$, then

$$\begin{aligned} T\sum_n \frac{e^{-i\omega_n\tau}}{i\omega_n - \omega} &= -\frac{i}{2\pi}\int_C dz \frac{e^{-z\tau}}{z-\omega} f(z), \\ &= -\frac{i}{2\pi}\int_{-\infty}^{\infty} dz e^{-z\tau} f(z)\left[\frac{1}{z+i0^+ - \omega} - \frac{1}{z - i0^+ - \omega}\right], \\ &= -e^{-\omega\tau} f(\omega). \end{aligned} \tag{3.58}$$

This result is well-defined because the Fermi factor provides convergence (asymptotically like $\exp[-\beta z]$) for $z \to \infty$ and the $\exp[-z\tau]$ term provides boundedness for $z \to -\infty$ when $\tau < 0$. Since $1 - f(z)$ has the same poles as $f(z)$ on the imaginary axis, with residues that have the opposite sign, and it behaves like $\exp[\beta z]$ for $z \to -\infty$, one finds

$$T\sum_n \frac{e^{-i\omega_n\tau}}{i\omega_n - \omega} = e^{-\omega\tau}[1 - f(\omega)], \tag{3.59}$$

for $\tau > 0$. The results in Eqs. (3.58) and (3.59) can then be substituted into Eq. (3.57) to get the final formula for the off-diagonal Green's function

$$G_{\alpha\beta ij}(\tau) = \begin{cases} -\frac{1}{\pi}\int d\omega \mathrm{Im}G_{\alpha\beta ij}(\omega)e^{-\omega\tau}[1-f(\omega)], & \tau > 0 \\ -\frac{1}{\pi}\int d\omega \mathrm{Im}G_{\alpha\beta ij}(\omega)e^{-\omega\tau}[-f(\omega)], & \tau < 0. \end{cases} \tag{3.60}$$

Now we note that we can restrict ourselves to the case $\tau_1 > \tau_2 > \tau_3 > \tau_4$ without loss of generality, because that is the ordering needed to get the relevant correlation functions. Then we employ Eq. (3.60) in Eq. (3.55) and use the fact that the summations over the spatial indices for the planes can be Fourier transformed, and then the momentum summation can be replaced by an integration over the two-dimensional DOS, to yield

$$
\begin{aligned}
F_{\alpha\beta}(\tau_1,\tau_2,\tau_3,\tau_4) = & \frac{a^2}{\pi^2} t_{\alpha\alpha+1} t_{\beta\beta+1} \int d\omega \int d\omega' \int d\epsilon^{\|} \rho^{2d}(\epsilon^{\|}) \\
& \times f(\omega)[1-f(\omega')] e^{-\omega(\tau_4-\tau_1)-\omega'(\tau_2-\tau_3)} \\
& \times \Big\{ \mathrm{Im}G_{\beta\alpha}(\epsilon^{\|},\omega)\mathrm{Im}G_{\alpha+1\beta+1}(\epsilon^{\|},\omega') \\
& + \mathrm{Im}G_{\beta+1\alpha+1}(\epsilon^{\|},\omega)\mathrm{Im}G_{\alpha\beta}(\epsilon^{\|},\omega') \\
& - \mathrm{Im}G_{\beta\alpha+1}(\epsilon^{\|},\omega)\mathrm{Im}G_{\alpha\beta+1}(\epsilon^{\|},\omega') \\
& - \mathrm{Im}G_{\beta+1\alpha}(\epsilon^{\|},\omega)\mathrm{Im}G_{\alpha+1\beta}(\epsilon^{\|},\omega') \Big\}. \qquad (3.61)
\end{aligned}
$$

Now we can evaluate the polarizations, and directly perform the analytic continuation. We Fourier transform the expression in Eq. (3.61) to get the Matsubara frequency representation. Then we replace $i\nu_l$ by $\nu + i0^+$, then we construct the transport coefficients on the real axis, and we finally take the limit $\nu \to 0$ to get the *dc* response. The factor $(\partial_\tau - \partial_{\tau'})/2$ gives a factor of $(\omega+\omega')/2$ which goes to $(\omega+\nu/2)$ after integrating over the delta function that arises in the analytic continuation. Setting $\nu = 0$ gives an extra power of ω in the integrand for each derivative factor in the response coefficient. The end result is

$$
\begin{aligned}
L_{ij\alpha\beta} = & \frac{a^2}{\pi} t_{\alpha\alpha+1} t_{\beta\beta+1} \int d\omega \left(-\frac{df(\omega)}{d\omega}\right) \omega^{i+j-2} \int d\epsilon^{\|} \rho^{2d}(\epsilon^{\|}) \\
& \{ \mathrm{Im}G_{\beta\alpha}(\epsilon^{\|},\omega)\mathrm{Im}G_{\alpha+1\beta+1}(\epsilon^{\|},\omega) + \mathrm{Im}G_{\alpha\beta}(\epsilon^{\|},\omega)\mathrm{Im}G_{\beta+1\alpha+1}(\epsilon^{\|},\omega) \\
& - \mathrm{Im}G_{\beta\alpha+1}(\epsilon^{\|},\omega)\mathrm{Im}G_{\alpha\beta+1}(\epsilon^{\|},\omega) - \mathrm{Im}G_{\alpha+1\beta}(\epsilon^{\|},\omega)\mathrm{Im}G_{\beta+1\alpha}(\epsilon^{\|},\omega) \}. \qquad (3.62)
\end{aligned}
$$

This is the generalized Jonson-Mahan theorem for inhomogeneous systems. Note that the equality of L_{12} with L_{21} is the Onsager reciprocal relation [Onsager (1931a); Onsager (1931b)].

With the expressions for the phenomenological coefficients that appear in Eqs. (3.49) and (3.50) determined, we now can move onto evaluating the transport in different cases of interest. The first point that needs to be emphasized is that the total number of electrons is always conserved in the system, so the charge current is conserved, and cannot change from plane to plane. There is no such conservation law for the heat current though, because the electrons can change the amount of heat that they carry depending on their local environment. Hence, it is the boundary conditions that we impose upon the heat current that determines how it behaves in a multilayered nanostructure. This point will become important as we analyze different experimental situations.

The first experiment we would like to analyze is the Peltier effect in a multilayered nanostructure. We imagine that the nanostructure is attached to a bath that maintains the entire structure at a fixed temperature, and we then turn on an external electric field. The Peltier effect is the ratio of the heat current to the charge current. A moment's reflection will show that the heat current is not necessarily conserved in this system, because we have to exchange heat with the reservoir to maintain a constant temperature profile. Hence, it isn't even obvious what ratio should be taken for the Peltier effect—the average heat current over the charge current, the total change in the heat current over the charge current, or the heat current transfered over the charge current. We now show how to determine all three of these results.

The starting point is the transport equations [(3.49) and (3.50)] with $T_\alpha = T$ independent of the plane number. As in the calculation of the current, we first determine the electric field by multiplying both sides of Eq. (3.49) by the inverse L_{11} matrix. Since the charge current is independent of the plane index, we find the electric field satisfies

$$E_\alpha = \frac{1}{e^2 a} \sum_\beta \left(L_{11}^{-1}\right)_{\alpha\beta} \langle j^{\mathrm{c,long}} \rangle. \tag{3.63}$$

Substituting this value of the electric field into Eq. (3.50) then yields

$$\langle j_\alpha^{\mathrm{Q,long}} \rangle = -\frac{1}{|e|} \sum_{\beta\gamma} L_{21\alpha\beta} \left(L_{11}^{-1}\right)_{\beta\gamma} \langle j^{\mathrm{c,long}} \rangle. \tag{3.64}$$

This is all we need to analyze the Peltier effect of a nanostructure. Note that the heat current generically will have α dependence, and hence will vary from plane to plane (see Fig. 3.15).

The first question we can ask is how much heat is lost or gained by the reservoir that is attached to the device to maintain isothermal conditions. This is determined by the ratio of the difference in the heat current at the right and the heat current at the left to the charge current. In equations,

$$\frac{\Delta\langle j^{\mathrm{Q,long}} \rangle}{\langle j^{\mathrm{c,long}} \rangle} = \frac{\langle j_R^{\mathrm{Q,long}} \rangle - \langle j_L^{\mathrm{Q,long}} \rangle}{\langle j^{\mathrm{c,long}} \rangle} = -\frac{1}{|e|} \sum_{\beta\gamma} \left(L_{21R\beta} - L_{21L\beta}\right) \left(L_{11}^{-1}\right)_{\beta\gamma}. \tag{3.65}$$

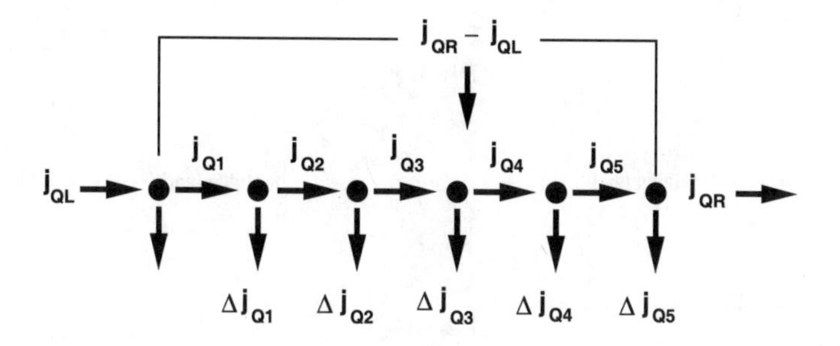

Fig. 3.15 Schematic diagram of the heat transfers in the Peltier effect. The dots refer to different planes. A heat current is incident from the left. As we go from one plane to another, the heat current changes, as heat is transfered to or from a reservoir to maintain the system at a constant temperature. For example, we can examine the total heat current transfered to the reservoirs ($j_{QR} - j_{QL}$), or we can examine the average heat current that flows through the device $\sum_\alpha j_{Q\alpha}/N$.

This would measure the net cooling or heating of the reservoir by the device as current flows. Similarly, we could measure the average heat flow carried through the device

$$\frac{\langle j_{\rm ave}^{\rm Q,long}\rangle}{\langle j^{\rm c,long}\rangle} = -\frac{1}{|e|}\frac{1}{N}\sum_{\alpha\beta\gamma} L_{21\alpha\beta}\left(L_{11}^{-1}\right)_{\beta\gamma}, \tag{3.66}$$

where N is the number of terms taken in the summation over the index α. This expression is analogous to the bulk Peltier effect, which measures the ratio of the heat to charge current flows (which are independent of position in a bulk system).

Next we examine the Seebeck effect and a thermal conductivity experiment. In both cases we work with an open circuit, so the total charge current vanishes $\langle j^{\rm c,long}\rangle = 0$. The Seebeck measurement is subtle, because we don't want to measure the voltage difference with probes at different temperatures, because there will be a contribution from the $\nabla T d\mu/dT$ terms to the voltage drop (and there may be a thermal link allowing heat to flow through the voltage probe). An actual experiment uses thermocouple probes, where one end of the probe is placed on the sample, and the other is placed in a constant T_0 bath. Two probes are needed to measure the voltage change and the temperature at two points along the sample. The net thermopower is measured relative to the thermopower of the metal used in one of the legs of the thermocouple (typically copper). For details, see [Domenicali (1954)] and [Nolas, *et al.* (2001)]; a simpler schematic picture of this issue is shown in Fig. 3.16. Alternatively, we can imagine the lead to the left placed in a bath at temperature T_0, the interface plane on the

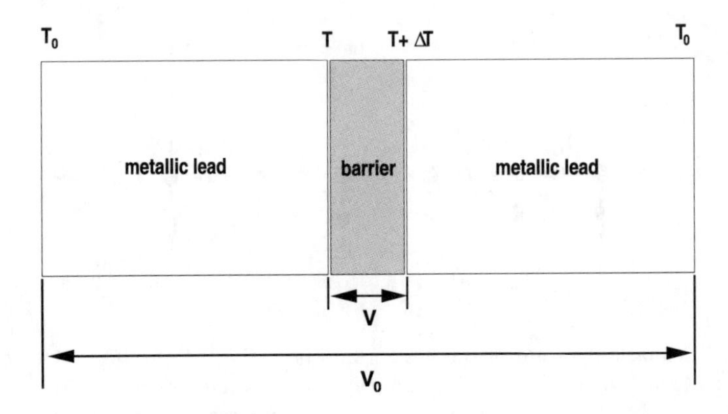

Fig. 3.16 Schematic diagram for how to measure the (relative) Seebeck effect. The metallic leads are composed of the same material, and a voltage probe is placed across the two ends which both are fixed at a temperature of T_0. The voltage across those probes is V_0. Since heat current will flow across the voltmeter if both ends are not at the same temperature, there is no way to directly measure the desired voltage V. But, since the change in voltage in the metallic lead in going from T_0 to T on the left hand side is exactly canceled by the change in voltage in going from T to T_0 on the right hand side, we find the difference between the voltage V and V_0 is just equal to the Seebeck coefficient of the metallic lead multiplied by ΔT. Hence, since a measurement uses V_0 instead of V, the Seebeck coefficient of the barrier is measured *relative to* the Seebeck coefficient of the metallic lead.

left held at temperature T, the interface on the right held at temperature $T+\Delta T$, and the lead to the right held at temperature T_0. The net effect on our analysis, if we assume the thermopower of copper can be neglected (or of the ballistic lead in the alternative picture), is that we neglect the $d\mu_\alpha/dT$ terms in our analysis (because the chemical potential at the probes is at a constant temperature when the potential difference is measured). With these caveats in mind, using Eq. (3.49), we find

$$E_\alpha = -\frac{1}{a|e|T}\sum_{\beta\gamma}\left(L_{11}^{-1}\right)_{\alpha\beta} L_{12\beta\gamma}(T_{\gamma+1} - T_\gamma). \tag{3.67}$$

Multiplying by a and summing over α yields the voltage drop across the device. We also need the temperature profile. Substituting Eq. (3.67) into Eq. (3.50), and noting that the heat current is conserved if the device is isolated and in the steady state (implying heat cannot be transfered out of any plane) because the system develops a temperature profile so that the heat current is conserved through the device. In this case, we can evaluate the temperature profile, which satisfies

$$T_{\alpha+1} - T_\alpha = -T \sum_\beta \left(M^{-1}\right)_{\alpha\beta} \langle J^{\mathrm{Q,long}} \rangle, \tag{3.68}$$

with the matrix M defined to be

$$M_{\alpha\beta} = L_{22\alpha\beta} - \sum_{\gamma\delta} L_{21\alpha\gamma} \left(L_{11}^{-1}\right)_{\gamma\delta} L_{12\delta\beta}. \tag{3.69}$$

Now we can sum Eq. (3.68) over α to get the temperature difference over the device. Hence the Seebeck effect becomes

$$\frac{\Delta V}{\Delta T} = -\frac{1}{|e|T} \frac{\sum_{\alpha\beta\gamma\delta}(L_{11}^{-1})_{\alpha\beta} L_{12\beta\gamma} M_{\gamma\delta}^{-1}}{\sum_{\alpha\beta} M_{\alpha\beta}^{-1}}. \tag{3.70}$$

Note that this is not equal to $1/T$ times the Peltier coefficient as in the bulk. Instead, we have a weighting of the L_{12} to L_{11} ratio by the matrix M, which is related to the thermal conductivity. This factor cancels in the bulk, where the M matrix depends on the difference of the spatial coordinates, and the $\mathbf{q} = 0$ response is independent of M because the common factor in the Fourier transform will cancel out (as can easily be proved by invoking the convolution theorem). If we do not measure ΔV via thermocouples at constant T, then the ΔV term is modified by a contribution from $d\mu_\alpha/dT$. We do not discuss that modification here, because it is not normally a technique used in measurements.

The thermal conductance is evaluated in a similar way, but does not require any subtlety in the measurement. We also work in an open circuit, and the heat current is conserved, because we isolate the system. Now we measure the ratio of the heat current to the temperature difference to find that the thermal conductance per unit area K satisfies

$$K = -\frac{\langle j^{\mathrm{Q,long}} \rangle}{\Delta T} = \frac{1}{T \sum_{\alpha\beta}(M^{-1})_{\alpha\beta}}, \tag{3.71}$$

and the thermal resistance-area product becomes

$$R_{\mathrm{th}} a^2 = T \sum_{\alpha\beta} \left(M^{-1}\right)_{\alpha\beta}. \tag{3.72}$$

The evaluation of the cooling efficiency of a refrigerator is more complicated than our above examples, because it corresponds to a situation where we have both a temperature gradient and charge current flow. It is instructive to first describe the standard "textbook" case of current flow for a bulk material before discussing the nanostructure case. But we simplify the bulk case to consider current flow through a single leg of a refrigerator, with the charge carriers being electrons. The modifications needed for analyzing a full refrigerator are straightforward, and described in detail in many texts (see chapter one of [Nolas, *et al.* (2001)] for an example).

In order to examine the heat transfer in a refrigerator, we need to include one term beyond linear response that gives rise to the Joule heating [Joule (1841)] in the volume of the leads, because this is one of the significant sources of irreversible heat production (heat is produced regardless of the direction of the current, unlike reversible thermoelectric effects where heat is produced or absorbed depending on the direction of the current). The equivalents of Eqs. (3.49) and (3.50) for the bulk material are

$$\langle j^{\mathrm{c}}\rangle = \frac{\sigma}{|e|}\frac{d\mu}{dT}\nabla T + \sigma E - S\sigma\nabla T, \tag{3.73}$$

$$\langle j^{\mathrm{Q}}\rangle = ST\frac{\sigma}{|e|}\frac{d\mu}{dT}\nabla T + ST\sigma E - \frac{1}{T}L_{22}\nabla T + V^{c}\langle j^{\mathrm{c}}\rangle, \tag{3.74}$$

where we have used the bulk conductivity σ and thermopower S. The extra term is the last term in Eq. (3.74), which represents the energy carried by an electron as it moves down the electrical potential "hill" (V^c is the position-dependent electrical potential of the external field). We imagine a slab of our material sandwiched between a hot and a cold heat reservoir (the hot reservoir at $z = L$ and the cold reservoir at $z = 0$). The slab of material has length L and a cross-sectional area A. The first step we take is to eliminate the electric field in favor of the charge current density using Eq. (3.73). Hence we set $E = \langle j^{\mathrm{c}}\rangle/\sigma - \nabla T(d\mu/dT)/|e| + S\nabla T$. The heat-current equation becomes

$$\langle j^{\mathrm{Q}}\rangle = (ST + V^{c})\langle j^{\mathrm{c}}\rangle - \kappa\nabla T, \tag{3.75}$$

where $\kappa = (L_{22} - L_{12}L_{21}/L_{11})/T$ is the thermal conductivity. It is conventional to assume that the σ, S, and κ coefficients are independent of T, which implies they have no variation with position z, induced by the temperature gradient.

The second step is to examine the heat production within the volume of the material. Since $\langle j^{\mathrm{Q}}\rangle$ represents the heat current, we can determine the heat balance by summing the heat current entering a given surface, and subtracting the heat current exiting the surface. What remains is the net heat production within the volume bounded by the surface. By using Gauss's law, the surface integral can be replaced by a volume integral, which can be summarized by the time rate of change of the heat production within a given volume element, called $\dot{Q}$, which is equal to $-\nabla \cdot \langle j^{\mathrm{Q}} \rangle$. In the steady state, the net heat production must vanish, otherwise, the local temperature would rise or fall to compensate for the heat being produced or absorbed. Hence, the steady-state response is characterized by a vanishing net heat production, or $-\nabla \cdot \langle j^{\mathrm{Q}} \rangle = 0$. For the form of the heat current in Eq. (3.75), we find (recall the continuity equation says $\nabla \cdot \langle j^{\mathrm{c}} \rangle = 0$ in the steady state)

$$\dot{Q} = -\nabla \cdot \langle j^{\mathrm{Q}} \rangle = \frac{\langle j^{\mathrm{c}} \rangle^2}{\sigma} - \frac{\langle j^{\mathrm{c}} \rangle}{|e|} \frac{d\mu}{dT} \nabla T - T \nabla S \cdot \langle j^{\mathrm{c}} \rangle + \nabla \kappa \cdot \nabla T + \kappa \nabla^2 T; \quad (3.76)$$

the first term on the right hand side is the Joule heat [Joule (1841)], the second term is the Thomson heat [Thomson (1851); Thomson (1854)] (after noting that $\nabla S = (dS/dT)\nabla T$), the third, fourth and fifth terms have no common name. In deriving Eq. (3.76), we substituted in for $-\nabla V^c = E$, using the representation for E in terms of the current and the temperature gradient. If we assume the coefficients are independent of T, then $\nabla S = \nabla \kappa = \nabla \mu = 0$, and we have just two terms in the heat balance. For the steady state, we set $\dot{Q} = 0$, or $\langle j^{\mathrm{c}} \rangle^2/\sigma = -\kappa \nabla^2 T$, which shows that the temperature profile curves as we approach either of the heat reservoirs. Solving this equation for the temperature profile yields

$$T(z) = T_c \left(\frac{L - z}{L} \right) + T_h \frac{z}{L} + \frac{\langle j^{\mathrm{c}} \rangle^2}{2\kappa\sigma} z(L - z), \quad (3.77)$$

where T_c is the temperature of the cold reservoir and T_h is the temperature of the hot reservoir.

The efficiency of a refrigerator is determined by the ratio of the heat transfered to the cold reservoir and the electrical energy required to drive the current through the device. The heat transfered is just $\langle j^{\mathrm{Q}} \rangle A$, while the electrical energy is $\langle j^{\mathrm{c}} \rangle \Delta V A$. The heat current is given in Eq. (3.75). We first note that $V^c = -Ez = -\langle j^{\mathrm{c}} \rangle z/\sigma - Sz\nabla T$, and then we take the derivative of the temperature in Eq. (3.77) with respect to z, to find a formula for ∇T (ignoring the $d\mu/dT$ term). Finally, we set $z = 0$ and $T = T_c$ to determine the heat transfer at the cold reservoir. The result is

$$\langle j^{\mathrm{Q}} \rangle A = \left[S T_c \langle j^{\mathrm{c}} \rangle - \kappa \frac{\Delta T}{L} - \frac{\langle j^{\mathrm{c}} \rangle^2}{2\sigma} L \right] A, \quad (3.78)$$

where $\Delta T = T_h - T_c$ is the difference in temperature between the two reservoirs. The potential difference is found by expressing the electric field in terms of the charge current and the temperature gradient, substituting in for the gradient according to the temperature profile, and then integrating over z from 0 to L. This yields

$$\Delta V = \frac{\langle j^{\rm c}\rangle}{\sigma}L + S\Delta T, \tag{3.79}$$

when we neglect $d\mu/dT$. The first term is the Ohm's law voltage drop, and the second is the thermoelectric voltage. The efficiency then becomes

$$\text{Eff.} = \frac{ST_c\langle j^{\rm c}\rangle - \frac{\kappa\Delta T}{L} - \frac{\langle j^{\rm c}\rangle^2 L}{2\sigma}}{\langle j^{\rm c}\rangle\left(\frac{\langle j^{\rm c}\rangle L}{\sigma} + S\Delta T\right)}. \tag{3.80}$$

Note that if we ignore the irreversible Joule heating and thermal conductivity contributions in the numerator, and the Ohm's law voltage in the denominator, then we recover the Carnot efficiency of $T_c/\Delta T$ [Carnot (1824)]. The irreversible processes always reduce the efficiency, often far below the Carnot limit.

There are many options that one can consider for optimizing the operation of the refrigerator. One can maximize the heat transfered to the cold reservoir by differentiating the numerator of Eq. (3.80) with respect to $\langle j^{\rm c}\rangle$ and setting it to zero. One can maximize the efficiency, by differentiating the ratio with respect to $\langle j^{\rm c}\rangle$ and setting it to zero. Finally, one can maximize the temperature difference between the hot and cold reservoirs. The choice depends on the application.

Having given a short tutorial on the bulk case, we now consider the nanostructure case in detail. The basic ideas are the same as in the bulk case, but now the technical details are much more complicated, because all of the transport coefficients are now matrices that vary with position. The starting point is the two phenomenological equations Eqs. (3.49) and (3.50), except we have to add the term $V^{\rm c}_\alpha\langle j^{\rm c,long}\rangle$ to the second equation, in order to ultimately describe the Joule heating properly. This term can be viewed as a nonlinear correction, but there are numerous other nonlinear corrections that are neglected. These other terms are likely to be small in comparison to the Joule heat, but there doesn't seem to be a complete analysis showing that other possible terms are unimportant. Since the heat transport equations are phenomenological in origin anyway, this does not cause any more fundamental problems. We also neglect the $d\mu/dT$ terms, since they are small in metals at low T, and because we want to simplify the analysis.

Our first step is to solve for the electric field, which is slightly more complicated than what we have done before:

$$E_\alpha = \frac{1}{e^2 a}\sum_\beta \left(L_{11}^{-1}\right)_{\alpha\beta} \langle j^{\mathrm{c,long}}\rangle - \frac{1}{|e|aT}\sum_\beta \left(L_{11}^{-1}L_{12}\right)_{\alpha\beta}(T_{\beta+1}-T_\beta). \quad (3.81)$$

Before we substitute this result into the heat-transport equation, we must note that $V_\alpha^c = -a\sum_{\beta\le\alpha} E_\beta$, which we write as $V_\alpha^c = -a\sum_\beta \theta_{\alpha\beta}E_\beta$, with the matrix θ defined to by $\theta_{\alpha\beta} = 1$ for $\beta \le \alpha$ and $\theta_{\alpha\beta} = 0$ for $\beta > \alpha$. Now we substitute the formula for the electric field into the formula for the heat transport, and collect all of the terms to give

$$\begin{aligned}\langle j_\alpha^{\mathrm{Q,long}}\rangle = &-\frac{1}{|e|}\sum_\beta \left(L_{21}L_{11}^{-1}\right)_{\alpha\beta}\langle j^{\mathrm{c,long}}\rangle - \frac{1}{T}\sum_\beta M_{\alpha\beta}(T_{\beta+1}-T_\beta)\\ &-\frac{1}{e^2}\sum_\beta \left(L_{11}^{-1}\right)_{\alpha\beta}\langle j^{\mathrm{c,long}}\rangle^2 + \frac{1}{|e|T}\sum_\beta \left(\theta L_{11}^{-1}L_{21}\right)_{\alpha\beta}(T_{\beta+1}-T_\beta),\end{aligned} \quad (3.82)$$

with the matrix M defined as before, $M = L_{22} - L_{21}L_{11}^{-1}L_{12}$. The terms in Eq. (3.82) are first the Peltier heat, then the heat from the thermal conductivity, followed by the Joule heat, and the last term appears to be related to the Thomson heat, but it isn't simple to extract the conventional Thomson heat form from this term (as it is currently written).

Our next step is to evaluate the steady-state heat transport. This was first worked out for inhomogeneous systems by Domenicali [Domenicali (1953)], which is reviewed in [Domenicali (1954)]. The strategy is identical to that used in the bulk: namely, we calculate the heat production within the volume by taking the negative divergence of the heat current. In the steady state, the net heat production must vanish, otherwise the temperature profile would change due to heating or cooling in the sample. For our discretized model of a multilayered nanostructure, setting the divergence of the heat current to zero, is equivalent to invoking heat-current conservation throughout the device. So the heat current does not change from plane to plane (which implies that the Peltier effect at a particular plane balances the thermal conduction of heat, the Joule heat produced, and the extra term which seems related to the Thomson heat when we are in the steady state). This allows us to invert Eq. (3.82), and solve for the temperature profile. The final result is lengthy, and we simplify it by introducing a new matrix,

$$N_{\alpha\beta} = -M_{\alpha\beta} + \frac{1}{|e|}\left(\theta L_{11}^{-1} L_{12}\right)_{\alpha\beta}. \tag{3.83}$$

In terms of N, the temperature profile is found from

$$T_{\alpha+1} - T_{\alpha} = T\sum_{\beta}\left(N^{-1}\right)_{\alpha\beta}\langle j^{\mathrm{Q,long}}\rangle + \frac{T}{|e|}\sum_{\beta}\left(N^{-1}L_{21}L_{11}^{-1}\right)_{\alpha\beta}\langle j^{\mathrm{c,long}}\rangle$$
$$+ \frac{T}{e^2}\sum_{\beta}\left(N^{-1}L_{11}^{-1}\right)_{\alpha\beta}\langle j^{\mathrm{c,long}}\rangle^2. \tag{3.84}$$

Summing Eq. (3.84) over α yields the total temperature change ΔT between the left and right ends of the multilayered nanostructure. The voltage difference can be found by substituting the temperature profile into Eq. (3.81), multiplying the field by a and summing over α. The end result is

$$\Delta V = \frac{1}{e^2}\sum_{\alpha\beta}\left(L_{11}^{-1} - L_{11}^{-1}L_{12}N^{-1}L_{21}L_{11}^{-1}\right)_{\alpha\beta}\langle j^{\mathrm{c,long}}\rangle$$
$$- \frac{1}{e^3}\sum_{\alpha\beta}\left(L_{11}^{-1}L_{12}N^{-1}L_{11}^{-1}\right)_{\alpha\beta}\langle j^{\mathrm{c,long}}\rangle^2$$
$$- \frac{1}{|e|}\sum_{\alpha\beta}\left(L_{11}^{-1}L_{12}N^{-1}\right)_{\alpha\beta}\langle j^{\mathrm{Q,long}}\rangle. \tag{3.85}$$

The efficiency is found from the ratio of $\langle j^{\mathrm{Q,long}}\rangle A$ to $\langle j^{\mathrm{c,long}}\rangle \Delta V A$. We have all of these results from the manipulations summarized above. But it isn't as useful to express the results in terms of the charge and heat currents, but instead, in terms of the charge current and the change in temperature ΔT. So we first express the heat current in terms of the temperature change, from Eq. (3.84), as

$$\langle j^{\mathrm{Q,long}}\rangle = \frac{\Delta T}{T\sum_{\alpha\beta}N_{\alpha\beta}^{-1}} - \frac{1}{|e|}\frac{\sum_{\alpha\beta}(N^{-1}L_{21}L_{11}^{-1})_{\alpha\beta}\langle j^{\mathrm{c,long}}\rangle}{\sum_{\alpha\beta}N_{\alpha\beta}^{-1}}$$
$$- \frac{1}{e^2}\frac{\sum_{\alpha\beta}(N^{-1}L_{11}^{-1})_{\alpha\beta}\langle j^{\mathrm{c,long}}\rangle^2}{\sum_{\alpha\beta}N_{\alpha\beta}^{-1}}. \tag{3.86}$$

Replacing $\langle j^{\mathrm{Q,long}}\rangle$ in favor of ΔT finally yields the efficiency as

$$\begin{aligned}
\text{Eff.} = \Bigg[& -\frac{1}{|e|}\frac{\sum_{\alpha\beta}(N^{-1}L_{21}L_{11}^{-1})_{\alpha\beta}\langle j^{\mathrm{c,long}}\rangle}{\sum_{\alpha\beta}N_{\alpha\beta}^{-1}} + \frac{\Delta T}{T\sum_{\alpha\beta}N_{\alpha\beta}^{-1}} \\
& - \frac{1}{e^2}\frac{\sum_{\alpha\beta}(N^{-1}L_{11}^{-1})_{\alpha\beta}\langle j^{\mathrm{c,long}}\rangle^2}{\sum_{\alpha\beta}N_{\alpha\beta}^{-1}}\Bigg] \Bigg/ \\
& \langle j^{\mathrm{c,long}}\rangle \Bigg[\frac{1}{e^2}\sum_{\alpha\beta}(L_{11}^{-1})_{\alpha\beta}\langle j^{\mathrm{c,long}}\rangle \\
& + \frac{1}{e^2}\Bigg\{\frac{\sum_{\alpha\beta}(L_{11}^{-1}L_{12}N^{-1})_{\alpha\beta}\sum_{\gamma\delta}(N^{-1}L_{21}L_{11}^{-1})_{\gamma\delta}}{\sum_{\alpha\beta}N_{\alpha\beta}^{-1}} \\
& - \sum_{\alpha\beta}(L_{11}^{-1}L_{12}N^{-1}L_{21}L_{11}^{-1})_{\alpha\beta}\Bigg\}\langle j^{\mathrm{c,long}}\rangle \\
& - \frac{\Delta T}{T|e|}\frac{\sum_{\alpha\beta}(L_{11}^{-1}L_{12}N^{-1})_{\alpha\beta}}{\sum_{\alpha\beta}N_{\alpha\beta}^{-1}} \\
& + \frac{1}{|e|^3}\Bigg\{\frac{\sum_{\alpha\beta}(L_{11}^{-1}L_{12}N^{-1})_{\alpha\beta}\sum_{\gamma\delta}(N^{-1}L_{11}^{-1})_{\gamma\delta}}{\sum_{\alpha\beta}N_{\alpha\beta}^{-1}} \\
& - \sum_{\alpha\beta}(L_{11}^{-1}L_{12}N^{-1}L_{11}^{-1})_{\alpha\beta}\Bigg\}\langle j^{\mathrm{c,long}}\rangle^2\Bigg]
\end{aligned} \tag{3.87}$$

The first term in the numerator is the generalization of the Peltier heat piece, the second term is the heat transported from the thermal conductivity, and the third is the Joule heat (but the factor of 1/2 in the bulk case isn't obvious from this matrix form), just like in the bulk case. The denominator, however, has some extra terms. The first is the Ohmic potential drop, the second is a term not found in the bulk, the third is the voltage due to the Seebeck effect, and the fourth is also a term not found in the bulk. The two extra terms both vanish in the bulk. While our analysis has been completely straightforward, the final results are quite complex!

A similar type of analysis can be employed to analyze power generators, but we will not go into the details of that here.

There are no additional computational issues for heat transport over and above the issues we have already seen with the charge transport.

The only point that must be emphasized is that for thermoelectric transport problems we need to have particle-hole asymmetry, so they invariably involve electronic charge reconstruction at the interfaces. The charge profile that results, does cause some significant technical challenges for determining the Green's functions on the real-frequency axis. But once they are known, then determining the transport coefficient matrices follows directly.

3.8 Superconducting Leads and Josephson Junctions

Superconductivity was discovered in the laboratory of Heike Kammerlingh-Onnes in 1911 [Onnes (1911)] (G. Flimm, G. Holst, and C. Dorsmann all participated in the experiments). The resistance of mercury was found to drop to essentially zero at about 4.2 K. We now know that the resistance is exactly zero in a superconductor, but the development of a theory that explains this took many years, and was finally achieved by John Bardeen, Leon Cooper, and Robert Schrieffer in 1957 [Bardeen, Cooper, and Schrieffer (1957)] and is called the BCS theory. It is assumed that the reader has some familiarity with the phenomena of superconductivity at the level of undergraduate solid state physics courses. Here we will develop some of the basic ideas for the many-body theory of superconductivity.

Our starting point is in the bulk, with an attractive Hubbard model ($U^{\mathrm{H}} = -|U|$). The Hubbard model interaction is an instantaneous on-site interaction; while it is normally taken to be repulsive, to describe the mutual repulsion of the like-charged electrons, there have been numerous studies of the attractive version of the model, which can lead to superconductivity due to strong local effective electron-electron attraction. This model is quite different from the conventional electron-phonon type models, which have retarded interactions arising from the fact that the phonon energy scales are much smaller than the electron energy scales. Nevertheless, we will see the attractive Hubbard model is one of the simplest models for superconductivity, and since it obeys all of the universal features of the electron-phonon-based models, it gives similar results. We study it in detail here.

As before, we begin with an examination of the equation of motion for the Green's functions. On the imaginary-time axis, the exact EOM given in Eq. (2.37), which we modify here for the attractive Hubbard model case, becomes

$$(-\partial_\tau + \mu)G_{ij\sigma}(\tau) + \sum_\delta t_{ii+\delta}G_{i+\delta j\sigma}(\tau) - \theta(\tau)|U|\langle n_{i-\sigma}(\tau)c_{i\sigma}(\tau)c^\dagger_{j\sigma}(0)\rangle$$
$$+\theta(-\tau)|U|\langle c^\dagger_{j\sigma}(0)n_{i-\sigma}(\tau)c_{i\sigma}(\tau)\rangle = \delta_{ij}\delta(\tau). \tag{3.88}$$

Unfortunately, there is no analytic formula for the two-particle operator averages that are multiplied by $|U|$. In the spirit of the BCS theory, we will evaluate the averages by using a variant of Wick's theorem, which is necessary to allow the system to become superconducting. This approach is the Hartree-Fock approximation for the ordered phase of the attractive Hubbard model (the first Hartree-Fock calculation of the phase diagram of the three-dimensional repulsive Hubbard model is [Penn (1966)]). As you may recall, Wick's theorem takes an operator average involving four Fermion operators, and writes it as a sum (with the appropriate sign) of the product of the averages of all possible operator pairs. If we assume we have no magnetic order, then operator averages of the form $\langle c^\dagger_{i\uparrow}(\tau)c_{j\downarrow}(0)\rangle$ vanish, because the electron cannot change its spin as it propagates in imaginary time from 0 to τ (and in space from site j to site i). One might have also set the operator average $\langle c^\dagger_{i\uparrow}(\tau)c^\dagger_{j\downarrow}(0)\rangle$ to zero as well, because the number of electrons is conserved. But the BCS theory tells us that we need to allow such averages to be nonzero in the superconducting state, since a superconductor involves a phase-coherent mixture of states with different electron number. Hence, in the Hartree-Fock theory for superconductivity, we approximate the $\sigma =\uparrow$ case of Eq. (3.88) by

$$(-\partial_\tau + \mu)G_{ij\uparrow}(\tau) + \sum_\delta t_{ii+\delta}G_{i+\delta j\uparrow}(\tau)$$
$$-\theta(\tau)|U|\langle n_{i\downarrow}\rangle\langle c_{i\uparrow}(\tau)c^\dagger_{j\uparrow}(0)\rangle - \theta(\tau)|U|\langle c^\dagger_{i\downarrow}(\tau)c^\dagger_{j\uparrow}(0)\rangle\langle c_{i\downarrow}(0)c_{i\uparrow}(0)\rangle$$
$$+\theta(-\tau)|U|\langle n_{i\downarrow}\rangle\langle c^\dagger_{j\uparrow}(0)c_{i\uparrow}(\tau)\rangle + \theta(-\tau)|U|\langle c^\dagger_{j\uparrow}(0)c^\dagger_{i\downarrow}(\tau)\rangle\langle c_{i\downarrow}(0)c_{i\uparrow}(0)\rangle$$
$$= \delta_{ij}\delta(\tau), \tag{3.89}$$

where we removed the τ dependence in some operator averages due to time-translation invariance. It is conventional to now define two additional Green's functions F and $\bar{F}$, which are called the anomalous Green's functions. They are defined by the following:

$$F_{ij} = -\langle \mathcal{T}_\tau c_{i\uparrow}(\tau)c_{j\downarrow}(0)\rangle \ ; \qquad \bar{F}_{ij} = -\langle \mathcal{T}_\tau c^\dagger_{i\downarrow}(\tau)c^\dagger_{j\uparrow}(0)\rangle. \tag{3.90}$$

These functions are connected by a "Hermitian" identity $\bar{F}_{ij}(\tau) = [F_{ji}(\tau)]^*$, which can be verified by direct calculation, and the use of time-translation invariance in the operator averages. Using these definitions, we find the Eq. (3.89) can be written as

$$(-\partial_\tau + \mu)G_{ij\uparrow}(\tau) + \sum_\delta t_{ii+\delta}G_{i+\delta j\uparrow}(\tau) + |U|\langle n_{i\downarrow}\rangle G_{ij}(\tau) \\ +|U|F_{ii}(0^-)\bar{F}_{ij}(\tau) = \delta_{ij}\delta(\tau). \tag{3.91}$$

This differential equation cannot be immediately solved because we do not know the explicit value for the anomalous Green's function $\bar{F}$. Hence, we need to derive an equation of motion for it as well. Starting from the definition in Eq. (3.90), and taking the derivative with respect to τ as usual, leads to the following equation for the attractive Hubbard model:

$$(-\partial_\tau - \mu)\bar{F}_{ij}(\tau) - \sum_\delta t^*_{ii+\delta}\bar{F}_{i+\delta j}(\tau) - |U|\langle n_{i\uparrow}\rangle \bar{F}_{ij}(\tau) \\ + |U|\bar{F}_{ii}(0^-)G_{ij\uparrow}(\tau) = 0, \tag{3.92}$$

where we used Hermiticity of the hopping matrix to replace $t_{i+\delta i}$ by $t^*_{ii+\delta}$. These two sets of equations [Eqs. (3.91) and (3.92)] now form a closed set of equations for the two Green's functions. It is customary to define the superconducting gap function Δ_i by $\Delta_i = |U|F_{ii}(0^-)$. Then $\Delta_i^* = |U|\bar{F}_{ii}(0^-)$.

If we assume there are no supercurrents flowing, even if we are in the superconducting state, our solution in the bulk will be homogeneous. This means that $\langle n_{i\sigma}\rangle$ and Δ_i will be independent of the lattice site i. If we have no external magnetic fields either, then the spin up and spin down Green's functions are identical too. In this case, we can immediately Fourier transform our results to momentum space (because the real-space summations are convolutions) and find

$$(-\partial_\tau + \mu + \frac{1}{2}|U|\langle n\rangle - \epsilon_{\mathbf{k}})G_{\mathbf{k}\uparrow}(\tau) + \Delta\bar{F}_{\mathbf{k}}(\tau) = \delta(\tau), \tag{3.93}$$

$$(-\partial_\tau - \mu - \frac{1}{2}|U|\langle n\rangle + \epsilon_{\mathbf{k}})\bar{F}_{\mathbf{k}}(\tau) + \Delta^* G_{\mathbf{k}\uparrow}(\tau) = 0. \tag{3.94}$$

The next step is to introduce the Matsubara frequencies just as before, by Fourier transforming the τ variable, and remembering that is is antiperiodic with period β. This yields

$$(i\omega_n + \mu + \frac{1}{2}|U|\langle n\rangle - \epsilon_{\mathbf{k}})G_{\mathbf{k}\uparrow}(i\omega_n) + \Delta\bar{F}_{\mathbf{k}}(i\omega_n) = 1, \tag{3.95}$$

$$(i\omega_n - \mu - \frac{1}{2}|U|\langle n\rangle + \epsilon_{\mathbf{k}})\bar{F}_{\mathbf{k}}(i\omega_n) + \Delta^* G_{\mathbf{k}\uparrow}(i\omega_n) = 0. \tag{3.96}$$

These equations can now be solved directly to find

$$G_{\mathbf{k}\uparrow}(i\omega_n) = -\frac{i\omega_n - \mu - \frac{1}{2}|U|\langle n\rangle + \epsilon_{\mathbf{k}}}{\omega_n^2 + \left(\mu + \frac{1}{2}|U|\langle n\rangle - \epsilon_{\mathbf{k}}\right)^2 + |\Delta|^2}, \tag{3.97}$$

$$\bar{F}_{\mathbf{k}}(i\omega_n) = \frac{\Delta^*}{\omega_n^2 + \left(\mu + \frac{1}{2}|U|\langle n\rangle - \epsilon_{\mathbf{k}}\right)^2 + |\Delta|^2}. \tag{3.98}$$

To arrive at a self-consistent solution, with a nonzero superconducting order parameter, we use the definition of the superconducting gap to note that we must have the following equation hold

$$|U|\bar{F}_{ii}(0^-) = |U|T\sum_n\sum_{\mathbf{k}}\bar{F}_{\mathbf{k}}(i\omega_n) = \Delta^*. \tag{3.99}$$

Since $\bar{F}$ has a Δ^* in the numerator [see Eq. (3.98)], there always is a solution with $\Delta^* = 0$, which corresponds to the normal-state solution. If the temperature is low enough, then we can satisfy the equation that results by canceling the Δ^* from the left and right hand sides, to find the BCS gap equation

$$1 = |U|T\sum_n\int d\epsilon\rho(\epsilon)\frac{1}{\omega_n^2 + \left(\mu + \frac{1}{2}|U|\langle n\rangle - \epsilon\right)^2 + |\Delta|^2}. \tag{3.100}$$

It can be shown, by comparing the free energies, that whenever a superconducting solution exists (*i.e.*, $\Delta \neq 0$), it is lower in energy than the normal-state solution [Bardeen, Cooper, and Schrieffer (1957)]. The superconducting transition temperature occurs when $\Delta(T)$ goes continuously to zero. Hence we can find T_c, by finding the temperature that satisfies Eq. (3.100) with $\Delta = 0$. See Prob. A.29 for an analysis of how to numerically solve for the T_c and the gap in the BCS model.

In this presentation of the BCS theory we have not discussed phonons, which are the usual objects which interact with electrons to produce superconductivity. Since phonon energy scales are much smaller than electron energy scales, these phonons are only active in scattering electrons within a thin energy shell about the Fermi surface (the thickness of the shell is on the order of the Debye temperature for the phonons). Hence, in most treatments for superconductivity, we can approximate the density of states by its constant value at the Fermi surface, and the integral over energy can be performed exactly. Once this is done, it turns out that all of the superconducting properties have universal relations between them. These relations are only approximate if one includes nonconstant DOS effects, or, more importantly, if one includes the retardation effects associated with the fact that phonons move much slower than electrons in a crystal, so the effective electron-electron interaction mediated by the phonons is retarded in time. Accounting for these deviations has been the source of much work in the 60s, 70s, and 80s. It is believed these effects are all well understood in low-temperature superconductors.

We will not discuss these kinds of effects further here, because the deviations are often too small to need to be taken into account when determining properties of devices like Josephson junctions. Instead, we continue with examining the properties of the attractive Hubbard model in the Hartree-Fock approximation, and we note that if we pick $|U|$ to be small enough that the superconducting T_c is much less than the hopping, then we are in the BCS regime, and we should recover all of the well-known universal relations.

In Fig. 3.17, we show plots of the superconducting transition temperature as a function of $|U|$, of the superconducting gap as a function of T, and of a universal plot of $\Delta(T)/\Delta(T \to 0)$ versus T/T_c. Note how all results essentially collapse onto a universal curve when plotted in this fashion. This is the BCS prediction for the behavior, and the Hartree-Fock approximation to the attractive Hubbard model is described well by the BCS theory

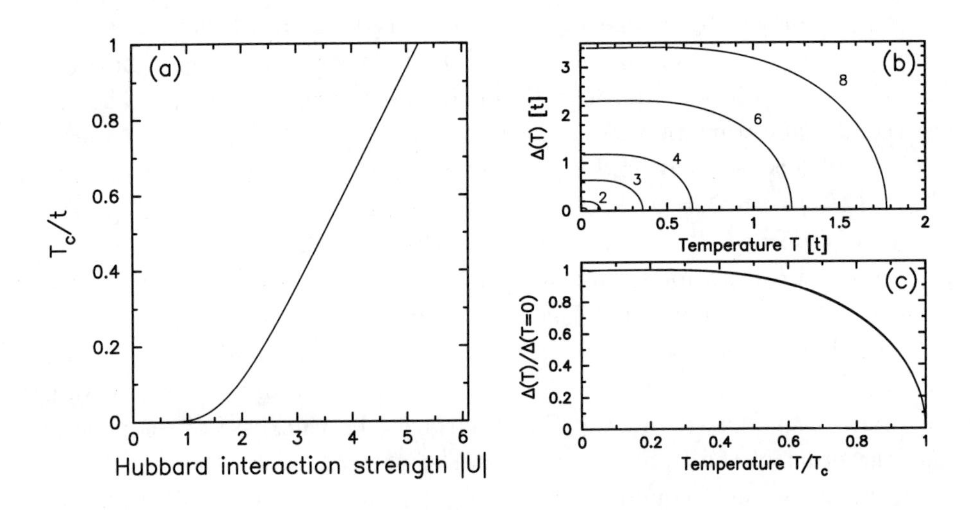

Fig. 3.17 (a) Transition temperature, (b) superconducting gap, and (c) universal plot of the relative gap versus the reduced temperature for the Hartree-Fock approximation to the Hubbard model on a simple-cubic lattice. In panel (b), seven values of $|U|$ are plotted: $|U| = 1$ (not visible), 1.5, 2, 3, 4, 6, and 8.

for a wide range of interaction strengths. The superconducting transition temperature (in panel a) seems to go to zero at a finite value of $|U|$, but this is actually not the case, as it depends on $|U|$ like $T_c \propto \exp[-C/|U|]$ for some constant C. The superconducting transition temperature also appears to increase without bound as $|U|$ becomes large. This is an actual artifact of the BCS theory. The correct behavior is for the transition temperature to have a maximum on the order of about one twentieth of the bandwidth, and then it decreases for larger values of $|U|$. This occurs because one enters the strong-coupling regime where the system forms pairs of electrons at high temperature, but they do not achieve the phase coherence needed for superconductivity until a much lower temperature is reached (see, for example, [Freericks, Jarrell and Scalapino (1993)]).

It turns out that there is a powerful and compact way of encapsulating the BCS equations into a 2×2 matrix form. This is called the Nambu-Gor'kov formalism [Nambu (1960); Gor'kov (1959)]. The starting point for this approach is to note that the pairing of electrons is between time-reversed states. So the state with spin up and momentum $\mathbf{k}$ is paired with the state with spin down and momentum $-\mathbf{k}$. These two states have the same energy because they form a so-called Kramers doublet [Kramers (1930)]; put in other words, because the Hamiltonian is time-reversal symmetric (when there is no magnetic field), the states that are paired together have the same energy when there is no current flowing. This motivates constructing a Fermionic spinor from a spin up electron and a spin down hole. We define the spinor $\psi_i(\tau) = [c_{i\uparrow}(\tau), c_{i\downarrow}^\dagger(\tau)]$ (not to be confused with a commutator) and form the 2×2 matrix Green's function

$$
\begin{aligned}
\mathbb{G}_{ij}(\tau) = -\langle \mathcal{T}_\tau \psi_i(\tau) \otimes \psi_j^\dagger(0)\rangle &= \begin{pmatrix} -\langle \mathcal{T}_\tau c_{i\uparrow}(\tau) c_{j\uparrow}^\dagger(0)\rangle & -\langle \mathcal{T}_\tau c_{i\uparrow}(\tau) c_{j\downarrow}(0)\rangle \\ -\langle \mathcal{T}_\tau c_{i\downarrow}^\dagger(\tau) c_{j\uparrow}^\dagger(0)\rangle & -\langle \mathcal{T}_\tau c_{i\downarrow}^\dagger(\tau) c_{j\downarrow}(0)\rangle \end{pmatrix} \\
&= \begin{pmatrix} G_{ij\uparrow}(\tau) & F_{ij}(\tau) \\ \bar{F}_{ij}(\tau) & -G_{ji\downarrow}(-\tau) \end{pmatrix}. \qquad (3.101)
\end{aligned}
$$

After defining the 2×2 Green's function matrix, we now need to determine its equation of motion. The EOMs for the $G_\uparrow$ and $\bar{F}$ functions appear in Eqs. (3.93) and (3.94). The EOMs for the other two Green's functions can be shown to be

$$\left(-\partial_\tau - \mu - \frac{1}{2}|U|\langle n\rangle + \epsilon_{\mathbf{k}}\right)[-G_{ji\downarrow}(-\tau)] + \Delta_i^* F_{ij}(\tau) = \delta(\tau) \quad (3.102)$$

$$\left(-\partial_\tau + \mu + \frac{1}{2}|U|\langle n\rangle - \epsilon_{\mathbf{k}}\right) F_{ij}(\tau) + \Delta_i\left[-G_{ji\downarrow}(-\tau)\right] = 0 \quad (3.103)$$

as derived in Prob. A.30. Now we perform a Fourier transformation to momentum space under the assumption that the superconducting gap is uniform in space. The four equations for the Green's functions can be summarized by the following matrix equation

$$\begin{pmatrix} -\partial_\tau + \mu + \frac{1}{2}|U|\langle n\rangle - \epsilon_{\mathbf{k}} & \Delta \\ \Delta^* & -\partial_\tau - \mu - \frac{1}{2}|U|\langle n\rangle + \epsilon_{\mathbf{k}} \end{pmatrix} \times \begin{pmatrix} G_{\mathbf{k}\uparrow}(\tau) & F_{\mathbf{k}}(\tau) \\ \bar{F}_{\mathbf{k}}(\tau) & -G_{-\mathbf{k}\downarrow}(-\tau) \end{pmatrix} = \begin{pmatrix} \delta(\tau) & 0 \\ 0 & \delta(\tau) \end{pmatrix}. \quad (3.104)$$

Fourier transforming from imaginary time to Matsubara frequencies transforms Eq. (3.104) into

$$\begin{pmatrix} i\omega_n + \mu + \frac{1}{2}|U|\langle n\rangle - \epsilon_{\mathbf{k}} & \Delta \\ \Delta^* & i\omega_n - \mu - \frac{1}{2}|U|\langle n\rangle + \epsilon_{\mathbf{k}} \end{pmatrix} \times \begin{pmatrix} G_{\mathbf{k}\uparrow}(i\omega_n) & F_{\mathbf{k}}(i\omega_n) \\ \bar{F}_{\mathbf{k}}(i\omega_n) & -G^*_{-\mathbf{k}\downarrow}(i\omega_n) \end{pmatrix} = \begin{pmatrix} 1 & 0 \\ 0 & 1 \end{pmatrix}. \quad (3.105)$$

Using $\mathbb{G}_{\mathbf{k}}(i\omega_n)$ for the matrix Green's function, and recalling the three Pauli spin matrices and the identity matrix

$$\tau_1 = \begin{pmatrix} 0 & 1 \\ 1 & 0 \end{pmatrix}, \quad \tau_2 = \begin{pmatrix} 0 & -i \\ i & 0 \end{pmatrix}, \quad \tau_3 = \begin{pmatrix} 1 & 0 \\ 0 & -1 \end{pmatrix}, \quad \mathbb{I} = \begin{pmatrix} 1 & 0 \\ 0 & 1 \end{pmatrix}, \quad (3.106)$$

allows us to write a compact form for the EOM

$$\left(i\omega_n \mathbb{I} + \left[\mu + \frac{1}{2}|U|\langle n\rangle - \epsilon_{\mathbf{k}} \right] \tau_3 + \mathrm{Re}\Delta\tau_1 - \mathrm{Im}\Delta\tau_2 \right) \mathbb{G}_{\mathbf{k}}(i\omega_n) = \mathbb{I}. \quad (3.107)$$

Finally, we solve for the Green's function by inverting the matrix on the left hand side of Eq. (3.107). The Green's function becomes

$$\mathbb{G}_{\mathbf{k}}(i\omega_n) = \frac{-i\omega_n \mathbb{I} + \left[\mu + \frac{1}{2}|U|\langle n\rangle - \epsilon_{\mathbf{k}}\right] \tau_3 + \mathrm{Re}\Delta\tau_1 - \mathrm{Im}\Delta\tau_2}{\omega_n^2 + \left[\mu + \frac{1}{2}|U|\langle n\rangle - \epsilon_{\mathbf{k}}\right]^2 + |\Delta|^2}, \quad (3.108)$$

which agrees with Eqs. (3.97) and (3.98) as it must.

We wish to go beyond the simple Hartree-Fock approximation to the attractive Hubbard model to describe additional scattering and how it can degrade the superconductivity. We will use local charge scattering via the Falicov-Kimball model for this purpose. We can solve the Falicov-Kimball model (plus the Hartree-Fock approximation to the attractive Hubbard model) exactly using the DMFT algorithm. To do so, we must first describe how to solve the Falicov-Kimball model impurity for spin-one-half electrons with additional anomalous dynamical mean fields that allow us to determine the anomalous Green's functions. This derivation is a straightforward generalization of the work that we already did for the spinless electron in the normal state, but it does require a careful analysis.

We start with a spin-one-half electron evolving in normal dynamical mean fields $\lambda_\uparrow$ and $\lambda_\downarrow$ and in anomalous dynamical mean fields α and $\bar{\alpha}$ (because the up spins do not interact directly with the down spins, we can solve this problem directly). The four evolution operators are then

$$S(\lambda_\uparrow) = \exp\left[-\int_0^\beta d\tau \int_0^\beta d\tau' \lambda_\uparrow(\tau,\tau') c_\uparrow^\dagger(\tau) c_\uparrow(\tau')\right] \tag{3.109}$$

$$S(\lambda_\downarrow) = \exp\left[-\int_0^\beta d\tau \int_0^\beta d\tau' \lambda_\downarrow(\tau,\tau') c_\downarrow^\dagger(\tau) c_\downarrow(\tau')\right] \tag{3.110}$$

$$S'(\alpha) = \exp\left[-\int_0^\beta d\tau \int_0^\beta d\tau' \alpha(\tau,\tau') c_\downarrow(\tau) c_\uparrow(\tau')\right] \tag{3.111}$$

$$\bar{S}'(\bar{\alpha}) = \exp\left[-\int_0^\beta d\tau \int_0^\beta d\tau' \bar{\alpha}(\tau,\tau') c_\uparrow^\dagger(\tau) c_\downarrow^\dagger(\tau')\right], \tag{3.112}$$

and the partition function to be evaluated is

$$\mathcal{Z}_{\text{imp}}(\lambda_\uparrow, \lambda_\downarrow, \alpha, \bar{\alpha}) = \text{Tr}_{c_\uparrow, c_\downarrow} \left\{ \mathcal{T}_\tau e^{-\beta(\mathcal{H}_{\text{imp}} - \mu\mathcal{N})} S(\lambda_\uparrow) S(\lambda_\downarrow) S'(\alpha) \bar{S}'(\bar{\alpha}) \right\}. \tag{3.113}$$

The Green's functions satisfy $G_\sigma(\tau,\tau') = -\delta \ln \mathcal{Z}_{\text{imp}}/\delta\lambda_\sigma(\tau',\tau)$, $F(\tau,\tau') = -\delta \ln \mathcal{Z}_{\text{imp}}/\delta\alpha(\tau',\tau)$, and $\bar{F}(\tau,\tau') = -\delta \ln \mathcal{Z}_{\text{imp}}/\delta\bar{\alpha}(\tau',\tau)$. Fourier transforming changes the functional derivatives to ordinary derivatives: $G_\sigma(i\omega_n) = -\partial \mathcal{Z}_{\text{imp}}/\partial\lambda_\sigma(i\omega_n)$, $F(i\omega_n) = -\partial \mathcal{Z}_{\text{imp}}/\partial\alpha(i\omega_n)$, and $\bar{F}(i\omega_n) = -\partial \mathcal{Z}_{\text{imp}}/\partial\bar{\alpha}(i\omega_n)$. Just like we did before in the normal case, we need to evaluate the EOMs for the Green's functions. The derivation is similar to what was done previously, and after Fourier transforming from imaginary time to Matsubara frequencies, we get

$$\begin{pmatrix} i\omega_n + \mu - \lambda_\uparrow(i\omega_n) & -\bar{\alpha}(i\omega_n) \\ -\alpha(i\omega_n) & i\omega_n - \mu + \lambda_\downarrow^*(i\omega_n) \end{pmatrix} \mathbb{G}(i\omega_n) = \mathbb{I}, \tag{3.114}$$

which can be directly solved to obtain the Green's function. This is performed in Prob. A.31. Once the Green's functions have been determined, we can then find the impurity partition function, which becomes

$$\mathcal{Z}_{\rm imp} = 4e^{\beta\mu}\prod_n \frac{[i\omega_n + \mu - \lambda_\uparrow(i\omega_n)][i\omega_n - \mu + \lambda_\downarrow^*(i\omega_n)] + \bar{\alpha}(i\omega_n)\alpha(i\omega_n)}{(i\omega_n)^2}. \tag{3.115}$$

We solve the problem for the Falicov-Kimball model in the same way as we did in the normal state. We form the full partition function by adding together the impurity partition function evaluated at μ and the impurity partition function evaluated at $\mu - U^{\rm FK}$ [with an additional weight of $\exp(-\beta E_f)$ for spinless localized electrons]. We will not go through the algebra for this here because the Falicov-Kimball model alone cannot support a superconducting state; it involves interactions with static particles, and Anderson's theorem [Anderson (1959b)] says that static interactions cannot lead to superconductivity. We will instead leap to the solution of the attractive Hubbard plus Falicov-Kimball model. To do this, we need to shift each λ_σ field by $-|U|\langle n\rangle/2$ and the α ($\bar{\alpha}$) fields are shifted by $-\Delta^*$ ($-\Delta$) respectively. The full partition function becomes

$$\begin{aligned}\mathcal{Z} = 4e^{\beta\mu}\prod_n \Big\{&[i\omega_n + \mu + \frac{1}{2}|U|\langle n\rangle - \lambda_\uparrow(i\omega_n)][i\omega_n + \mu + \frac{1}{2}|U|\langle n\rangle - \lambda_\downarrow(i\omega_n)]^* \\ &+ [\alpha(i\omega_n) - \Delta^*][\bar{\alpha}(i\omega_n) - \Delta]\Big\}\Big/(i\omega_n)^2 \\ + 4e^{\beta(\mu - U^{\rm FK})}e^{-\beta(E_f - \mu_f)}\prod_n \Big\{&[i\omega_n + \mu + \frac{1}{2}|U|\langle n\rangle - \lambda_\uparrow(i\omega_n) - U^{\rm FK}] \\ &\times[i\omega_n + \mu + \frac{1}{2}|U|\langle n\rangle - \lambda_\downarrow(i\omega_n) - U^{\rm FK}]^* \\ &+ [\alpha(i\omega_n) - \Delta^*][\bar{\alpha}(i\omega_n) - \Delta]\Big\}\Big/(i\omega_n)^2.\end{aligned} \tag{3.116}$$

The Green's functions next follow by taking derivatives with respect to the appropriate dynamical mean fields. First we need to define the localized electron filling $w_1 = 1 - w_0$, which satisfies

$$\begin{aligned} w_0 = (1 - w_1) = \Bigg[&1 + e^{-\beta(E_f + U^{\rm FK} - \mu_f)} \\ &\times \prod_n \frac{|i\omega_n + \mu + \frac{1}{2}|U|\langle n\rangle - U^{\rm FK} - \lambda(i\omega_n)|^2 + |\alpha(i\omega_n) - \Delta^*|^2}{|i\omega_n + \mu + \frac{1}{2}|U|\langle n\rangle - \lambda(i\omega_n)|^2 + |\alpha(i\omega_n) - \Delta^*|^2}\Bigg]^{-1}.\end{aligned} \tag{3.117}$$

In Eq. (3.117), we have used the fact that the self-consistent solution satisfies $\lambda_\uparrow = \lambda_\downarrow = \lambda$ and $\bar{\alpha} = \alpha^*$. Now we take the appropriate derivatives with respect to the dynamical mean fields to find the Green's functions, which become

$$G_\uparrow(i\omega_n) = w_0 \frac{[i\omega_n + \mu + \frac{1}{2}|U|\langle n\rangle - \lambda(i\omega_n)]^*}{|i\omega_n + \mu + \frac{1}{2}|U|\langle n\rangle - \lambda(i\omega_n)|^2 + |\alpha(i\omega_n) - \Delta^*|^2} \tag{3.118}$$
$$+ w_1 \frac{[i\omega_n + \mu + \frac{1}{2}|U|\langle n\rangle - U^{\rm FK} - \lambda(i\omega_n)]^*}{|i\omega_n + \mu + \frac{1}{2}|U|\langle n\rangle - U^{\rm FK} - \lambda(i\omega_n)|^2 + |\alpha(i\omega_n) - \Delta^*|^2}$$

$$F(i\omega_n) = w_0 \frac{-\alpha(i\omega_n)^* + \Delta}{|i\omega_n + \mu + \frac{1}{2}|U|\langle n\rangle - \lambda(i\omega_n)|^2 + |\alpha(i\omega_n) - \Delta^*|^2} \tag{3.119}$$
$$+ w_1 \frac{-\alpha(i\omega_n)^* + \Delta}{|i\omega_n + \mu + \frac{1}{2}|U|\langle n\rangle - U^{\rm FK} - \lambda(i\omega_n)|^2 + |\alpha(i\omega_n) - \Delta^*|^2}$$

$$\bar{F}(i\omega_n) = w_0 \frac{[-\alpha(i\omega_n) + \Delta^*]}{|i\omega_n + \mu + \frac{1}{2}|U|\langle n\rangle - \lambda(i\omega_n)|^2 + |\alpha(i\omega_n) - \Delta^*|^2} \tag{3.120}$$
$$+ w_1 \frac{[-\alpha(i\omega_n) + \Delta^*]}{|i\omega_n + \mu + \frac{1}{2}|U|\langle n\rangle - U^{\rm FK} - \lambda(i\omega_n)|^2 + |\alpha(i\omega_n) - \Delta^*|^2}$$

$$-G_\downarrow(i\omega_n)^* = w_0 \frac{-[i\omega_n + \mu + \frac{1}{2}|U|\langle n\rangle - \lambda(i\omega_n)]}{|i\omega_n + \mu + \frac{1}{2}|U|\langle n\rangle - \lambda(i\omega_n)|^2 + |\alpha(i\omega_n) - \Delta^*|^2} \tag{3.121}$$
$$+ w_1 \frac{-[i\omega_n + \mu + \frac{1}{2}|U|\langle n\rangle - U^{\rm FK} - \lambda(i\omega_n)]}{|i\omega_n + \mu + \frac{1}{2}|U|\langle n\rangle - U^{\rm FK} - \lambda(i\omega_n)|^2 + |\alpha(i\omega_n) - \Delta^*|^2}$$

To complete the DMFT algorithm, we need to extract the self-energy from the impurity Green's function (this is the self-energy of the attractive Hubbard model plus the Falicov-Kimball model, with the Hubbard piece evaluated in the Hartree-Fock approximation). We begin with the definition of the effective medium in the Nambu-Gor'kov notation:

$$\mathbb{G}_0^{-1}(i\omega_n) = \begin{pmatrix} i\omega_n + \mu - \lambda(i\omega_n) & -\alpha(i\omega_n)^* \\ -\alpha(i\omega_n) & i\omega_n - \mu + \lambda(i\omega_n)^* \end{pmatrix}, \tag{3.122}$$

and the 2×2 matrix self-energy $\bar{\Sigma}$ then satisfies

$$\begin{aligned} \bar{\Sigma}(i\omega_n) &= \mathbb{G}_0(i\omega_n)^{-1} - \mathbb{G}(i\omega_n)^{-1}, \\ &= \begin{pmatrix} \Sigma(i\omega_n) & \Phi(i\omega_n) \\ \Phi^*(i\omega_n) & -\Sigma^*(i\omega_n) \end{pmatrix} \\ &= \begin{pmatrix} -\frac{1}{2}|U|\langle n\rangle + \Sigma^{\rm FK}(i\omega_n) & \Delta + \Phi^{\rm FK}(i\omega_n) \\ \Delta^* + \Phi^{\rm FK*}(i\omega_n) & \frac{1}{2}|U|\langle n\rangle - \Sigma^{\rm FK*}(i\omega_n) \end{pmatrix}. \end{aligned} \tag{3.123}$$

In Eq. (3.123), we have separated out the Hartree-Fock contributions from the Falicov-Kimball model contributions for clarity. Finally, the local Green's function for the lattice is determined from the appropriate Hilbert transformation

$$\mathbb{G}(i\omega_n) = \int d\epsilon \rho(\epsilon) \left[i\omega_n \mathbb{I} + (\mu - \epsilon)\tau_3 - \bar{\Sigma}(i\omega_n) \right]^{-1}. \tag{3.124}$$

(See Prob. A.32 for efficient ways to evaluate this Hilbert transform.) The full DMFT algorithm starts with the self-energy matrix set equal to a convenient starting value. Unlike in the normal state, where we often set the self-energy to zero, here we need to ensure that the off-diagonal piece of the self-energy Φ is nonzero to begin with, or we will not stabilize a superconducting solution; the off-diagonal piece will iterate to zero if we are above T_c. Eq. (3.124) is then used to find the local Green's function matrix. The Dyson equation in Eq. (3.123) is then employed to find the the effective medium $\mathbb{G}_0$. Next the impurity Green's function is determined from Eqs. (3.118–3.121), and the new self-energy follows from the Dyson equation Eq. (3.123). The loop is iterated until the self-energy stops changing.

Note that we no longer have a simple way of determining the transition temperature, as we did before with the BCS gap equation [in Eq. (3.100)], because we do not have any way of determining the precise proportionality of the α fields with the gap Δ (in other words we need to know $\lim_{\Delta \to 0} \alpha(i\omega_n)/\Delta^*$ in order to perform calculations above T_c). This is needed to cancel the factor of Δ that we canceled from both sides of the BCS gap equation. Instead, we now need to solve for the gap as a function of temperature, and find the T_c from the point where the gap goes to zero [which is easiest to do by linearly extrapolating $\Delta^2(T)$ with a linear extrapolation as $T \to T_c$]. Such a procedure is much more laborious, because there is "critical slowing down" of the iterative solutions near T_c. Alternatively, one could calculate the pair-field susceptibility and find the temperature where it diverges, but this is beyond the scope of the book.

In Fig. 3.18, we plot the reduction of the superconducting T_c for the half-filled Hubbard-Falicov-Kimball model, with an attractive interaction of $|U| = 4$ and various U^{FK} (with fixed $w_1 = 0.5$). In panel (a), we show T_c versus U^{FK} which is reduced initially by a quadratic dependence on U^{FK}; this arises from the fact that the self-energy for the Falicov-Kimball model has an imaginary part that grows like $(U^{\mathrm{FK}})^2$ initially. Once we hit the insulating phase around $U^{\mathrm{FK}} \approx 4.92$, the T_c drops rapidly, going to zero near 5.5. In panel (b), we show a plot of the superconducting gap; the

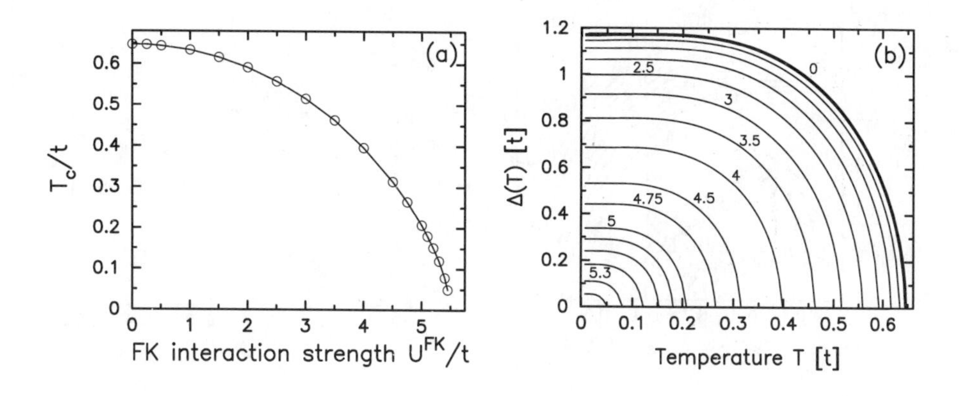

Fig. 3.18 (a) Transition temperature and (b) superconducting gap versus temperature for the $|U| = 4$ attractive Hubbard model with $w_1 = 0.5$ and various values of U^{FK}. Note how the transition temperature is reduced, and that the dependence is initially quadratic in U^{FK}. Once we reach the insulating phase (around $U^{\mathrm{FK}} \approx 4.92$) the T_c rapidly drops to zero. In the right panel, the superconducting gap is plotted. These gap functions all satisfy the universal BCS form when plotted in reduced variables. We chose U^{FK}=0, 0.25, 0.5, 1, 1.5, 2, 2.5, 3, 3.5, 4, 4.5, 4.75, 5, 5.1, 5.2, 5.3, 5.4, and 5.45.

shape stays essentially the same as T_c is reduced. In fact, if those results were plotted in terms of the reduced gap and reduced temperature, they would show the universal BCS form.

In Fig. 3.19, we show a more experimentally relevant plot of the gap reduction for a fixed scattering strength of the defects, but with a varying concentration w_1. In panel (a), we show the results for $U^{\mathrm{FK}} = 3$, and in panel (b), we show the results for $U^{\mathrm{FK}} = 6$. The inset to panel (a) is the superconducting transition temperature. These kinds of curves can be produced by exposing superconductors to varying doses of ion damage, or by introducing different concentrations of nonmagnetic impurities. While it is well-known in the theory of superconductivity that magnetic impurities are strong pair breakers [Anderson (1959b)], nonmagnetic impurities can have the same effect if they push the system close to a metal-insulator transition. Indeed, the T_c reduction is initially linear in the concentration of defects, similar to the magnetic-impurity prediction [Abrikosov and Gor'kov (1960)], but the decrease slows as we reach half-filling ($w_1 = 0.5$), where the T_c reduction is maximal. In panel (b), the T_c is suppressed all the way to zero because the system undergoes a density-driven metal-insulator transition (at $w_1 = 0.5$), and the scattering is too strong as the insulating phase is approached to support the superconductivity. Note how the dip in T_c is very

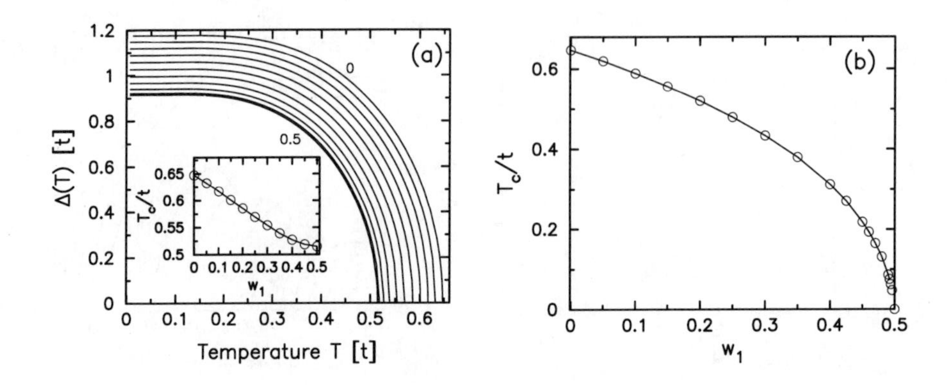

Fig. 3.19 (a) Superconducting gap versus temperature for the $|U| = 4$ attractive Hubbard-Falicov-Kimball model with $U^{\mathrm{FK}} = 3$ and various values of w_1 in steps of 0.05; the conduction electron filling is fixed at 1.0. The maximal depression occurs at $w_1 = 0.5$ and curves are identical with $w_1 \to 1 - w_1$ due to particle-hole symmetry. These curves are closer to what is seen in experiment, as defects typically have a given interaction strength, with their concentration being varied. Inset is the behavior of T_c, which initially decreases linearly with the concentration of defects. This behavior was first proposed by [Abrikosov and Gor'kov (1960)] for the case of magnetic impurities; nonmagnetic impurities, like what we have here are not supposed to change T_c if the DOS is unchanged [Anderson (1959b)], but our scattering is strong enough to change the DOS, thereby leading to a reduction in T_c. The rate of decrease of T_c must slow down as $w_1 \to 0.5$, because the T_c is a minimum at that point. (b) Superconducting T_c for $U^{\mathrm{FK}} = 6$. In this case, the T_c is suppressed to zero for large enough defect concentration (compare with Fig. 3.18). The gap function continues to display the universal form, and so it is not included.

sharp as w_1 approaches 0.5; it appears to decrease like $T_c \approx \sqrt{0.5 - w_1}$, but there may be small deviations at the lowest temperatures, which are difficult to determine accurately. This is the same kind of behavior as seen with magnetic impurities [Tinkham (1975)], but here it is driven by the proximity to the metal-insulator transition, rather than the breaking of time-reversal invariance, which arises from the spin-flip scattering off of a magnetic impurity.

One of the most important properties of a superconductor is that it can carry current without losses. We would like to calculate the current that a superconductor can carry within our many-body formalism. The basic physical idea is that there is a superconducting gap, so we can modify the distribution of electrons in the ground state, and still maintain the superconductivity, if we do not change the energy by more than the condensation energy resulting from the presence of the gap. To carry current, we need to shift the center of the distribution of electrons (in momentum space) from the origin, to a finite wavevector $\mathbf{Q}$. Then the electrons will have a net flow in the $\mathbf{Q}$ direction, because the distribution has a nonzero total momentum. This implies that we are pairing the electronic states $\mathbf{k}+\mathbf{Q}/2$ and spin-up with the states $-\mathbf{k}+\mathbf{Q}/2$ and spin down. Since these states are not related by time-reversal symmetry any more, they will have somewhat different energies, and it is more difficult to pair them together (when the difference in energy is larger than the superconducting gap, the two states won't gain energy through pairing, and will become normal). When $\mathbf{Q}$ becomes too large, the system will lose its superconductivity. The maximal value of the current density as a function of $\mathbf{Q}$ is called the critical current density of the superconductor.

It turns out that a uniform current will flow in the bulk if the superconducting gap behaves like

$$\Delta_j = e^{i\mathbf{Q}\cdot\mathbf{R}_j}\Delta, \tag{3.125}$$

with Δ real (the uniform current flow is a theoretical construct, actual materials often have inhomogeneous current profiles). At this point we take this statement as an *ansatz*, and we will show later that indeed it does produce a uniform current. The presence of a phase gradient (the phase grows linearly with position) for the superconducting gap function breaks the translational invariance of the system. This is to be expected, because we now have a direction defined by the direction that the current is flowing. It turns out that the normal Green's functions retain their translational invariance, but the anomalous Green's functions do not. Our strategy is to find the correct phase factor that we need to multiply the anomalous Green's functions by to restore translational invariance. We will produce this factor by introducing a canonical transformation to the Fermionic creation and annihilation operators, whose net result will be to introduce a complex phase onto the hopping matrix elements. Once translational invariance has been restored, we will then be able to solve the problem directly by going to momentum space.

It is not easy to determine what phase factor is needed to restore translational invariance to the Green's functions in the presence of a phase gradient on the superconducting gap. We now go through this derivation step by step. We take the EOMs in Eqs. (3.91) and (3.92), and Fourier transform to Matsubara frequencies:

$$\sum_l [(i\omega_n + \mu + \frac{1}{2}|U|\langle n\rangle)\delta_{il} + t_{il}]G_{lj\uparrow}(i\omega_n) + \Delta_i \bar{F}_{ij}(i\omega_n) = \delta_{ij}, \quad (3.126)$$

$$\sum_l [(i\omega_n - \mu - \frac{1}{2}|U|\langle n\rangle)\delta_{il} - t^*_{il}]\bar{F}_{lj}(i\omega_n) + \Delta^*_i G_{ij\uparrow}(i\omega_n) = 0. \quad (3.127)$$

Solve Eq. (3.127) for $G_{ij\uparrow}$

$$G_{ij\uparrow}(i\omega_n) = -\frac{1}{\Delta^*_i}\sum_l [(i\omega_n - \mu - \frac{1}{2}|U|\langle n\rangle)\delta_{il} - t^*_{il}]\bar{F}_{lj}(i\omega_n), \quad (3.128)$$

and substitute the result into Eq. (3.126). We write the anomalous Green's function with a double Fourier transform, because we do not know that it is translation invariant when there is a current flowing. The formula is

$$\bar{F}_{ij}(i\omega_n) = \sum_{\mathbf{k}\mathbf{k}'} e^{i\mathbf{k}\cdot\mathbf{R}_i + i\mathbf{k}'\cdot\mathbf{R}_j}\bar{F}_{\mathbf{k}\mathbf{k}'}(i\omega_n). \quad (3.129)$$

Substituting this formula into the expression allows the summation over the spatial indices to be carried out, and after some significant algebra, one finds

$$\sum_{\mathbf{k}\mathbf{k}'} e^{i(\mathbf{k}+\mathbf{Q})\cdot\mathbf{R}_i + i\mathbf{k}'\cdot\mathbf{R}_j}\bar{F}_{\mathbf{k}\mathbf{k}'}(i\omega_n)$$
$$\times [-(i\omega_n + \mu + \frac{1}{2}|U|\langle n\rangle - \epsilon_{\mathbf{k}+\mathbf{Q}})(i\omega_n - \mu - \frac{1}{2}|U|\langle n\rangle + \epsilon_{\mathbf{k}}) + \Delta^2] = \Delta\delta_{ij}. \quad (3.130)$$

If we shift the i and j spatial components by a nearest-neighbor translation vector, the right hand side is unchanged, because the constant Δ is independent of i and j, hence we learn that the factor $\exp[i\delta\cdot(\mathbf{k}+\mathbf{Q}+\mathbf{k}')] = 1$. Since we restrict our momenta to the Brillouin zone, we learn that $\mathbf{k}' = -\mathbf{k} - \mathbf{Q}$. So we find

$$\sum_{\mathbf{k}} e^{i(\mathbf{k}+\mathbf{Q})\cdot(\mathbf{R}_i - \mathbf{R}_j)}\bar{F}_{\mathbf{k},-\mathbf{k}-\mathbf{Q}}(i\omega_n)$$
$$\times [-(i\omega_n + \mu + \frac{1}{2}|U|\langle n\rangle - \epsilon_{\mathbf{k}+\mathbf{Q}})(i\omega_n - \mu - \frac{1}{2}|U|\langle n\rangle + \epsilon_{\mathbf{k}}) + \Delta^2] = \Delta\delta_{ij}. \quad (3.131)$$

The only way to have the left hand side become a Kronecker delta function is for the summand to be just the exponential factor. In other words, we must have

$$\bar{F}_{\mathbf{k},-\mathbf{k}-\mathbf{Q}}(i\omega_n) = \frac{\Delta}{-(i\omega_n + \mu + \frac{1}{2}|U|\langle n\rangle - \epsilon_{\mathbf{k}+\mathbf{Q}})(i\omega_n - \mu - \frac{1}{2}|U|\langle n\rangle + \epsilon_{\mathbf{k}}) + \Delta^2}. \tag{3.132}$$

Now we can construct the real-space anomalous Green's function from the Fourier transform, or

$$\bar{F}_{ij}(i\omega_n) = \sum_{\mathbf{k}} e^{i\mathbf{k}\cdot(\mathbf{R}_i - \mathbf{R}_j) - \mathbf{Q}\cdot\mathbf{R}_j} \times \frac{\Delta}{-(i\omega_n + \mu + \frac{1}{2}|U|\langle n\rangle - \epsilon_{\mathbf{k}+\mathbf{Q}})(i\omega_n - \mu - \frac{1}{2}|U|\langle n\rangle + \epsilon_{\mathbf{k}}) + \Delta^2}. \tag{3.133}$$

This final formula does indeed show that the anomalous Green's function is not translation invariant. But if we define a new anomalous Green's function

$$\bar{\mathcal{F}}_{ij}(i\omega_n) = e^{i\frac{\mathbf{Q}}{2}\cdot(\mathbf{R}_i + \mathbf{R}_j)} \bar{F}_{ij}(i\omega_n), \tag{3.134}$$

then it is easy to see that $\bar{\mathcal{F}}_{ij}$ is translationally invariant. Now we use Eq. (3.128) to see that the normal Green's function $G_{ij\uparrow}$ is translationally invariant; nevertheless, we define a new normal Green's function via

$$\mathcal{G}_{ij\uparrow}(i\omega_n) = e^{i\frac{\mathbf{Q}}{2}\cdot(\mathbf{R}_i - \mathbf{R}_j)} G_{ij\uparrow}(i\omega_n). \tag{3.135}$$

Both Eqs. (3.134) and (3.135) follow if we introduce new electron creation and annihilation operators

$$\tilde{c}_{i\sigma} = e^{-\frac{i}{2}\mathbf{Q}\cdot\mathbf{R}_i} c_{i\sigma}, \quad \tilde{c}^{\dagger}_{i\sigma} = e^{\frac{i}{2}\mathbf{Q}\cdot\mathbf{R}_i} c^{\dagger}_{i\sigma}, \tag{3.136}$$

and express the Green's functions in terms of the tilde operators. Since all of the anticommutation relations of the tilde operators are unchanged after the phases in Eq. (3.136) are introduced, this transformation is called a canonical transformation.

The canonical transformation of the creation and annihilation operators restores full translational symmetry to the system. Hence, our strategy for finding the critical current is to (i) transform the Hamiltonian to be expressed in terms of the tilde operators, (ii) Fourier transform to solve the problem in momentum space, and (iii) invert the canonical transformation to find the original Green's functions. The current can then be determined from these original Green's functions.

The first step is to express the attractive Hubbard plus Falicov-Kimball model in terms of the tilde operators. Since $\tilde{n}_{i\sigma} = n_{i\sigma}$ (because the two phase factors cancel), the only change in the form of the Hamiltonian is for the hopping matrix, because $\tilde{c}_{i\sigma}^\dagger \tilde{c}_{i+\delta\sigma} = \exp[-i\mathbf{Q}\cdot\delta/2]c_{i\sigma}^\dagger c_{i+\delta\sigma}$. Hence, if we define $\tilde{t}_{ii+\delta} = \exp[i\mathbf{Q}\cdot\delta/2]t_{ii+\delta}$, the Hamiltonian in the tilde operators is

$$\tilde{\mathcal{H}} - \mu\tilde{\mathcal{N}} = -\sum_{\delta\sigma}\tilde{t}_{ii+\delta}\tilde{c}_{i\sigma}^\dagger\tilde{c}_{i+\delta\sigma} - |U|\sum_i \tilde{n}_{i\uparrow}\tilde{n}_{i\downarrow} - \mu\sum_i(\tilde{n}_{i\uparrow}+\tilde{n}_{i\downarrow})$$
$$+ E_f\sum_i w_i + U^{\mathrm{FK}}\sum_i w_i(\tilde{n}_{i\uparrow}+\tilde{n}_{i\downarrow}). \tag{3.137}$$

The only difference from the original Hamiltonian is that the hopping matrix is now complex-valued (although it remains Hermitian). This achieves our first goal of transforming the phase gradient of the order parameter onto the hopping matrix elements. The hopping matrix is translationally invariant, so we can diagonalize the problem by going to momentum space. Because of the extra phase factors on the hopping matrix, the eigenvalues of $-\tilde{t}_{ij}$ are $\epsilon_{\mathbf{k}+\mathbf{Q}/2}$, while the eigenvalues of $-\tilde{t}_{ij}^*$ are $\epsilon_{\mathbf{k}-\mathbf{Q}/2}$. The DMFT algorithm is essentially the same as before, because the impurity part of the problem is unchanged. The only modification is that the Hilbert transform becomes

$$\tilde{\mathbb{G}}(i\omega_n) = \sum_{\mathbf{k}}\left[\begin{pmatrix} i\omega_n + \mu - \epsilon_{\mathbf{k}+\mathbf{Q}/2} & 0 \\ 0 & i\omega_n - \mu + \epsilon_{\mathbf{k}-\mathbf{Q}/2}\end{pmatrix} - \Sigma(i\omega_n)\right]^{-1}. \tag{3.138}$$

Note that we cannot replace the sum over momentum by an integral over the DOS any more because the integrand does not depend solely on $\epsilon_{\mathbf{k}}$. But if we work on a simple cubic lattice, and note that if we choose $\mathbf{Q}$ in the z-direction, then

$$\epsilon_{\mathbf{k}\pm\mathbf{Q}/2} = -2t[\cos(\mathbf{k}_x) + \cos(\mathbf{k}_y)]$$
$$- 2t\cos(\mathbf{k}_z)\cos(\mathbf{Q}_z/2) \pm 2t\sin(\mathbf{k}_z)\sin(\mathbf{Q}_z/2). \tag{3.139}$$

The summation over $\mathbf{k}_x$ and $\mathbf{k}_y$ can be performed by integrating over the two-dimensional DOS; we then have one final integration over $\mathbf{k}_z$. This last integral over $\mathbf{k}_z$ can be performed via contour integration and the residue theorem rather than calculating it numerically; see Prob. A.33 for details.

Once the Green's functions for the tilde variables have been determined, we find the Green's functions for the original variables by introducing the appropriate phase factors. The current density is finally found by evaluating the expectation value of the current operator (it is the same current operator that we used in the normal state, and we evaluate it, neglecting vertex corrections, as before). The result is

$$\mathbf{j}_z = \frac{4|e|t}{\hbar a^2} T \sum_n \mathrm{Im} G_{ii+\delta_z}(i\omega_n) = \frac{4|e|t}{\hbar a^2} T \sum_n \mathrm{Im}\left[e^{i\mathbf{Q}a/2} \mathcal{G}_{ii+\delta_z}(i\omega_n)\right], \tag{3.140}$$

where there is an additional factor of 2 arising from the sum over spin states. The nearest-neighbor Green's function is found from

$$G_{ii+\delta_z}(i\omega_n) = \sum_{\mathbf{k}} e^{i(\mathbf{k}_z + \frac{\mathbf{Q}_z}{2})} \mathcal{G}_{\mathbf{k}}(i\omega_n). \tag{3.141}$$

Calculating this nearest-neighbor Green's function requires evaluating another integral similar to the one used for the generalized Hilbert transform [see Prob. A.33 (d)]. Note that the current will vanish when $\mathbf{Q} = 0$ or when $\Delta = 0$. Finally, we increase $\mathbf{Q}$ until the system cannot support superconducting order any more. The maximal current density is the critical current density. Note that the direct calculation of the current in this case shows that it is uniform, so the *ansatz* that we keep the magnitude of the gap fixed, and vary just the phase of the gap through space, is verified *a posteriori.*

In Fig. 3.20, we plot current-phase-gradient relation for the bulk supercurrent. In panel (a), we fix the density of charge scatterers and vary the strength of the scattering, while in panel (b) we fix the strength of the scatterers and vary their concentration. Note how the charge scatterers initially have a small effect on the current (when the scattering strength is small), but then they suppress it all the way to zero as the scattering increases. The filled circles show the critical current for each case. The location of the critical current migrates to the left as the scattering increases. The results are similar when we increase the concentration of scatterers, all with the same strength of scattering, but the reduction appears to be linear in the concentration for most concentrations. An interesting early review of critical currents in superconductors can be found in [Bardeen (1962)].

The problem is only slightly more complicated for nanostructures. In this case, we do not know the phase profile for the BCS gap, and it need not be described by a simple phase gradient. We will actually determine the phase profile via self-consistently solving the problem. We start with a given current in the bulk, which is described by a uniform phase gradient.

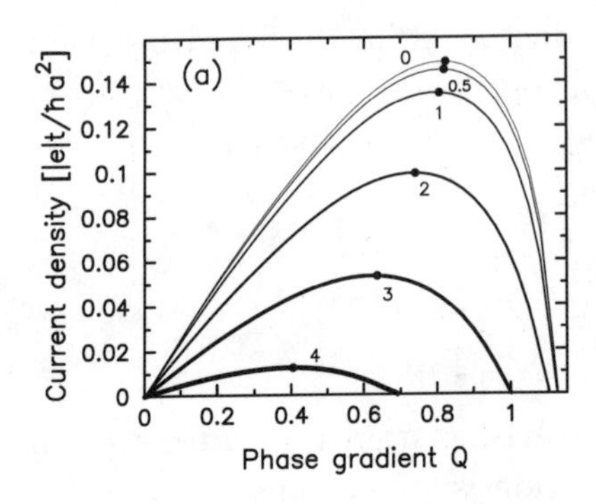

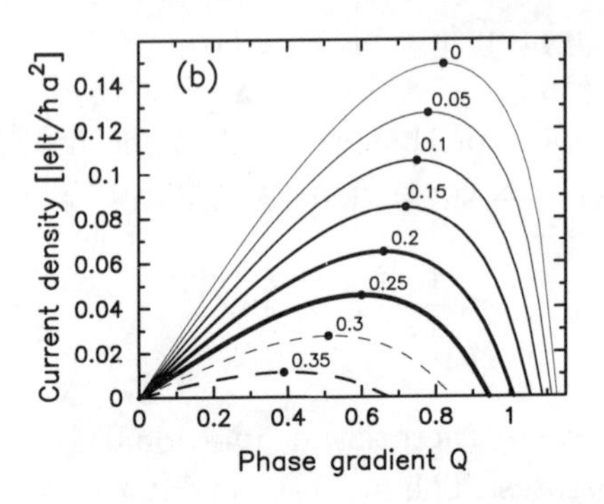

Fig. 3.20 (a) Current as a function of the phase gradient at $T = 0.3$ for the $|U| = 4$ attractive Hubbard model with $w_1 = 0.5$ and various values of U^{FK}. Note how the critical current is initially insensitive to the charge scatterers, but then is reduced dramatically as the system moves closer to the metal-insulator transition. The solid circles show the critical current for each case. (b) Current as a function of phase gradient at $T = 0.3$ for the $|U| = 4$ attractive Hubbard model with $U^{\mathrm{FK}} = 6$ and various values of w_1. Note how the critical current is reduced essentially linearly with the concentration of charge scatterers for most cases.

We fix the phase gradient to this value by choosing the change in phase of F_α to be equal to $\mathbf{Q}$ at the first self-consistent plane of the nanostructure. We describe the general anomalous Green's functions via an overall phase $F_\alpha = |F_\alpha| \exp[i\phi_\alpha]$. In the bulk, we would have $\phi_\alpha = \mathbf{Q}_z\alpha$, but for the nanostructure we will see deviations of this form. As before, we transform the phase that accumulates on the anomalous Green's functions onto a phase of the hopping matrix elements. This is accomplished by making the same canonical transformation as before

$$\tilde{c}_{\alpha i\sigma} = e^{-i\phi_\alpha/2} c_{\alpha i\sigma}, \quad \tilde{c}^\dagger_{\alpha i\sigma} = e^{i\phi_\alpha/2} c^\dagger_{\alpha i\sigma}; \tag{3.142}$$

this produces the change in the hopping $t_{\alpha\alpha+1} \to \exp[-i(\phi_\alpha - \phi_{\alpha+1})] t_{\alpha\alpha+1}$. Because the total charge is conserved, even if we have an inhomogeneous system, one important consistency check is to verify that the current density does not change from one plane to the next. This method of computation works with a current-biased junction, just like what would be done in an experiment. As the current is increased, we will find a point where the current density through the device reaches a maximum. Once this occurs, we have achieved the critical current density of the device. It is always smaller than the critical current density in the bulk.

Note that we need to use a generalization of the quantum zipper algorithm, within the Nambu-Gor'kov formalism, to solve for the Green's functions, but such a modification is rather straightforward to carry out [Miller and Freericks (2001)]. The key issue is that the quantum zipper algorithm will work with 2×2 matrix functions for the recursion to the left or to the right. One needs to be particularly careful about the phases on the hopping matrix elements as well, because those matrix elements are complex when we perform the canonical transformation to the tilde variables.

The Nambu-Gor'kov generalization of the quantum zipper algorithm begins with the following equation for the Matsubara frequency Green's function in the mixed basis (in terms of the tilde operators):

$$\tilde{\mathbb{G}}_{\alpha\beta}(\mathbf{k}^{\|}, i\omega_n) = \\ \begin{pmatrix} \ddots & \mathbb{B}_{\alpha-1} & 0 & \vdots & \vdots & \\ \ldots\, i\omega_n\mathbb{I} - \mathbb{A}_{\alpha-1}(\mathbf{k}^{\|}) & & \mathbb{B}_{\alpha} & 0 & 0 & \ldots \\ \ldots & \mathbb{B}^{\dagger}_{\alpha} & i\omega_n\mathbb{I} - \mathbb{A}_{\alpha}(\mathbf{k}^{\|}) & \mathbb{B}_{\alpha+1} & 0 & \ldots \\ \ldots & 0 & \mathbb{B}^{\dagger}_{\alpha+1} & i\omega_n\mathbb{I} - \mathbb{A}_{\alpha+1}(\mathbf{k}^{\|}) & \mathbb{B}_{\alpha+2} & 0 \\ & 0 & 0 & \ddots & \ddots & \ddots \end{pmatrix}^{-1}_{\alpha\beta} \tag{3.143}$$

where the matrices $\mathbb{A}_{\alpha}(\mathbf{k}^{\|})$ are the total planar energies for a particular plane, given by

$$\mathbb{A}_{\alpha}(\mathbf{k}^{\|}) = \begin{pmatrix} \epsilon^{\|}_{\alpha\mathbf{k}^{\|}} + \Sigma_{\alpha}(i\omega_n) - \mu & \Phi_{\alpha}(i\omega_n) \\ \Phi^*_{\alpha}(i\omega_n) & -\epsilon^{\|}_{\alpha\mathbf{k}^{\|}} - \Sigma^*_{\alpha}(i\omega_n) + \mu \end{pmatrix} \tag{3.144}$$

and the matrices $\mathbb{B}_{\alpha}$ are the hopping terms (which couple the $\alpha - 1$st and αth planes),

$$\mathbb{B}_{\alpha} = \begin{pmatrix} -t_{\alpha-1\alpha} & 0 \\ 0 & t^*_{\alpha-1\alpha} \end{pmatrix}. \tag{3.145}$$

The planar Green's function, $\tilde{\mathbb{G}}_{\alpha\alpha}(\mathbf{k}^{\|}, i\omega_n)$, is readily evaluated as a combination of continued fractions just as we did before, except now we need to keep track of the additional 2×2 matrix structure of the Nambu-Gor'kov approach. We define the matrix-valued right function, $\mathbb{R}_{\alpha}(\mathbf{k}^{\|}, i\omega_n)$, and the matrix-valued left function, $\mathbb{L}_{\alpha}(\mathbf{k}^{\|}, i\omega_n)$, from their respective recursion relations

$$\mathbb{R}_{\alpha}(\mathbf{k}^{\|}, i\omega_n) = i\omega_n\mathbb{I} - \mathbb{A}_{\alpha}(\mathbf{k}^{\|}) - \mathbb{B}_{\alpha+1}\mathbb{R}^{-1}_{\alpha+1}(\mathbf{k}^{\|}, i\omega_n)\mathbb{B}^{\dagger}_{\alpha+1} \tag{3.146}$$

$$\mathbb{L}_{\alpha}(\mathbf{k}^{\|}, i\omega_n) = i\omega_n\mathbb{I} - \mathbb{A}_{\alpha}(\mathbf{k}^{\|}) - \mathbb{B}^{\dagger}_{\alpha}\mathbb{L}^{-1}_{\alpha-1}(\mathbf{k}^{\|}, i\omega_n)\mathbb{B}_{\alpha}\,. \tag{3.147}$$

The recursive calculation continues to infinity, but once it has been extended to planes in the uniform bulk region, the coefficients for each plane become constant. The effect of a constant spatial phase gradient in Δ is equivalent to a constant phase factor in the hopping integral, $t_{\alpha\alpha+1}$, that does not change between planes in the bulk (when we work with the tilde operators instead of the original operators). Hence, by equating all $\mathbb{R}_\alpha(\mathbf{k}^\parallel, i\omega_n)$ with $\mathbb{R}_\infty(\mathbf{k}^\parallel, i\omega_n)$ for α outside of the self-consistent region and $\mathbb{L}_\alpha(\mathbf{k}^\parallel, i\omega_n)$ with $\mathbb{L}_{-\infty}(\mathbf{k}^\parallel, i\omega_n)$ for α in the bulk limit, an exact terminator function can be calculated as the solution of a complex quadratic matrix equation:

$$\mathbb{R}_\infty(\mathbf{k}, i\omega_n)\mathbb{B}_\infty^{\dagger-1}\mathbb{R}_\infty(\mathbf{k}^\parallel, i\omega_n) + \left[\mathbb{A}_\infty(\mathbf{k}^\parallel) - i\omega_n\mathbb{I}\right]\mathbb{B}_\infty^{\dagger-1}\mathbb{R}_\infty(\mathbf{k}^\parallel, i\omega_n) + \mathbb{B}_\infty = \mathbb{O} \tag{3.148}$$

$$\mathbb{L}_{-\infty}(\mathbf{k}^\parallel, i\omega_n)\mathbb{B}_{-\infty}^{-1}\mathbb{L}_{-\infty}(\mathbf{k}^\parallel, i\omega_n) + \left[\mathbb{A}_{-\infty}(\mathbf{k}^\parallel) - i\omega_n\mathbb{I}\right]\mathbb{B}_{-\infty}^{-1}\mathbb{L}_{-\infty}(\mathbf{k}^\parallel, i\omega_n) + \mathbb{B}_{-\infty}^\dagger = \mathbb{O}. \tag{3.149}$$

Note that the same terminator function is used for all sites in the intermediate layers, and the functions $\mathbb{R}_\alpha$ and $\mathbb{L}_\alpha$ calculated for one site are also used in the calculation for the next site. These matrix quadratic equations are solved iteratively for the bulk.

The continued fractions form the local planar Green's functions, according to

$$\tilde{\mathbb{G}}_{\alpha\alpha}(\mathbf{k}^\parallel, i\omega_n) = \Big\{i\omega_n\mathbb{I} - \mathbb{A}_\alpha(\mathbf{k}^\parallel) - \mathbb{B}_\alpha^\dagger\mathbb{L}_{\alpha-1}^{-1}(\mathbf{k}^\parallel, i\omega_n)\mathbb{B}_\alpha - \mathbb{B}_{\alpha+1}\mathbb{R}_{\alpha+1}^{-1}(\mathbf{k}^\parallel, i\omega_n)\mathbb{B}_{\alpha+1}^\dagger\Big\}^{-1} \tag{3.150}$$

which, using Eqs. (3.146-3.147), can be simplified to

$$\tilde{\mathbb{G}}_{\alpha\alpha}(\mathbf{k}^\parallel, i\omega_n) = \left[\mathbb{R}_\alpha(\mathbf{k}^\parallel, i\omega_n) + \mathbb{L}_\alpha(\mathbf{k}^\parallel, i\omega_n) - i\omega_n\mathbb{I} + \mathbb{A}_\alpha(\mathbf{k}^\parallel)\right]^{-1} \tag{3.151}$$

The Green's functions connecting neighboring planes, α and $\alpha \pm 1$, which are required to calculate the current flow, satisfy two equivalent forms

$$\begin{aligned}\tilde{\mathbb{G}}_{\alpha\alpha+1}(\mathbf{k}^\parallel, i\omega_n) &= -\tilde{\mathbb{G}}_{\alpha\alpha}(\mathbf{k}^\parallel, i\omega_n)\mathbb{B}_{\alpha+1}\mathbb{R}_{\alpha+1}^{-1}(\mathbf{k}^\parallel, i\omega_n) \\ &= -\mathbb{L}_\alpha^{-1}(\mathbf{k}^\parallel, i\omega_n)\mathbb{B}_{\alpha+1}\tilde{\mathbb{G}}_{\alpha+1\alpha+1}(\mathbf{k}^\parallel, i\omega_n) \end{aligned}\tag{3.152}$$

$$\begin{aligned}\tilde{\mathbb{G}}_{\alpha\alpha-1}(\mathbf{k}^\parallel, i\omega_n) &= -\tilde{\mathbb{G}}_{\alpha\alpha}(\mathbf{k}^\parallel, i\omega_n)\mathbb{B}_\alpha^\dagger\mathbb{L}_{\alpha-1}^{-1}(\mathbf{k}^\parallel, i\omega_n) \\ &= -\mathbb{R}_{\alpha+1}^{-1}(\mathbf{k}^\parallel, i\omega_n)\mathbb{B}_\alpha^\dagger\tilde{\mathbb{G}}_{\alpha-1\alpha-1}(\mathbf{k}^\parallel, i\omega_n). \end{aligned}\tag{3.153}$$

The current per unit cell area, $J_{\alpha\alpha+1}$, which flows along each link between two neighboring planes, α and $\alpha+1$, in the z-direction is finally given by:

$$J_{\alpha\alpha+1} = \frac{4|e|t}{\hbar a^2} T \sum_n \int_{-\infty}^{\infty} d\epsilon^{\|} \rho^{2d}(\epsilon^{\|}) \mathrm{Im}\left[e^{i(\phi_\alpha - \phi_{\alpha+1})} \mathcal{G}_{\alpha\alpha+1}(\epsilon^{\|}, i\omega_n) \right]; \tag{3.154}$$

the phase factor is needed because we express the current in terms of the original electron creation and annihilation operators. A stringent convergence check for self-consistency, when there is a phase difference between the bulk superconductors, is that the current flow is constant from one plane to the next (*i.e.*, $J_{\alpha\alpha+1}$ is independent of α).

The approach given here is a generalization of the Bogoliubov-DeGennes equations [Bogoliubov, Tolmachev and Shirkov (1958); de Gennes (1966)] to allow for strong electron correlations. In addition, by employing the local approximation and inhomogeneous DMFT, we can solve the resulting equations much more efficiently on a computer.

There are some additional numerical issues that arise in these calculations. It can become difficult to determine the phase profile as we near the critical current. A worthwhile strategy is to slowly increase the current bias (*i. e.*, the phase gradient $\mathbf{Q}$ at the first plane), and to use the phase profile from a previous calculation as the starting point for the next calculation. This "simulated annealing" type strategy often aids in being able to stabilize the iterative nature of the solutions. Second, when the critical current density of the device becomes too small, then one can run into numerical problems which appear to arise from loss of precision error. In many of our calculations, we found it difficult to stabilize solutions when the critical current density was smaller than 10^{-12} times the critical current density in the bulk.

A complete discussion of solutions of these equations in the case of a Josephson junction is given in Chapter 5.

3.9 Finite Dimensions and Vertex Corrections

Our theory has assumed that the vertex corrections vanish for inhomogeneous DMFT if they vanish in the bulk. Strictly speaking this is not true. For example, the vanishing of the vertex corrections for the conductivity, arises solely from the fact that the velocity operator is odd in parity, while the irreducible vertex is even in parity. When combined in the Bethe-Salpeter equation to form the conductivity response function, this difference in parity causes the vertex-correction terms to vanish. When we consider an inhomogeneous multilayered device, this parity argument no longer holds for longitudinal transport. The charge vertex generically varies from plane to plane, so it does not have a well-defined parity, and the current operator is expressed in a real-space format, which also does not have a well-defined parity. Hence, we can no longer explicitly show that the vertex corrections vanish. We have assumed in this book that the modifications due to the effect of the charge vertices will be small enough that they can be neglected. It would be quite interesting to examine this problem directly and see how large these vertex corrections can be even if the vertex functions remain local.

Our exposition in this chapter has used the local approximation throughout, where the self-energy and the irreducible vertex function are both local. In any finite-dimensional system, this does not hold, so our approach is only approximate. For the three-dimensional cases we have studied, the expectation is that the self-energy and irreducible vertex continue to have limited momentum dependence, so the local approximation does capture most of the important physical behavior in these systems. One can ask, however, what can be done to reintroduce the small momentum dependence into the self-energy and the vertex functions. There has been a great body of work on this problem in the bulk. The basic idea is to generalize the self-consistent impurity problem to a self-consistent cluster problem which allows some level of momentum dependence to be restored. One such technique is the dynamical cluster approximation [Hettler, *et al.* (1998); Hettler, *et al.* (2000); Jarrell and Krishnamurthy (2001)] (DCA), which restores momentum dependence in a coarse-grained fashion. Using the DCA, provides a self-consistent, systematic means to take into account finite-dimensional effects, and it does so in a manner that always maintains the correct analyticity of the Green's functions, self-energies, and response functions. While the DCA has only rarely been applied to three-dimensional systems [Kent, *et al.* (2005)], it has been applied to many two-dimensional problems [Aryanpour, Hettler and Jarrell (2002); Macridin, Jarrell and Maier (2004)], and it does show the expected behavior—the self-energy picks up mild momentum dependence, and the irreducible vertex does affect the transport properties, but does not change qualitative features, only quantitative details.

One could imagine generalizing our inhomogeneous DMFT approach to include some momentum dependence via using the DCA to determine the local Green's functions on each plane. This can be most easily done if the self-energy is allowed to have momentum dependence only in the planar directions. Then the sum over $\mathbf{k}_x$ and $\mathbf{k}_y$ cannot be replaced by a sum over the two-dimensional DOS, but this only creates a more complicated numerical algorithm, because we need to replace a one-dimensional integral over the $2d$-bandstructure by a two-dimensional integral over the Brillouin zone. Then the impurity solver for each plane would need to be replaced by the DCA approach. If it is necessary to include long-range spatial correlations in the inhomogeneous (longitudinal) direction, then our whole method of approach to this problem would have to be redeveloped, because the quantum zipper algorithm could no longer be applied.

One obvious system that would be interesting to study with this technique is the class of high-temperature superconductors. The copper-oxygen planes are strongly correlated, but the planes are often only weakly coupled together through the material. Hence, a DCA approach to handle each plane, coupled with the inhomogeneous DMFT approach for coupling the planes together, could yield an interesting theory for these systems. This problem has not yet been attempted by anyone.

An alternate approach that allows some momentum dependence to be restored to these systems would be to calculate the perturbative contributions to the momentum dependence of the self-energy and irreducible vertices. This technique can be problematic, because the simplest way to do this usually breaks causality for some frequencies, hence the theory becomes unphysical in some regimes. Nevertheless, as a quick way to see the size of finite-dimensional effects, this approach may be a useful one to apply to some problems.

Chapter 4

Thouless Energy and Normal-State Transport

4.1 Heuristic Derivation of the Generalized Thouless Energy

Semiclassical approaches often lead to interesting ideas for analyzing quantum-mechanical behavior. The Thouless energy is one such idea that has proved to be remarkably important as a quantum-mechanical energy scale. The idea for the emergence of such an energy scale originated with work of Thouless in the 1970s [Edwards and Thouless (1972); Thouless (1974)]. In this work, which was first numerical, the idea of examining how the quantum-mechanical levels were spaced in energy (for the disorder problem) and how that level spacing related to the localization transition, was first introduced. This notion of using a quantum-mechanical level spacing, denoted by Δ_{E} (and not to be confused with the superconducting gap), to determine properties related to transport is a useful concept, which has now been adopted into a new object called the Thouless energy. The Thouless energy for a diffusive conductor is typically defined as the quantum-mechanical energy scale that can be extracted from the dwell time of the quasiparticle within a region of thickness L of the conducting material. In diffusive transport, the motion of the particles is random, so the average (or dwell) time that each particle spends inside of a region is proportional to the square of the size of the region, with the proportionality being the inverse of the diffusion constant $\mathcal{D}$. This motivates the definition of an energy scale via

$$E_{\mathrm{Th}} = \frac{\hbar}{t_{\mathrm{dwell}}} = \frac{\hbar\mathcal{D}}{L^2}. \tag{4.1}$$

When the resistance of a disordered material is examined via numerical means, it is found that a dimensionless version of the resistance r_{N} (where we divide by the quantum of resistance for spin-one-half particles $R_Q = h/2e^2$) can be interpreted as the ratio of two energy scales: the Thouless energy E_{Th} and the level spacing Δ_E. The formula is

$$r_N = \frac{R_N}{R_Q} = \frac{1}{2\pi}\frac{\Delta_E}{E_{Th}}. \tag{4.2}$$

The localization transition occurs near the region where $r_N \approx 1$.

The inverse level spacing can be interpreted as the change in the number of electrons in the material per unit of energy. Since the total number of electrons (determined by adjusting the chemical potential μ) satisfies

$$N(\mu) = V \int d\omega\, A(\omega - \mu) f(\omega - \mu), \tag{4.3}$$

the inverse level spacing becomes $\Delta_E^{-1} = dN/d\mu$. The local DOS $A(\omega - \mu)$ is independent of the chemical potential because $A(\omega)$ has the chemical potential located at $\omega = 0$, so by subtracting μ from the argument, we remove the chemical-potential dependence. Thus the inverse level spacing satisfies

$$\Delta_E^{-1} = V \int d\omega\, A(\omega) \left[-\frac{df(\omega)}{d\omega}\right]. \tag{4.4}$$

Since the derivative of the Fermi-Dirac distribution is sharply peaked around $\omega = 0$ for low T, we can replace the DOS factor in the integrand by $A(0)$, for metals at low T, and then the integral over ω yields 1 (recall that the DOS is measured from the chemical potential, so $A(0)$ is the DOS at the Fermi energy E_F). The final formula for the inverse level spacing in metals is then

$$\Delta_E^{-1} = VA(0). \tag{4.5}$$

Using Eqs. (4.1) and (4.2) then produces a resistance equal to

$$R_N = \frac{h}{2e^2}\frac{L^2}{h\mathcal{D}}\Delta_E = \frac{L}{2e^2\mathcal{D}A(0)A}, \tag{4.6}$$

which is the common form one finds for a diffusive conductor whose volume is $V = LA$ (this form is often derived from the so-called Einstein relation; see the next section for details).

A similar line of reasoning could be used to define the dwell time in a ballistic conductor (which suffers no scattering) to determine a ballistic Thouless energy, but we do not go through such a path, because we instead want to focus on a more general form for the Thouless energy that can be inferred from the reasoning we have already developed. The idea is to note that if we take Eq. (4.4) as the fundamental definition for the inverse level spacing, even if we have a strongly correlated metal or insulator, then we can use numerical (or experimental) results for the resistance of different size materials to determine the Thouless energy via the relation developed in Eq. (4.2) [Freericks (2004a); Freericks (2004b)]:

$$\begin{aligned} E_{\rm Th} &= \frac{R_{\rm Q}}{2\pi}\frac{1}{R_{\rm N}\Delta_{\rm E}^{-1}} \\ &= \frac{\hbar}{2e^2 R_{\rm N} AL \int d\omega\, A(\omega)[-df(\omega)/d\omega]}. \end{aligned} \tag{4.7}$$

4.2 Thouless Energy in Metals

There are typically two kinds of metals that are discussed in reference to the Thouless energy—(i) ballistic metals that have no scattering and (ii) diffusive metals that can be described by semiclassical diffusion. We have already seen that the Thouless energy decreases like $1/L^2$ for a diffusive conductor. In a ballistic metal, the decrease is instead like $1/L$, because the dwell time in the region of size L is proportional to $L/v_{\rm F}$ with $v_{\rm F}$ some appropriate average Fermi velocity for the quasiparticles. One can also infer these dependences from Eq. (4.7) in the following way: for a diffusive conductor, the resistance grows linearly with the thickness L of the conductor (Ohm's law), so the Thouless energy decreases like $1/L^2$, while for a ballistic conductor the resistance is independent of L, so the Thouless energy decreases like $1/L$. However, it is interesting to find more specific results for the Thouless energy in these two limits because we can use them to compare with the body of work in the field, and ensure that our generalized Thouless energy approach makes sense.

We start with noninteracting electrons in three dimensions and we will consider the transport through a macroscopic region of cross-sectional area A and length L. Free electrons have an energy dispersion $\epsilon_{\mathbf{k}} = \hbar^2 k^2/2m$. It is a simple exercise to show that the DOS at the Fermi energy satisfies $A(0) = 2mk_{\rm F}/h^2$ and the Fermi velocity is $\mathbf{v}_{\rm F} = \nabla_{\mathbf{k}}\epsilon_{\mathbf{k}}/\hbar = \hbar\mathbf{k}_{\rm F}/m$, with $\mathbf{k}_{\rm F}$ the Fermi wavevector and m the effective mass of the quasiparticle. This yields an inverse level spacing of $\Delta_{\rm E}^{-1} = 2mALk_{\rm F}/h^2$.

In a ballistic conductor, we need to imagine the material of size $L \times A$ as sandwiched between two semi-infinite metallic leads. If there is a transmissivity $\mathcal{T}$ at the lead-metal interface, then the resistance is given by $R_{\rm N} = 4\pi R_{\rm Q}/k_{\rm F}^2 A\mathcal{T}$, which is the inverse of the number of conducting channels available at the Fermi surface (modified by the transmissivity). This is called the Sharvin resistance [Sharvin (1965)] or contact resistance between the lead and the metal. Using the inverse level spacing, and the generalized Thouless energy formula, then gives the ballistic Thouless energy as

$$E_{\rm Th}^{\rm ballistic} = \frac{\hbar v_{\rm F}\mathcal{T}}{4L}. \tag{4.8}$$

This behaves like $1/L$ as we know it should.

In a diffusive conductor, where we also sandwich the material of size $L \times A$ between two semi-infinite metallic leads, we have both a contact resistance plus the diffusive contribution from the diffusive conductor of length L. If we start from the bulk form for the resistivity, we use a simple Drude law [Drude (1900a); Drude (1900b)] to find the resistivity as a function of a mean-free-path $\ell = v_{\rm F}\tau$, with τ the relaxation time for the scattering process. The simple result is that

$$\rho_{\rm dc} = \frac{m}{\rho_e e^2 \tau} = \frac{3\pi R_{\rm Q}}{k_{\rm F}^2 \ell}, \tag{4.9}$$

with $\rho_e = k_{\rm F}^2/3\pi$ the electron density. If we assume a contact resistance given by $R_{\rm c} = 4\pi R_{\rm Q}/k_{\rm F}^2 A\mathcal{T}$ as derived above, then the total resistance becomes

$$R_{\rm N} = \left[\frac{4}{\mathcal{T}} + \frac{3L}{\ell}\right]\frac{\pi R_{\rm Q}}{k_{\rm F}^2 A}, \tag{4.10}$$

and the associated Thouless energy is

$$E_{\rm Th}^{\rm diffusive} = \frac{\hbar v_{\rm F}\ell}{3\left(1+\frac{4\ell}{3\mathcal{T}L}\right)L^2} = \frac{\hbar\mathcal{D}}{\left(1+\frac{4\ell}{3\mathcal{T}L}\right)L^2}, \tag{4.11}$$

where we used the fact that one can define the diffusion constant via $\mathcal{D} = v_{\rm F}\ell/3$ (this follows from the fact that the *dc* conductivity is $\sigma_{\rm dc} = k_{\rm F}^2\ell/3\pi R_{\rm Q}$, which can also be written as $\sigma_{\rm dc} = 2e^2\mathcal{D}A(0)$ via the Einstein relation—solving for $\mathcal{D}$ yields the desired result). This result differs slightly from the expected $1/L^2$ behavior due to the inclusion of the contact resistance. If the term $4\ell/2\mathcal{T}L$ is small, then we recover the $1/L^2$ behavior, but if it is non-negligible, we expect to see deviations.

Using the techniques described in Chapters 2 and 3, we can now generate both the resistance of a multilayered nanostructure and the bulk DOS, so we can evaluate all the terms in Eq. (4.7) and determine the Thouless energy of different devices. The numerical calculations to carry this out are complex, and a project to do so is described in Prob. A.37. For simplicity, we assume that the hopping between all lattice sites is the same, namely equal to t, which we use as our unit of energy; the lattice sites are taken to be the locations on a simple-cubic lattice. We examine the case of ballistic metal semi-infinite leads (no scattering) and a central barrier region that is described by the Falicov-Kimball model with an interaction strength U^{FK} and with a concentration of defects equal to $w_1 = 1/2$; these terms vary only from one plane to another, and are fixed within any given plane. We work at half-filling for both the leads and the barrier, so the chemical potential lies at $\mu = 0$ for all T and there is no electronic charge reconstruction because the chemical potentials match identically between the two different materials. Hence we only have three parameters to vary: (i) the strength of the interaction U^{FK}; (ii) the length of the barrier $L = Na$; and (iii) the temperature T. Nevertheless, this is a rich parameter space to examine, because there is a metal-insulator transition occurring at $U^{\mathrm{FK}} \approx 4.92t$. Note that it is only charge transport that is interesting to study for this case because the particle-hole symmetry implies that the thermopower vanishes, so there are no thermoelectric effects.

Let us begin with a review of the bulk charge transport on a simple-cubic lattice at half filling in the Falicov-Kimball model. The *dc* conductivity continuously goes to zero at the metal-insulator transition, which occurs near $U^{\mathrm{FK}} = 4.92$. In the metallic phase, the conductivity is a fairly flat function of T for low temperature, because the imaginary part of the self-energy does not have strong frequency dependence. If we concentrate our examination in this low-T region, then we can extract a mean-free-path for the electrons via the Einstein relation (using the interacting DOS). The result of this exercise is shown in Fig. 4.1. When the mean free path drops below one lattice spacing (at approximately $U^{\mathrm{FK}} = 3$) we have reached the so-called Ioffe-Regel limit [Ioffe and Regel (1960)] which is supposed to be the minimum possible metallic conductivity. Nevertheless, in the Falicov-Kimball model, the conductivity continues to continuously go to zero at $U^{\mathrm{FK}} \approx 4.92$. The diffusive picture for transport holds up to this critical transition point, but beyond that point, use of the formula for the mean free path may make no sense anymore.

In Fig. 4.2, we plot the relative value of the resistance of a strongly scattering metal $U^{\mathrm{FK}} = 2$ (panel a) and for an anomalous metal $U^{\mathrm{FK}} = 4$ (panel b) for different thicknesses of the barrier. One can see that as the barrier is made thicker, the shape of the curves becomes similar, and

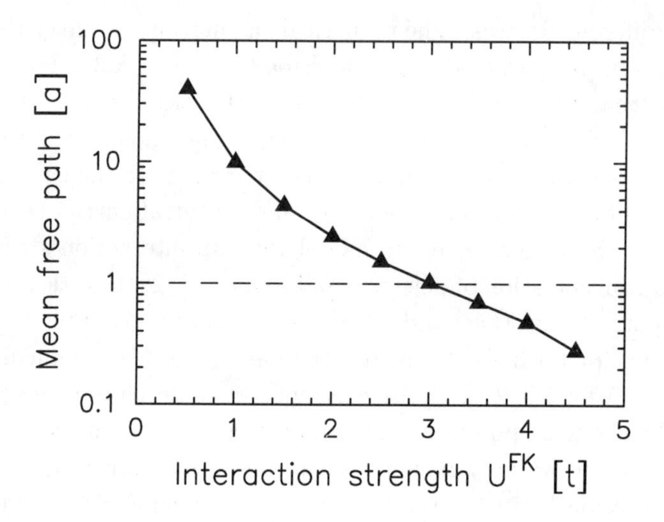

Fig. 4.1 Mean free path as determined from the Einstein relation for the bulk conductivity in the Falicov-Kimball model on a simple-cubic lattice at half filling. The temperature is $T = 0.01$. Although the mean free path drops below the Ioffe-Regel limit at $U^{\mathrm{FK}} \approx 3$, the conductivity (and mean free path) continuously goes to zero at the metal-insulator transition. *Adapted with permission from* [Tahvildar-Zadeh, Freericks, and Nikolić (2006)].

actually it reproduces the bulk resistivity shape, as one might expect. Since the resistivity is decreasing as T increases in both cases, these are very strongly scattering metals, but the decrease is minimal for $U^{\mathrm{FK}} = 2$, and at low T one can approximate the resistivity by just a constant, while it is a significant reduction for the $U^{\mathrm{FK}} = 4$ case, which arises due to the fact that the DOS has a large dip near the chemical potential, so more states are able to participate in the transport as the temperature is raised. In addition, we note that the cases with thin barriers always have conventional metallic behavior even in panel (b) where the mean free path is less than a lattice spacing, but the crossover to the anomalous metallic behavior occurs much sooner for the stronger scattering material.

Plots of the resistance-area-per-unit-cell product are shown in Fig. 4.3 as a function of the interaction strength U^{FK}. Note how there is virtually no indication of the metal-insulator transition in the thin barrier cases, but it becomes more obvious, with the resistance increasing sharply, as the thickness increases. In all cases there is a finite intercept at $U^{\mathrm{FK}} = 0$ due to a nonzero contact resistance. The numerical calculations become quite challenging to accurately determine the resistance in the Mott insulating

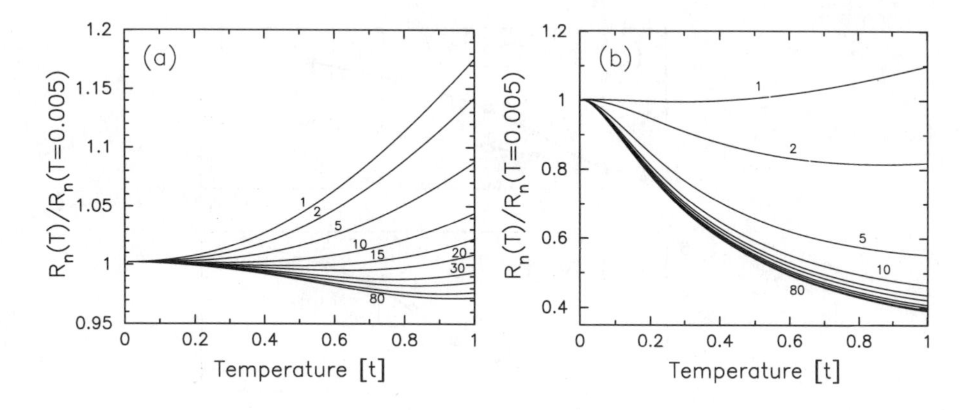

Fig. 4.2 Panel (a): relative resistance versus temperature for different thickness devices and $U^{\mathrm{FK}} = 2$. Panel (b): relative resistance versus temperature for different thickness devices and $U^{\mathrm{FK}} = 4$. We plot barrier thicknesses corresponding to 1, 2, 5, 10, 15, 20, 30, 40, 60, and 80 planes.

phase when the barrier is thick. We are unable to accurately determine R_n for interaction strengths larger than 6 at $N = 20$. This arises due to the fine structure that builds up in the self-energy on different planes, as illustrated in Fig. 4.4. This figure depicts how the self-energy develops a sharp peak reminiscent of the delta-function peak seen in the bulk Mott insulator, but here the peak has a finite width and height. This structure becomes very fine, and is challenging to determine accurately.

The numerical issues associated with a self-energy, like the one depicted in Fig. 4.4, arise from two separate issues: (i) first there is a huge variation in the self-energy over a short energy scale (the change is over almost nine orders of magnitude for a range of frequency about $0.1t$) and (ii) extracting the self-energy when the Green's function is small is subject to loss of precision since the DMFT algorithm requires taking the difference of two large numbers to yield the self-energy. While these issues are also present in the bulk (especially on the hypercubic lattice, where the "Mott-insulator" is really always a pseudogap), one can derive analytic expressions to determine the self-energy in regions where the numerics fails. For a nanostructure, it is much more difficult to try to construct such an analytic expression due to the inhomogeneity, hence calculations become infeasible when the numerics fails. One way to see that the numerics is failing is to note that the first visible error is usually that the exponential decay of the DOS at the chemical potential will stop as the number of planes increases (often, you will actually see a peak start to develop). Whenever this occurs, it is likely

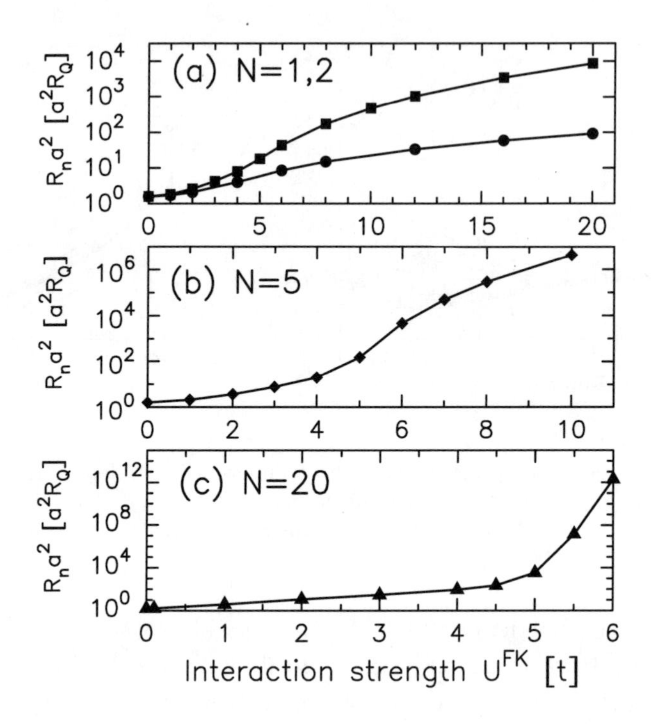

Fig. 4.3 Resistance as a function of U^{FK} for four different barrier thicknesses: (a) $N = 1$ (circles) and $N = 2$ (squares); (b) $N = 5$; and (c) $N = 20$. Note how there is essentially no indication of the metal-insulator transition for a single plane junction, but it becomes clear for the $N = 5$ and $N = 20$ cases. The temperature is $T = 0.01$. *Adapted with permission from* [Freericks, Nikolić and Miller (2001)] (© 2001 the American Physical Society).

that the numerical code is starting to have precision issues, so this should always be checked with moderately thick insulating barriers, to insure that the numerical precision is adequate. Sometimes increasing the number of quadrature steps, by reducing the grid spacing, can improve the precision enough that the results become viable, but at some point they undoubtedly fail.

Given the resistance and the bulk DOS, we are now able to calculate the Thouless energy. In the ballistic case, the Thouless energy trivially will be given by Eq. (4.8) because the resistance is given by the contact resistance for any thickness. Hence we focus on diffusive junctions that have contributions from both the contact resistance and the diffusive resistance from the interior of the barrier. We show results for two cases $U^{\mathrm{FK}} = 2$ and 4 in Fig. 4.5. These plots are of E_{Th} versus L on a semilogarithmic plot and

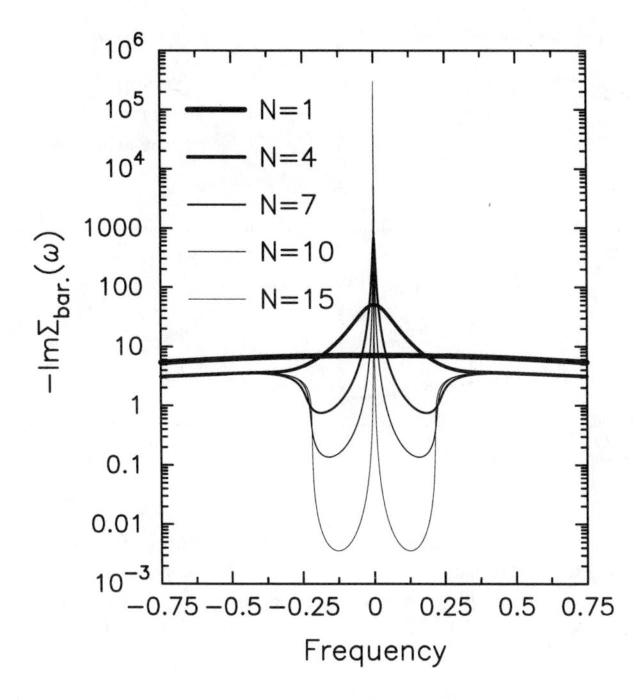

Fig. 4.4 Semilogarithmic plot of the imaginary part of the self-energy on the central plane of the barrier with five different thicknesses ($N = 1$, 4, 7, 10, and 15). The value of the interaction energy is $U^{\mathrm{FK}} = 6$. These results do not depend on temperature. *Reprinted with permission from* [Freericks (2004b)] (© 2004 the American Physical Society).

of $E_{\mathrm{Th}}L^2$ versus L on a linear plot. The former shows how the Thouless energy decreases, while the latter shows whether the $1/L^2$ dependence holds at large L; this $1/L^2$ dependence appears sooner for $U^{\mathrm{FK}} = 4$, because the contact resistance is a smaller relative contribution to the resistance in this case. The curves, which are plotted for different temperatures, show how the diffusion constant varies with T (it appears to increase almost linearly with T at high T for $U^{\mathrm{FK}} = 4$), but the overall dependence on T is rather weak, as expected for metals.

There is one other point to make about the "depression" of the diffusion constant in the top panels of Fig. 4.5. If the formula in Eq. (4.11) held exactly, then by using the mean free paths plotted in Fig. 4.1 we could determine precisely what the behavior for different cases should look like. But when we have very thin junctions, their resistance is relatively insensitive to the size of the interactions (see Fig. 4.3), which tells us they are

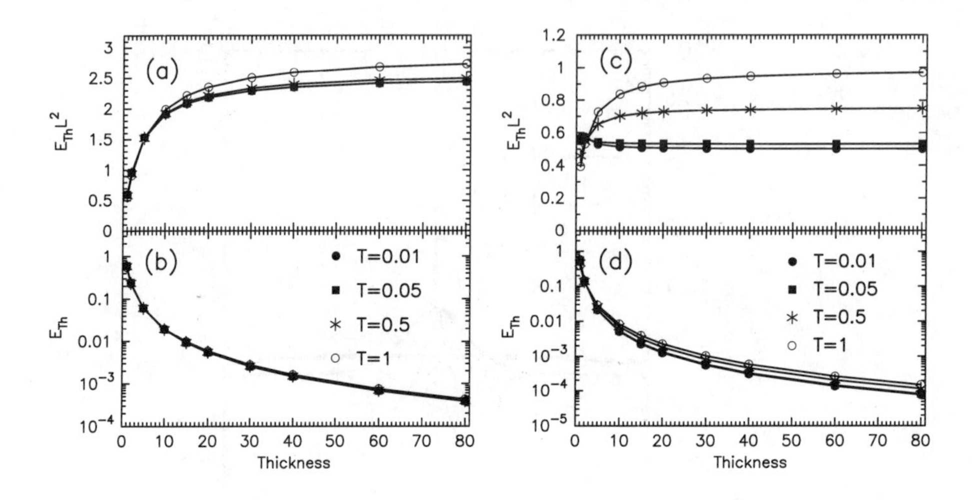

Fig. 4.5 Left panels: Thouless energy versus L for four different temperatures and $U^{\rm FK} = 2$. Right panels: the same but for $U^{\rm FK} = 4$. The top panels are $E_{\rm Th}L^2$ versus L, while the bottom panels are $E_{\rm Th}$ versus L on a semilog plot. *Reprinted with permission from* [Freericks (2004b)] (© 2004 the American Physical Society).

less resistive than one would predict from Eq. (4.10). This implies that the Thouless energy will be enhanced for the smallest thicknesses, and hence the diffusion constant won't vary as much as might be expected. This becomes more evident as the strength of the scattering increases.

These results show that the Thouless energy is well-defined and easily extracted from data for the resistance versus temperature. In actual junctions or devices, the contact resistance can create a significant modification of the $1/L^2$ dependence, but this is recovered in the large L limit. We will see in the next chapter how the Thouless energy plays a significant role in describing the behavior of Josephson junctions.

4.3 Thouless Energy in Insulators

In insulators, the Thouless energy has significantly stronger temperature dependence than in metals. This is because the inverse level spacing, as defined in Eq. (4.4), vanishes as $T \to 0$ in an insulator. Since the bulk *dc* resistivity becomes infinite, this may not seem like a problem, but the resistance of a device will not be infinite due to quantum-mechanical tunneling, which will allow electrons to flow through the junction even at $T = 0$. Hence, the Thouless energy will diverge for an insulator as $T \to 0$. Note

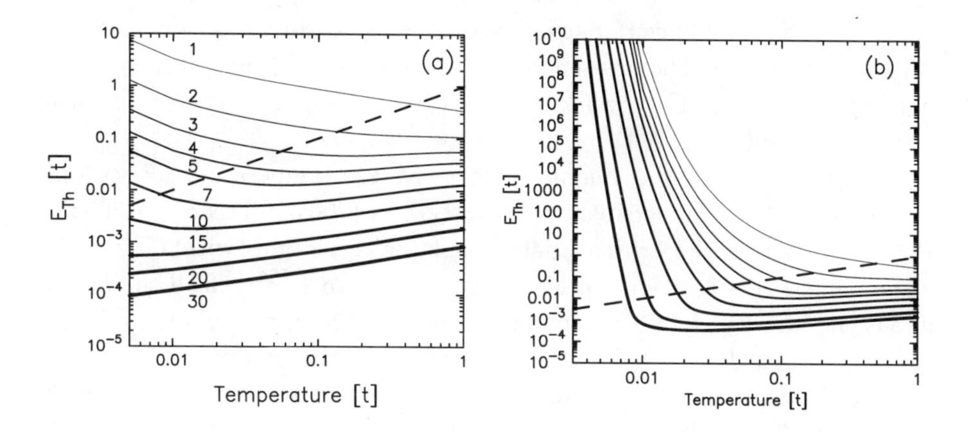

Fig. 4.6 Left panel: Thouless energy versus T for different thicknesses and $U^{\mathrm{FK}} = 5$. The dashed line is the curve $E_{\mathrm{Th}} = T$. *Reprinted with permission from* [Freericks (2005)]. Right panel: Thouless energy versus T for different thicknesses and $U^{\mathrm{FK}} = 6$. The dashed line is the curve $E_{\mathrm{Th}} = T$. Running from top to bottom, the curves are for $N = 1$, 2, 3, 4, 5, 7, 10, 15, and 20. *Reprinted with permission from* [Freericks (2004b)] (© 2004 the American Physical Society).

that it is easy to understand why the resistivity becomes infinite, but the resistance remains finite. The resistivity is defined for the bulk material, and the Kubo formula always involves an integral of an effective relaxation time multiplied by a derivative of a Fermi function. That derivative becomes a delta function at $T = 0$, so the conductivity is determined by the relaxation time at $\omega = 0$. If the relaxation time vanishes, then the resistivity will be infinite. By definition, the relaxation time does vanish in any insulator. On the other hand, when we investigate the resistance of a nanostructure, we need to evaluate a similar Kubo formula in real space, but here none of the integrands vanish, because the DOS is always finite at the chemical potential everywhere in the nanostructure. This arises due to the normal-state proximity effect, that allows the metallic wavefunctions to leak into the insulator. If the insulator was not attached to metallic leads, then the resistance need not be finite, but whenever it is attached to metallic leads, the resistance must be finite (although it can become extremely large in magnitude). Hence, we always see a finite resistance to the nanostructure that has an insulator sandwiched between two metallic leads.

If we want to interpret the Thouless energy in terms of a tunneling time (analogous to the dwell time), this would say that the tunneling time goes to zero as $T \to 0$, but it isn't clear that such an interpretation can be easily made and quantitative theories for the tunneling times of particles across barriers have been controversial. In any case, we do not need to try to interpret the Thouless energy in terms of a tunneling time, because our approach allows us to work directly with the Thouless energy to learn interesting results.

We show plots of the Thouless energy versus temperature for various thicknesses of the barrier in Fig. 4.6 [Freericks (2004b); Freericks (2005)]. Panel (a) is near the critical metal-insulator transition, with $U^{\mathrm{FK}} = 5$, while panel (b) is a small gap Mott insulator, with $U^{\mathrm{FK}} = 6$. The dashed line is the line $E_{\mathrm{Th}} = T$. The points of intersection of the two lines define the place where $E_{\mathrm{Th}} = T$ for a given thickness of the barrier.

Note how the more insulating the barrier becomes, the stronger the T dependence is. In the case $U^{\mathrm{FK}} = 5$, the temperature is not low enough to clearly see the divergence as $T \to 0$, but for the case with $U^{\mathrm{FK}} = 6$ the

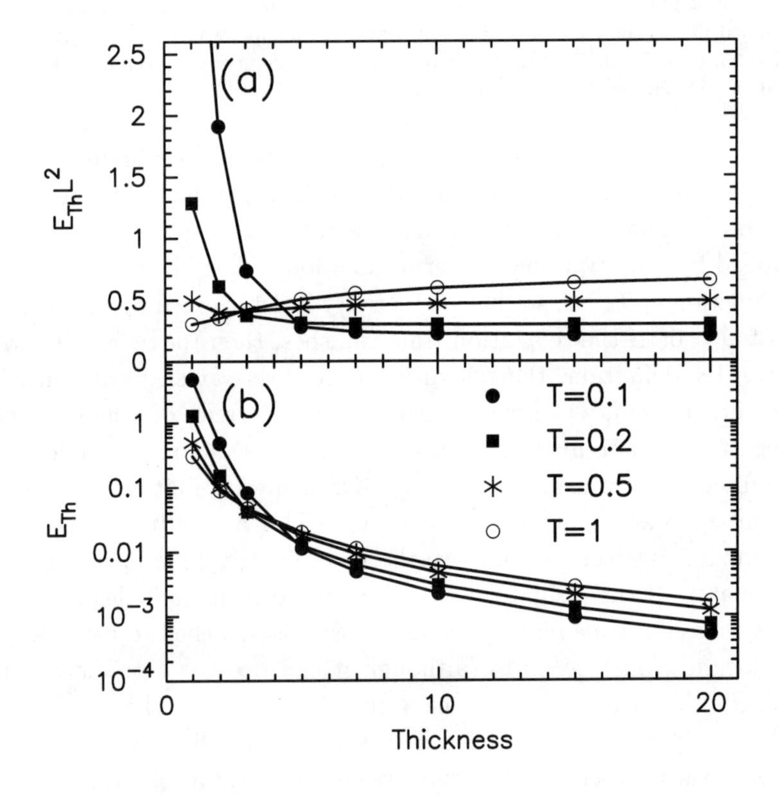

Fig. 4.7 (a) $E_{\mathrm{Th}}L^2$ and (b) E_{Th} versus L for $U^{\mathrm{FK}} = 6$ and various temperatures. Note how we see a $1/L^2$ behavior at high temperature, indicating that the thermally activated transport is diffusive inside a Mott-insulating barrier at high temperature.

divergence is obvious. Furthermore, the thicker the barrier is, the lower the temperature where $E_{\mathrm{Th}} = T$ and the lower the crossover temperature where the Thouless energy changes its slope from increasing to decreasing with T. Hence, the effective energy scale for transport is lower for a thicker barrier. While this may sound counter-intuitive, we will see in the next section, that this makes good sense as we unravel the crossover from tunneling to incoherent (Ohmic) transport in a device.

We plot the analog of Fig. 4.5 in Fig. 4.7 for $U^{\mathrm{FK}} = 6$, which is a small-gap Mott insulator. Note how at high temperature and for thick junctions, the Thouless energy behaves like $1/L^2$, indicating diffusive transport. For thinner junctions at low temperature, where the transport is via tunneling, the "diffusion constant" varies significantly with thickness initially. This is because the transport is not diffusive in that regime. It is interesting to see that as the device makes the transition from tunneling to incoherent, thermally activated, Ohmic transport, the character of the transport changes from quantum-mechanical tunneling to semiclassical diffusion. We will see a more dramatic illustration of this in the next section.

4.4 Crossover from Tunneling to Incoherent Transport in Devices

When the barrier is an insulator, we expect the system to display tunneling at low temperature, which crosses over to thermally activated transport at high temperature. We have already seen that the thermally activated transport is diffusive, when we examined the Thouless energy at high temperatures. Now we examine the resistance directly. Plotted in Fig. 4.8 is the resistance versus temperature on a log-log plot. Panel (a) shows the small-gap insulator $U^{\mathrm{FK}} = 6$ and panel (b) shows the near critical Mott-insulator $U^{\mathrm{FK}} = 5$. In the top panel, we can clearly see tunneling exhibited by the flat steps of equal size as the thickness increases. These equally spaced steps indicate the system has a resistance that grows exponentially with the thickness, which is a hallmark of the tunneling regime. Furthermore, the resistance depends only weakly on T in this regime, and since tunneling is a quantum effect, it does not display strong T dependence. As the temperature is made larger, the curves turn over, and all lie nearly on top of each other. This is the Ohmic linear scaling regime, where the resistance scales like $R_n \approx \rho_{\mathrm{dc}} L/A$. The thermally activated regime displays diffusive transport that can be described by Ohm's law. The magenta dashed curve

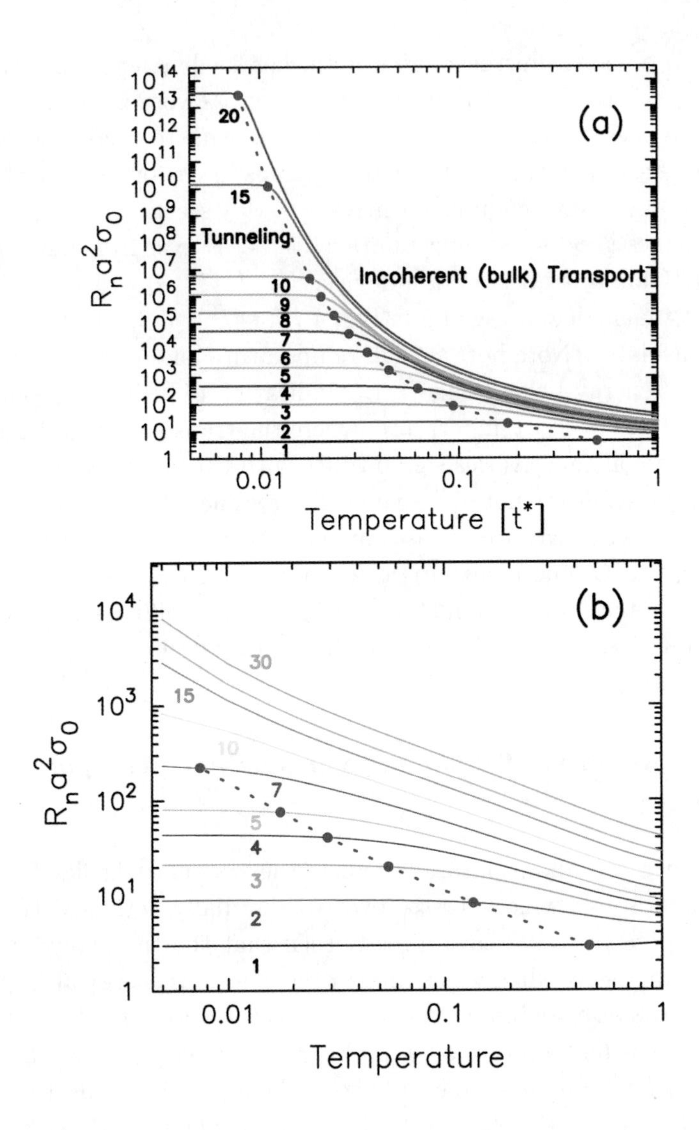

Fig. 4.8 Top panel: Resistance versus T on a log-log plot for different thicknesses of the barrier and $U^{\mathrm{FK}} = 6$. The dashed line with solid circles is the curve corresponding to points where $E_{\mathrm{Th}} = T$. *Reprinted with permission from* [Freericks (2004b)] (© 2004 the American Physical Society). Bottom panel: The same plot but for $U^{\mathrm{FK}} = 5$. *Reprinted with permission from* [Freericks (2005)]. The constant σ_0 satisfies $\sigma_0 = 4e^2/ha^2$.

with solid circles marks the points where the Thouless energy is equal to the temperature. We can clearly see that this criterion properly determines the crossover from tunneling to activated, incoherent transport.

Surprisingly, the crossover point does not seem to have any simple relationship to the barrier height; it also may seem strange that the crossover occurs at a lower temperature for thicker barriers. But these results can be easily understood with a little thought. The tunneling phenomenon acts like a "quantum short" through the device, which we can view as a conducting channel in parallel with the thermally activated channel. At low T, the thermally activated channel is too resistive, so all of the current flows via tunneling. As the T rises, the thermally activated resistance drops because the resistivity depends exponentially on the inverse of the temperature. As this resistivity drops, the resistance of the thermally activated conduction channel also drops, and once it becomes lower than the resistance due to tunneling, the majority of the current is carried by the thermally activated carriers. Since the tunneling resistance grows exponentially with the thickness, this crossover resistance is higher for thicker junctions, so they crossover at a lower temperature than the thinner junctions. The dependence on the barrier height is complicated because the barrier height enters into the activated behavior for the resistance, and it also plays a role in determining the magnitude of the tunneling resistance, but there is no simple way to determine precisely the functional dependence of the crossover temperature on the barrier height.

In panel (b), we see curves corresponding to the near critical Mott insulator. Not too surprisingly, this device does not behave in any simple fashion. The resistance does show flat plateaus at low T, but they are not equally spaced as the thickness increases, so they do not depend exponentially on the thickness. But they are not linear either, and this regime is really a new regime that probably does not have any simple analytic behavior to describe it because the DOS depends so strongly on energy in the region close to the chemical potential. At high temperatures, there is a crossover to the linear-scaling Ohmic regime, with the transport being diffusive. The Thouless energy does provide a reasonable "ballpark" estimate for this crossover, but it is not as good at predicting the location when we are too close to the metal-insulator transition; of course, if we push to the metallic side of the transition, there is no crossover.

In Fig. 4.9, we see a different perspective on this issue, with a plot of the resistance versus thickness at a fixed temperature $T = 0.01$ for $U^{\mathrm{FK}} = 2$, 4, 5, and 6. Panel (a) is a semilog plot and panel (b) is a linear plot. This allows us to easily find the Ohmic scaling regime and the exponential scaling regime. Both the strongly scattering metal $U^{\mathrm{FK}} = 2$ and the anomalous metal $U^{\mathrm{FK}} = 4$ display Ohmic scaling, plus a contact resistance, for all

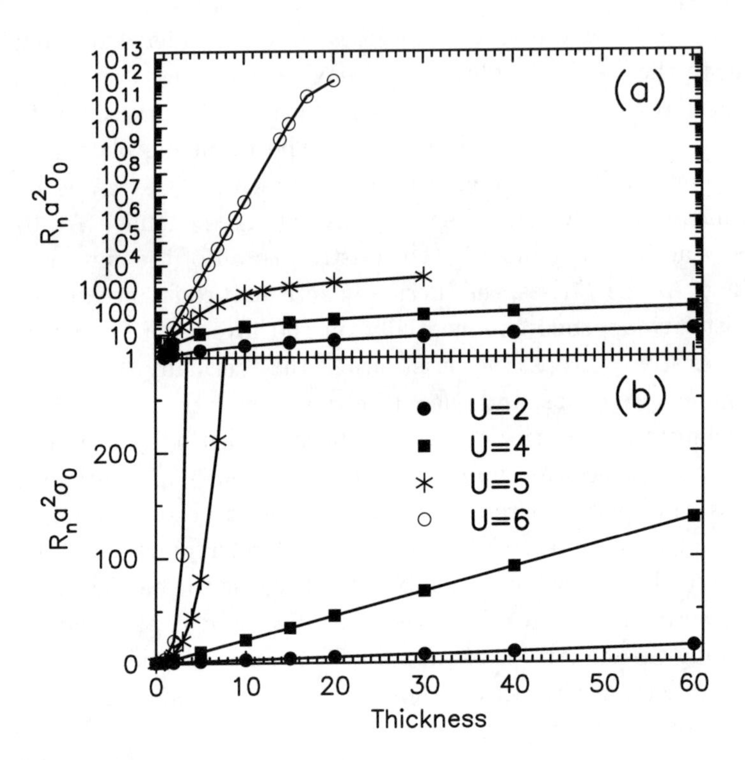

Fig. 4.9 Resistance versus thickness at $T = 0.01$ for $U^{\mathrm{FK}} = 2$, 4, 5, and 6. Panel (a) is a semilogarithmic plot, which can show exponential dependence of the resistance on thickness, while panel (b) is a linear plot, which can show Ohmic linear scaling of the resistance. The metallic phases ($U^{\mathrm{FK}} = 2$ and 4) obey perfect linear scaling, plus a small contact resistance. The small-gap Mott-insulator ($U^{\mathrm{FK}} = 6$) has exponential dependence of the resistance on L, while the near-critical insulator ($U^{\mathrm{FK}} = 5$) has behavior that increases faster than linear, but not fast enough to be exponential. *Reprinted with permission from* [Freericks (2004b)] (© 2004 the American Physical Society).

thicknesses. The slope of the curve agrees perfectly with the resistivity that is calculated from the bulk Kubo response function. The small-gap Mott insulator has perfect exponential dependence of the resistance on the thickness until we reach a thickness of about 20, where it starts to turn over. Indeed, the Thouless energy predicts the crossover thickness (sometimes called the Thouless length) occurs near 20 for $T = 0.01$. The near-critical insulator ($U^{\mathrm{FK}} = 5$) does not obey either simple scaling form—the resistance increases faster than linearly, but slower than exponential in the thickness.

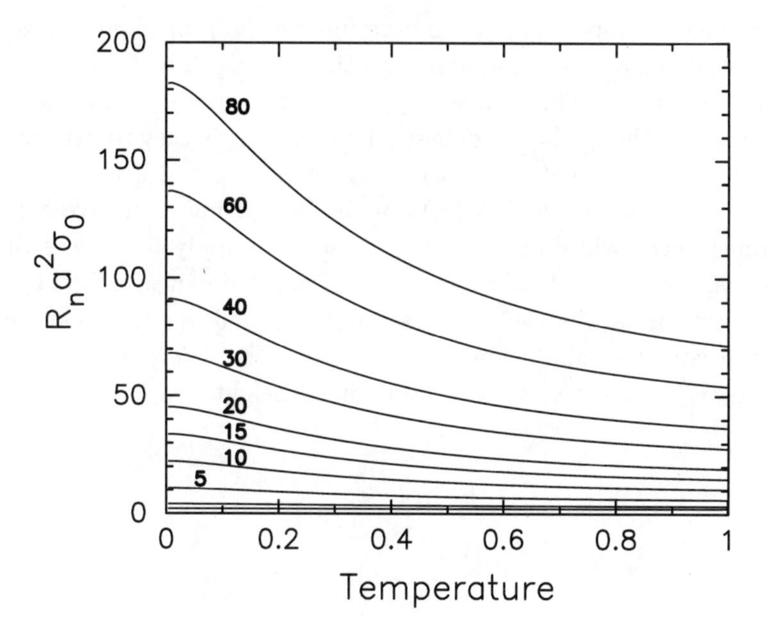

Fig. 4.10 Resistance versus temperature for a diffusive metal with $U^{\mathrm{FK}} = 4$. The different curves correspond to different thicknesses. Unlike the insulating case in Fig. 4.8, here we see no signal of tunneling behavior; instead, the curves all have a similar shape, indicating the linear scaling of the resistance at each T for diffusive transport. *Reprinted with permission from* [Freericks (2004b)] (© 2004 the American Physical Society).

Our final results are shown in Fig. 4.10, where we plot the resistance versus temperature for different thicknesses in a diffusive metal. Unlike the insulator case in Fig. 4.8, here we see the linear scaling of the resistance with thickness, as all curves share a similar shape (especially as the barrier is made thicker). This result is analogous to what we saw in Fig. 4.2, where the curves all had a similar shape, especially once the barrier was thick enough.

The results of this section clearly show that the Thouless energy, which can be thought of as the energy scale associated with charge transport, plays an important role in understanding the different behavior in multilayered nanostructures. In particular, we can use the Thouless energy to determine whether we have ballistic or diffusive transport by examining how it changes with thickness at large thickness and we can use it to determine the crossover from tunneling to incoherent transport in a junction (either as a function of thickness for fixed temperature, or as a function of temperature for fixed thickness). We will see in the next chapter, that it also plays an important role in determining the properties of Josephson junctions, especially for junctions with thick enough barriers.

The generalized Thouless energy should be a useful diagnostic tool for characterizing the quality of different junctions. It is easy to determine the Thouless energy if one can measure the resistance and if one has a good idea of the shape of the bulk DOS of the barrier material. We hope that experimentalists will find this concept useful to apply to devices that are being manufactured. It also can be used as an engineering guide, if one wants tunneling at a given temperature, the Thouless energy can immediately tell you at what thickness you should expect there to be a crossover to diffusive transport given a certain barrier height.

Chapter 5

Josephson Junctions and Superconducting Transport

5.1 Introduction to Superconducting Electronics Devices

Silicon-based transistors and circuits currently run at up to 4 GHz in commercially available computers, but are not expected to be able to be clocked much faster than 10 GHz. Other semiconductors may be able to push the clock speed a few times faster, but we are rapidly approaching the maximal speeds for semiconductor-based electronics. Superconducting electronics, on the other hand, which are based on Josephson junctions as the fundamental circuit element [Josephson (1962)], have inherent switching speeds that are dramatically higher. Niobium-based junctions have been used to create a T-flip-flop circuit that runs at 770 GHz [Chen *et al.* (1999)]. Other superconductors, like MgB_2, may be able to operate at even higher speeds. While it is true that superconducting electronics will need significant development to progress to even 100 GHz speeds in complicated circuits (tens to hundreds of thousands of junctions on a chip) like analog to digital converters (which are still much simpler than a microprocessor), the theoretical ceiling on clock speed is set much higher for superconductor-based technologies, and with enough time devoted to the development of advanced circuits, it may be possible to create ultrafast digital electronics from Josephson junction-based chips.

A Josephson junction consists of a multilayered sandwich that has superconducting leads to the right and the left, and a central barrier region that can be an insulator (SIS junction), a normal metal (SNS junction), a strongly correlated metal or insulator (SCmS or SCiS junction), or combinations of these (the SINIS junction has been examined recently). The central region of the Josephson junction is not inherently superconducting, so the superconductivity is reduced as we move from the superconductor into the barrier region. But the superconductivity cannot go to zero too rapidly, because that would cost too much in energy, so it gradually goes to zero, resulting in superconducting correlations that leak from the left superconductor into the right superconductor; from a technical standpoint, the gap function can go discontinuously to zero, but the pair-field Green's function decays much more slowly. This physical process allows superconducting pairs of electrons to travel through the device, and it typically creates a complicated nonlinear current-voltage characteristic; nonlinear behavior is often the most important aspect of a device for applications (recall the transistor has a nonlinear current-voltage relation, because current is exponentially suppressed to flow in one direction through a *pn*-junction; the nonlinear current-voltage relation of a Josephson junction is even more complex—see Fig. 1.1).

One of the most important aspects of a Josephson junction is how rapidly it switches from a state where current can flow with no voltage across the junction (it is driven by a gradient of the phase of the superconducting order parameter through the device) to a resistive state where

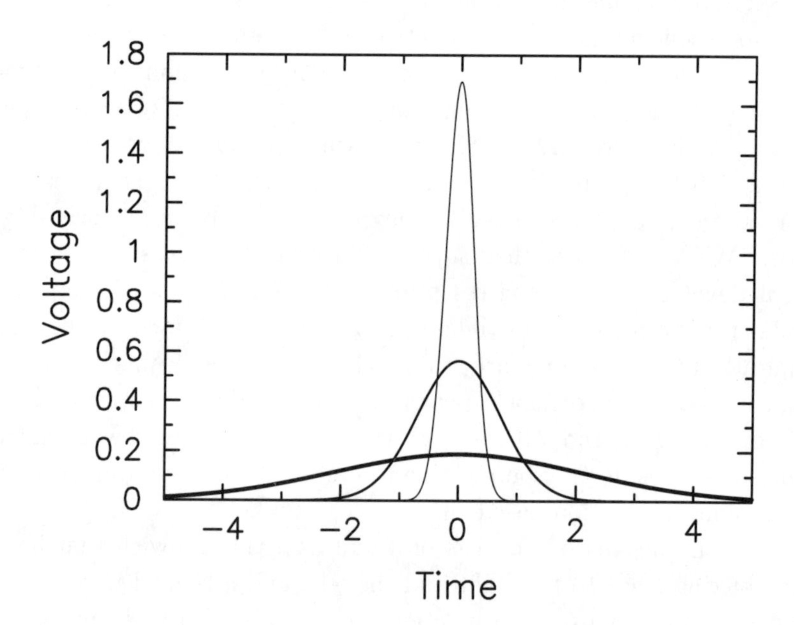

Fig. 5.1 Schematic plot of voltage pulses with different values of the characteristic voltage, given by the figure-of-merit $I_c R_n$. Note how the width of the pulse narrows as the characteristic voltage increases because the area under the curve is the same for all curves (and equal to the flux quantum).

there is current flowing with a voltage across the junction, and then back to the no voltage state. As this switching behavior occurs, we can examine a plot of the voltage versus time (see a schematic in Fig. 5.1). It turns out that the integral of this voltage pulse is equal to a flux quantum $\phi_0 = h/2e$. Hence, the higher the peak of the voltage pulse, the narrower the pulse, and thereby, the faster the switching speed. So the height of the voltage pulse will determine the switching speed of a Josephson junction, and we want to maximize this to get the narrowest pulse. The peak voltage, also called the characteristic voltage, is equal to the product of the critical current at zero voltage I_c and the normal-state resistance R_n. This product, I_cR_n, is also called the figure-of-merit of a Josephson junction.

The figure-of-merit has some simple limiting values in well-known cases. For a thin insulating barrier (called a SIS tunnel junction), the product is $I_cR_n = \pi\Delta(0)/2|e|$, which was first worked out by Ambegaokar and Baratoff [Ambegaokar and Baratoff (1963)]. This result is independent of the barrier height, and only depends on the size of the superconducting gap, which is around 1 meV for low-temperature superconductors. The fact that the figure-of-merit is independent of the barrier material is easy to understand: both the critical current and the resistance will depend on the barrier height, but they do so in a reciprocal fashion, leading to an I_cR_n product that is independent of the barrier height. At low temperature, one can also show that I_cR_n satisfies $I_cR_n = \pi\Delta(0)/|e|$ for thin metal-barrier junctions (SNS junctions) [Kulik and Omelyanchuk (1977)]. This is also independent of the properties of the barrier, but is more complicated to derive, because it relies on evaluating properties of the proximity effect in the normal metal.

Current Josephson-based devices are manufactured from niobium-aluminum-oxide-niobium trilayers [Rowell, Gurvitch and Geerk (1981); Gurvitch, Washington and Huggins (1983)]. The growth process involves first depositing a Nb layer, followed by a thin Al layer. Then the device is exposed to oxygen, which forms a disordered, nonstoichiometric AlO_x barrier layer, and finally it is capped with an upper layer of Nb. The aluminum layer needs to be thin, but it does not need to be controlled too stringently, because the proximity effect will make a thin aluminum layer superconducting, since it sits on top of the niobium. The thickness of the barrier is determined by the amount of oxygen that the aluminum is exposed to, but if the barrier is too thin, then it does not form a uniform barrier, and transport is dominated by so-called pinholes, which are hot-spots in the barrier that conduct electrons more easily than other regions (see Fig. 1.8). The barrier height of pure sapphire (Al_2O_3) is about 4.5 eV, but the barrier height observed in actual Josephson junctions is often much smaller, on the order of $1-2$ eV, and this is because the aluminum oxide is not stoichiometric, due to it being oxygen-deficient (see Fig. 1.9).

These Nb-AlO_x-Nb Josephson junctions are tunnel junctions, which have a multivalued, hysteretic current-voltage curve [see Fig. 1.1 (a) for an example]. But the current-voltage characteristic needs to be single valued (corresponding to what is called a nonhysteretic junction) for use in rapid single-flux quantum logic (RSFQ) [Mukhanov, Semenov and Likharev (1987); Likharev (2000)], which is the fastest operating logic for a Josephson junction device [see Fig. 1.1 (b)]. A hysteretic Josephson junction can be converted to a nonhysteretic junction by adding a shunt resistor in parallel with the Josephson junction. This is what is used for current Nb-based junctions that are employed on chips to make electronics devices. The area of the junction in currently available technology is on the order of a square micron. If we make the area smaller, we can get into a regime called the self-shunted junction regime, where the junction has a nonhysteretic current-voltage curve without requiring an external shunt resistor.

To understand what parameters are required for self-shunted junctions, we perform some simple back-of-the-envelope calculations. For typical Josephson junctions, we need to run on the order of 1 mA of current through each junction to prevent errors in the circuit. If the junction area is 0.1 μ^2, then the junction will have a current density of 10^6 kA/cm^2. The I_cR_n product tends to be about 1 mV, so the resistance is about 1 Ω. If the barrier thickness is between $1-10$ nm, then the resistivity ranges from $\rho_{\rm dc}$ =1-10 mΩ-cm; this is near the metal-insulator transition, which typically occurs around 1 mΩ-cm. Hence, as the area of the Josephson junction is made smaller, it is natural to consider junctions made from materials that lie close to the metal-insulator transition.

One example of a class of self-shunted junctions that have shown promise for Josephson technology is tantalum-deficient Ta_xN barriers [Kaul, *et al.* (2001)]. This material is a normal-metal barrier, that can be tuned to pass through a metal-insulator transition at Ta_3N_5 [Yu, *et al.* (2002)]. These junctions illustrate good Josephson properties at low temperature, but the properties in available prototype junctions change too rapidly with temperature (as the temperature is raised), which is problematic for circuits, since the switching speed is related to the I_cR_n product and timing errors can quickly force a RSFQ circuit to cease working.

5.2 Superconducting Proximity Effect

When a superconductor is placed close to a nonsuperconducting material, we have an inhomogeneous many-body problem to solve. The superconductivity is reduced within the superconductor as we approach the interface, while superconducting correlations leak into the nonsuperconducting material, and decay with some characteristic length scale. The healing length of the superconducting gap in the superconductor, from its reduced value at the interface, to its bulk value far from the interface is called the

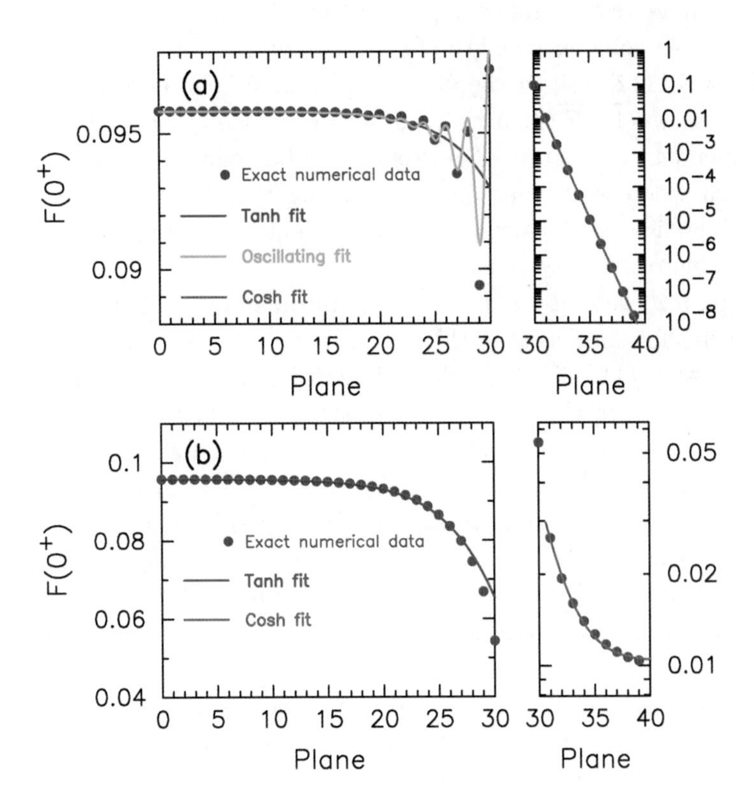

Fig. 5.2 Plot of the anomalous Green's function at equal time (the pair-field amplitude) as a function of the plane number for a Josephson junction with 20 planes in the barrier. The superconducting part (up to plane 30) has $U^{\mathrm{H}} = -2$ and is at half-filling. The barrier (plane 31 to 40) is described by a Falicov-Kimball model with $U^{\mathrm{H}} = 0$, $w_1 = 0.5$, and at half-filling as well. The top plots (a) are for $U^{\mathrm{FK}} = 6$ while the bottom plots (b) are for $U^{\mathrm{FK}} = 2$. The superconducting planes are plotted on a linear scale (on the left), while the barrier planes are plotted on a log scale. The blue dots are the calculated numerical data, and the curves are various fits described in the text. *Adapted with permission from* [Tahvildar-Zadeh, Freericks and Nikolić (2006)].

Ginzburg-Landau coherence length because they were the first to create a phenomenological model that describes how superconductivity is depressed near the surface of a superconductor [Ginzburg and Landau (1950)]. This healing length should be related to the superconducting coherence length ξ_0, which is the average distance between paired electrons in the superconductor (and can be calculated in the bulk). The BCS theory says the coherence length satisfies $\xi_0 = \hbar v_\mathrm{F}/\pi\Delta(0)$, where v_F is an appropriate average Fermi velocity, averaged over the Fermi surface. If we average the modulus of the Fermi velocity over the Fermi surface on a simple-cubic lattice with nearest-neighbor hopping at half-filling, we find $v_\mathrm{F} \approx 3.15at/\hbar$, which yields a coherence length of $\xi_0 \approx 5.1a$ when $U^\mathrm{H} = -2$ [$\Delta(0) = 0.198t$ and $T_c = 0.1112$]. Since the Ginzburg-Landau coherence length satisfies $\xi_\mathrm{GL} = 0.74\xi_0/\sqrt{1 - T/T_c}$ near T_c, a fit of the healing length near T_c allows an independent extraction of ξ_0 from the inhomogeneous solution; we find $\xi_0 \approx 5.2a$ when we fit our data, indicating that the coherence length is around 5 lattice spacings when $U^\mathrm{H} = -2$. This is a short coherence length superconductor.

In Fig. 5.2, we plot the pair-field amplitude (value of the local anomalous Green's function as $t \to 0^+$) as a function of position at a temperature $T = 0.01 \approx T_c/11$. The Josephson junction has 30 self consistent planes to the right and to the left, and 20 planes in the barrier; only half of the junction is shown because the results are symmetric for the other half. The top panels are for $U^\mathrm{FK} = 6$ and the bottom for $U^\mathrm{FK} = 2$, with $w_1 = 0.5$ and half-filling. The blue dots are the numerical results found by solving the inhomogeneous DMFT equations in the superconducting phase. The different curves are different types of fits, according to different fitting forms. But before we discuss the fitting, it is useful to examine the numerical data in more detail. Note how the gap is suppressed more in the normal-metal barrier (bottom panels) than in the correlated insulator barrier (top panels) for the superconducting part of the junction (left). In the normal-metal case, there are only small oscillations, but in the insulating case, the pair-field amplitude has significant oscillations. It is believed these oscillations arise from Fermi wavelength effects, since a semi-infinite superconducting lead terminated by the vacuum should have oscillations that decay as we move away from the surface. For the normal-metal barrier, it makes sense to try the phenomenological Ginzburg-Landau form for fitting the pair-field amplitude via

$$F_\alpha(0^+) = F_{\rm bulk}(0^+)\tanh\left(\frac{|\alpha+\alpha_0|a}{\sqrt{2}\xi_{\rm GL}}\right), \tag{5.1}$$

where we view α_0 and $\xi_{\rm GL}$ as phenomenological fit parameters. This form is supposed to only hold near T_c, but it has an exponential decay from the bulk value far from the interface for all T, and hence is a reasonable first try. We can see the fit (red curve) is quite good for the SNS junction, with significant deviations only near the interface, while it fits the average well (if the oscillations are smoothed out) for the SCiS junction. When we have oscillations, we modify the fit in Eq. (5.1) by multiplying by $a + b\sin[k(\alpha+\bar{\alpha})a]/|\alpha+\alpha'|^\nu$, which provides a simple decaying oscillatory form on top of the exponential healing. One can see the fit (green curve) is quite good in the insulator case now, with deviations also occurring near the interface only. The period of the oscillations is about two lattice spacings, which is not easy to understand, because the Fermi surface has a range of wavevectors over the Fermi surface. The decay of the amplitude is also ill understood—it decays like the inverse of the eighth power of the distance from the interface. Note that $\xi_{\rm GL}$ is essentially independent of the strength of the correlations in the barrier and is about $5a$.

Within the barrier, the decay is governed by the normal-metal coherence length ξ_N, and is expected to be exponential in the distance from the interface. Because the correlations grow in a symmetric way as we approach the second interface to the right, the simplest functional form to use for a fit is

$$F_\alpha(0^+) = F\cosh\left(\frac{\alpha a}{\xi_N}\right). \tag{5.2}$$

This form works almost equally well in metallic or insulating barriers. The coherence length is a function of temperature, and it also varies with the thickness of the barrier. It is an increasing function of the thickness for thin barriers, but then approaches a limit as the system becomes thick enough. This thick-barrier limit is what we call the normal-metal coherence length of the junction. As can be seen in the right panels of Fig. 5.2, ξ_N depends strongly on $U^{\rm FK}$ (it is about $6.5a$ for $U^{\rm FK} = 2$ and $0.6a$ for $U^{\rm FK} = 6$). We do not need to include any fitting with oscillations within the barrier itself. Previous results showed one additional oscillation within the barrier [Freericks, Nikolić and Miller (2001); Freericks, Nikolić and Miller (2002)], but that oscillation appears to be a bug in an earlier version of the code.

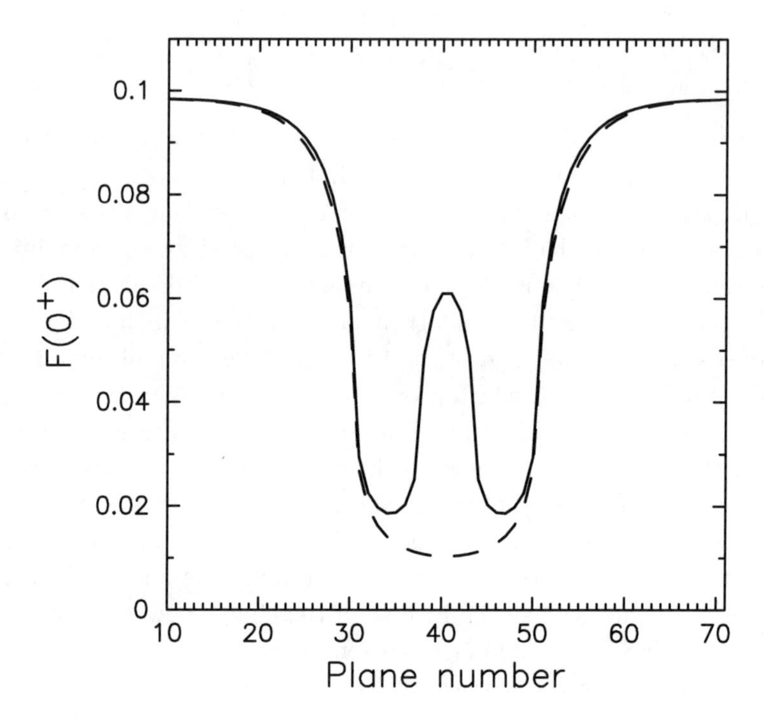

Fig. 5.3 Plot of the anomalous Green's function at equal time as a function of the plane number for a SNSNS Josephson junction with 30 self-consistent planes in the left lead, 7 planes of normal metal, 6 planes of superconductor, 7 planes of normal metal, and then 30 self-consistent planes in the right lead. The dashed line shows the result for a 20 plane SNS junction. The temperature is $T = 0.01$. *Reprinted with permission from* [Freericks, Nikolić and Miller (2002)] (©World Scientific Publishing Co. Pte. Ltd., Singapore).

We can get more complicated results if we introduce additional inhomogeneities on length scales smaller than the respective coherence lengths. For example, in Fig. 5.3, we plot the pair-field amplitude for a SNSNS junction, which has the middle 6 planes of the barrier replaced by superconductors [Freericks, Nikolić and Miller (2002)]. One can see the expected growth of the pair-field amplitude as we enter the superconductor in the center of the device. Since, it will turn out that the critical current of the Josephson junction depends on how large a phase gradient can be placed over the central plane of the barrier, one would expect a higher critical current for the SNSNS junction, because the superconducting correlations are enhanced. Indeed, it increases by over a factor of two.

Finally, we can ask what happens to the Josephson junction if there is an electronic charge reconstruction at the interfaces. This can arise from a chemical potential mismatch between the superconductor and the barrier. If we take clean metals for both the superconductor and the normal metal, so the screened dipole layers at the interfaces provide significant scattering near the interfaces, we can consider the junction to be similar to a SINIS junction, which can be thought of as a SIS junction that has the insulating barrier split in half and filled in the center by a normal metal to try to make the junction thicker, and hence less susceptible to pinholes, and possibly could produce junctions with smaller spreads in parameters across a chip; unfortunately, it seems like the reproducibility of each insulating barrier is difficult to achieve if the insulators are grown artificially. But the screened-dipole layer approach might be more reproducible, because it relies on intrinsic properties of the materials, which would most likely create symmetric, well-defined barriers at the interfaces.

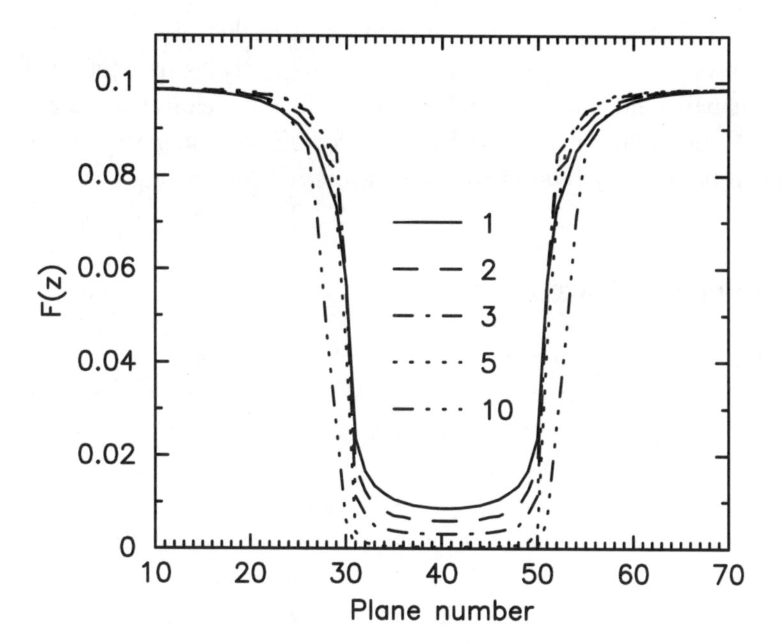

Fig. 5.4 Plot of the anomalous Green's function at equal time as a function of the plane number for a SINIS Josephson junction with 30 self-consistent planes in the left lead, 20 planes of normal metal, and 30 self-consistent planes in the right lead. The different curves correspond to different Fermi level mismatches (ΔE_{F} labels the different curves). The electronic charge reconstruction is plotted in Fig. 3.11. *Reprinted with permission from* [Freericks, Nikolić and Miller (2002)] (©World Scientific Publishing Co. Pte. Ltd., Singapore).

An example of a SINIS junction, generated by electronic charge reconstruction, is plotted in Fig. 5.4 [Freericks, Nikolić and Miller (2002)]. The first point to note, is that as the charge redistribution increases, due to a larger mismatch, the pair-field amplitude is significantly reduced within the normal-metal barrier. In the superconductor, we see the inverse proximity effect is initially reduced in magnitude, as we saw in the correlated insulator barriers, but as the Fermi-energy mismatch is increased, the inverse proximity effect becomes larger. This occurs because we are starting to see significant effects of the charge reconstruction, which extends further and further into the superconducting lead as the mismatch is made larger (more correctly, the magnitude of the scattering which arises from the change in the charge density becomes larger, since the screening length is unchanged).

In this section, we have shown the wide variety of different length scales that are associated with Josephson junctions. They display both proximity effects and inverse proximity effects, as well as Fermi-wavelength-driven oscillations. We have focused solely on length scales associated directly with the pair-field amplitude here. We will see below that there are a number of other important length scales for other properties, but many of them can be directly related to the length scales described here.

5.3 Josephson Current

In the bulk, we applied a constant phase gradient to generate a supercurrent when we were in the superconducting state. We saw that there was a fairly wide range of Q values where the current increased as the phase gradient increased, but then over a fairly narrow range, the current suddenly dropped and then vanished (see Fig. 3.20) once the superconductor could sustain no more current (because the phase gradient was too large). Since the phase variation from plane to plane is nonuniform in a Josephson junction, it is difficult to characterize the change in the phase by a single phase gradient (which one would we choose). Instead, we examine the total phase change over the barrier region of the Josephson junction. We define the barrier to "begin" halfway between the last superconducting plane on the left and the first barrier plane on the left (this is at position 30.5 for our calculations) and to "end" halfway between the last barrier plane on the right and the first superconducting plane on the right. Then we simply accumulate all of the phase change across the entire barrier, and use that total phase change, called the phase across the Josephson junction, to characterize the current flowing through the junction. It is easy to see that the phase gradient must change from plane-to-plane, because the pair-field amplitude changes from plane-to-plane. Since the current passing through each plane must be a constant value through the device, the phase gradient must vary to compensate for the changes in the pair-field amplitude through the junction. We expect the phase gradient to be the largest at the region where the pair-field amplitude is the smallest, and, once that plane can no longer sustain a larger phase gradient across it, the current stops increasing, and we reach the critical current of the device. In most cases we consider, this maximal phase gradient occurs at the central plane of the barrier.

In Fig. 5.5, we show the current-phase relation for a number of different thickness SNS junctions. The superconductor has $U^{\mathrm{H}} = -2$ and is at

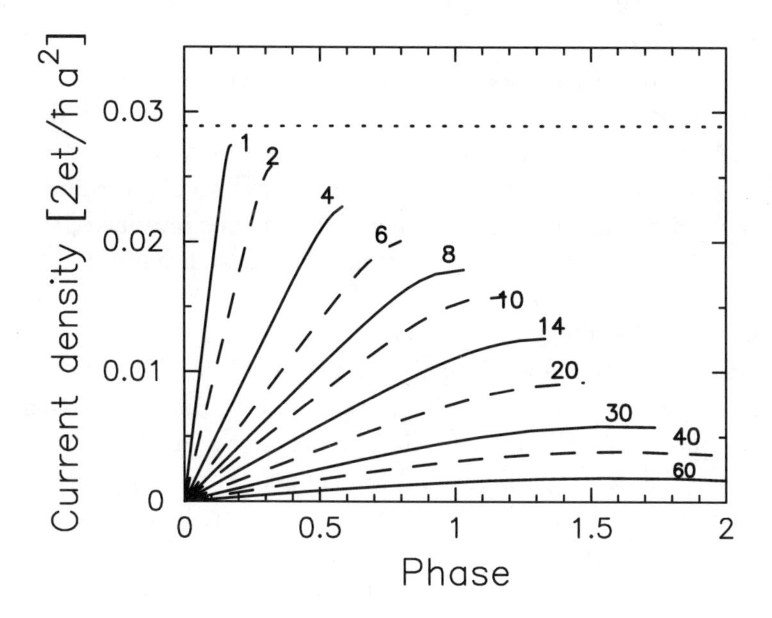

Fig. 5.5 Current-phase relationship for a series of different thickness SNS junctions. The barrier is a ballistic metal with no scattering, while the leads also have no scattering, but they do have a superconducting attraction of $U^{\mathrm{H}} = -2$. All parts of the device are at half-filling, and the temperature is $T = 0.01 \approx T_c/11$. The different curves are for different thicknesses of the barrier. One can see that as the barrier is made thicker, the current drops, and the curve becomes more sinusoidal. In fact, if we plot $J(\phi)/J_c$, versus the phase ϕ, we find simple sinusoidal behavior once the thickness is larger than about 25 planes. The dashed line is the critical current density in the bulk. *Adapted with permission from* [Freericks, Nikolić and Miller (2002)] (original figure ©World Scientific Publishing Co. Pte. Ltd., Singapore).

half-filling. The normal-metal barrier has no Hubbard or Falicov-Kimball interaction and is also at half-filling. The thin barrier ($N = 1$) can sustain a critical current density nearly as high as the critical current density in the bulk. The shape of the curve is far from sinusoidal. As the thickness increases, the current density drops, and the curves become more sinusoidal in shape. Once we reach a thickness of about 25 planes, we recover sinusoidal behavior for all larger thicknesses. Pure sinusoidal behavior is expected from the [Josephson (1962)] result of an SIS junction, but we see it here for thick enough SNS junctions as well. The deviations from the sinusoidal shape arise mainly from the self-consistency of the solutions. When the critical current density is close enough to that of the bulk, we need to examine the solutions with full self-consistency to achieve high accuracy. When scattering is added into the barrier, we also recover the sinusoidal behavior for thicker barriers, but the thickness where this occurs becomes much less than 25 planes.

It is also interesting to examine what the actual phase change looks like over each of the planes of the junction. Since the phase change is dominated by the gradient term, we subtract it from the results we plot, and show just the phase-deviation $\delta\phi_\alpha = \phi_\alpha - \alpha Q$. Since the phase returns to the bulk result of a constant phase gradient far from the interfaces, the phase-deviation plot must become flat as we move far from the interface in either direction. But it does not need to go to zero as $\alpha \to \infty$. Instead, it can go to a constant. The difference between the right and the left phase-deviation functions represents the additional phase accumulated over the barrier, in addition to the phase corresponding to the phase gradient multiplied by the thickness of the barrier. Note that in previous work [Freericks, Nikolić and Miller (2001); Freericks, Nikolić and Miller (2002)], too large of a gradient was subtracted, so that the phase-deviation function went to zero far to the left and far to the right (or one could view that as a slightly different definition of the phase deviation function).

The phase-deviation function for a ballistic normal metal and for a weakly scattering diffusive normal metal SNS junction is plotted in Fig. 5.6 (a) and (b). Note how the deviation goes to a constant far from the interfaces, as expected, because we need to recover the bulk limit there. But there is always an additional total phase shift due to the barrier that needs to be added on to determine the total phase across the junction. As the interaction strength increases, the contribution from the phase-deviation function becomes more and more important. When it completely dominates the phase change, then we can neglect the contribution from the

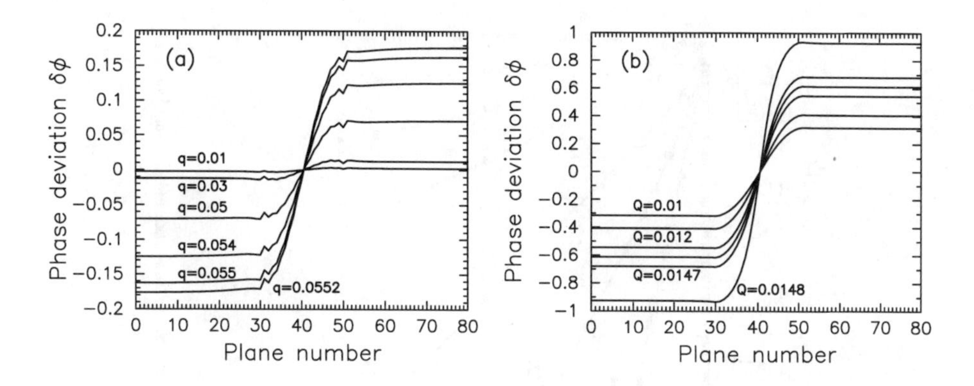

Fig. 5.6 Phase-deviation plot for a twenty-plane barrier SNS junction with (a) no scattering and (b) scattering given by $U^{\mathrm{FK}} = 1$ and $w_1 = 0.5$. The curves are clustered around the critical current of the junction, where Q is maximal (recall the value of the critical current density is determined by the value of Q that we introduce in the leads, because current is conserved through the junction). Note how the general shape of the phase deviation is similar for all cases, and how there is some sharp features present in the SNS junction with no scattering. Only a small amount of scattering is needed to completely smooth out all of that extra structure. The slopes of the phase deviation function are largest at the central plane, indicating that the total phase gradient is largest there, as discussed in the text. *Panel (a) adapted with permission from* [Freericks, Nikolić and Miller (2002)] (original figure ©World Scientific Publishing Co. Pte. Ltd., Singapore).

gradient, which is quite small; this is how Josephson performed his original analysis [Josephson (1962)]. In addition, note the sharp structure present at the interface for the barrier that has no scattering (a), and how it is washed out when there is scattering in the barrier (b). The slope of the phase deviation curve is the largest at the central plane of the barrier, which says that the total phase gradient is largest there. This is to be expected, because the pair-field amplitude is the smallest there.

The critical current density for a ballistic metal barrier ($U^{\mathrm{FK}} = 0$), a diffusive metal barrier ($U^{\mathrm{FK}} = 1$ and 2), a strongly correlated metal near the metal-insulator transition ($U^{\mathrm{FK}} = 4$), a near-critical Mott insulator $U^{\mathrm{FK}} = 5$, and a small-gap Mott insulator $U^{\mathrm{FK}} = 6$, is shown in Fig. 5.7. We use a semi-logarithmic plot to explicitly show the exponential decay of the critical current density with the thickness of the junction. The behavior is qualitatively the same for all different kinds of barriers, but the decay rate increases dramatically as the scattering increases. Note how the exponential decay of the critical current only sets in once the barrier is thick enough, as there is some curvature to the curves for thin barriers. Furthermore, the BCS theory predicts that the decay of the critical current with the thickness

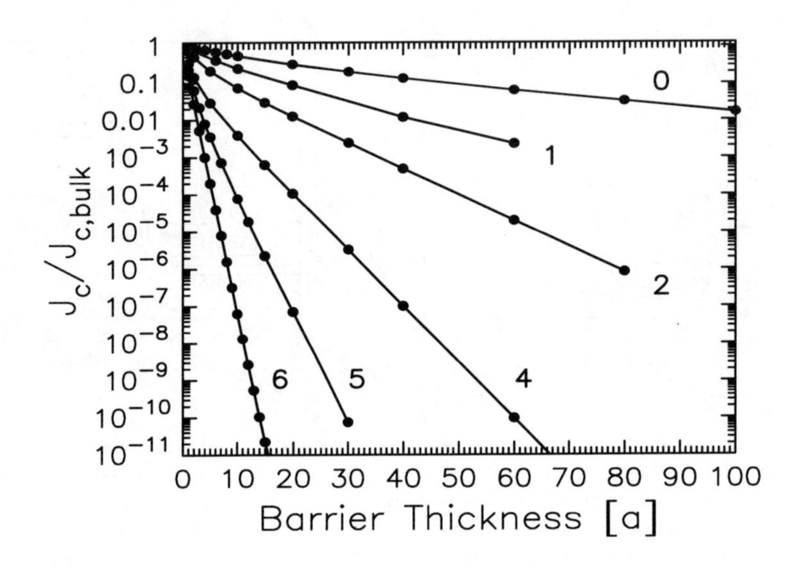

Fig. 5.7 Critical current density (normalized by the bulk critical current density) for barriers described by the Falicov-Kimball model with $w_1 = 0.5$ and different U^{FK} values as a function of the thickness of the barrier (with $T = 0.01$). Note how in all cases (ballistic metal barrier, diffusive metal barrier, and insulating barrier) the critical current density decreases exponentially with the thickness of the barrier. The rate of decrease increases substantially after the metal-insulator transition. The superconductor is at half-filling with $U^{\text{H}} = -2$. The barrier is described by the Falicov-Kimball model at half-filling with $w_1 = 0.5$ and U^{FK} values shown as the labels to the different curves. *Adapted with permission from* [Tahvildar-Zadeh, Freericks and Nikolić (2006)].

of the barrier is also governed by the normal-metal coherence length ξ_N, which describes the exponential decay of the pair-field amplitude when there is no current flowing. Comparing our numerical results for these two different calculations, verifies that they are essentially identical. The numerical values for ξ_N, as extracted from the above plot, are as follows: (i) $U = 0$, $\xi_N = 29.1$; (ii) $U = 1$, $\xi_N = 11$; (iii) $U = 2$, $\xi_N = 6.26$; (iv) $U = 4$, $\xi_N = 2.85$; (v) $U = 5$, $\xi_N = 1.4$; (vi) $U = 5.5$, $\xi_N = 0.79$; and (vii) $U = 6$, $\xi_N = 0.62$.

In the quasiclassical theory of superconductivity, the decay length for a diffusive SNS junction is supposed to be governed by the thermal diffusion length, which is defined to be $\xi_T = \sqrt{\hbar\mathcal{D}/2\pi T}$, where $\mathcal{D}$ is the diffusion constant. We can extract the diffusion constant by using the Einstein relation, as described in Chapter 4, if we know the bulk value of the resistivity. Performing this analysis for these different barriers, produces good agreement with the length scales extracted directly from the data. The approach does not work well once $U > 4.92$, where the bulk system has a metal-insulator transition. The final results for the thermal diffusion length are: (i) $U = 1$, $\xi_T = 12.8$; (ii) $U = 2$, $\xi_N = 6.37$; and (iii) $U = 4$, $\xi_N = 2.82$. Agreement with the above data is quite good (we cannot perform such an analysis for the ballistic metal case, or for the insulating cases at low temperature).

So the Josephson current that flows through the junction is described well by quasiclassical notions when the barrier is a diffusive metal, but the inhomogeneous DMFT approach allows us to also investigate properties of insulators, and to study systems close to the strongly correlated (Mott) metal-insulator transition.

In Fig. 5.8, we plot the critical current density as a function of barrier thickness for a SINIS junction where both materials are at half-filling, but the Fermi level of the barrier is shifted by ΔE_f as shown in the labels of the figure. This shift causes there to be a screened-dipole layer, which corresponds to the electronic charge reconstruction at the interface. Note how the presence of this charge scattering causes a significant reduction of the critical current, which then settles down to an exponential dependence

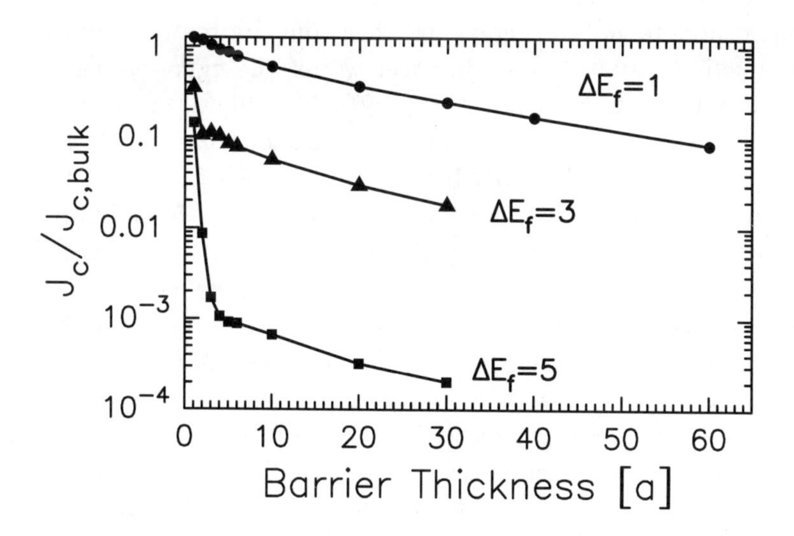

Fig. 5.8 Critical current density (normalized by the bulk critical current density) for SINIS junctions where the "insulating" layers are generated by an electronic charge reconstruction (at half-filling) due to a Fermi-level mismatch. Note how the critical current is initially reduced dramatically due to the scattering off the screened dipole layers, but then it has an exponential dependence on the barrier thickness once the barrier is thick enough. *Adapted with permission from* [Nikolić, Freericks and Miller (2002a)] (original figure © 2002 the American Physical Society).

on the barrier thickness once the barrier is thick enough. There is no scattering in the barrier, and the charge screening length is about three lattice spacings.

5.4 Figure-of-Merit for a Josephson Junction

The critical current is one piece of the figure-of-merit for a Josephson junctions. The other piece is the normal-state resistance. The critical current can be calculated on the imaginary axis (using Matsubara frequencies), but in the ordered superconducting state. The normal-state resistance requires a real-axis calculation, but there is no superconducting order. In fact, we have already calculated the resistance as a function of temperature when we examined the Thouless energy in Chapter 4. So we simply combine the two results to get the I_cR_n product. Note that we actually calculate I_c/A and R_nA for the junctions, but the area factors cancel when we evaluate the products.

Since the I_cR_n product is a voltage, it is common to multiply by the electric charge to get an energy, and then compare that energy to the superconducting gap at $T = 0$. In other words, the figure-of-merit is often reported in the combination $|e|I_cR_n/\Delta(0)$. We will do this here as well.

The figure-of-merit is plotted in Fig. 5.9 for four different thicknesses as a function of U^{FK} at $T = 0.01$. In panel (a), we plot the results for $N = 1$ (top) and $N = 2$ (bottom), in panel (b), we have $N = 5$, and in panel (c), we have $N = 20$. For the single-plane barrier, the I_cR_n product is highest for the metallic junctions, and drops as the scattering increases. Once U^{FK} is large enough, it becomes constant. The dashed line is the Ambegaokar-Baratoff prediction [Ambegaokar and Baratoff (1963)], and we can see our results drop below their results. This is because the gap is suppressed as we approach the interface, which results in a small reduction of the figure-of-merit. The $N = 2$ case has even more significant drop as scattering is added to the barrier, until we hit the metal-insulator transition (around $U^{\mathrm{FK}} \approx 4.92$), and it rises to end up near the constant value of the $N = 1$ case. In panel (b), we have a moderately thick barrier, and the results there are even more interesting. The figure-of-merit is suppressed significantly in the metallic phase, but then turns around dramatically and increases in the insulating phase, going above the value seen for the thin barriers. The thick barrier case in panel (c) behaves similarly, but to a higher extreme, as we see a large suppression on the metallic sign, and a hint of an increase

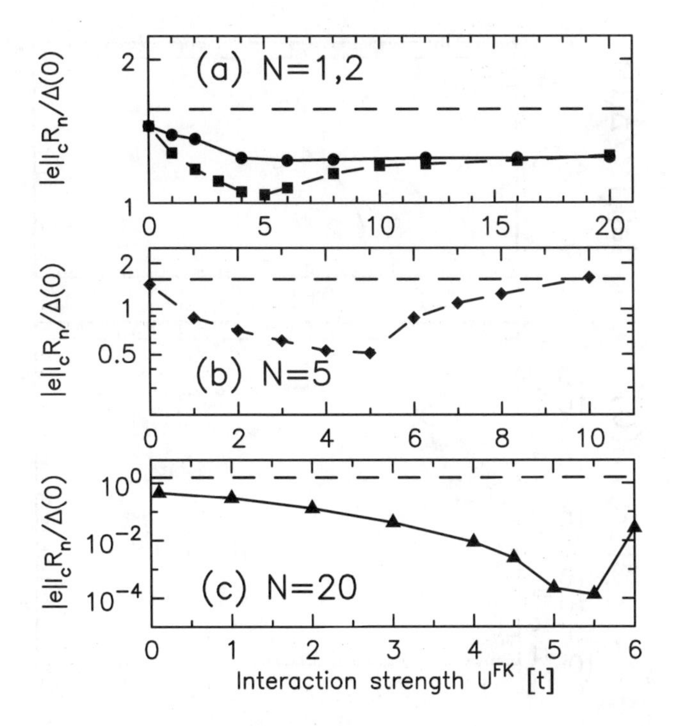

Fig. 5.9 Figure of merit at $T = 0.01$ as a function of U^{FK} for four different thickness Josephson junctions: (a) $N = 1$ and $N = 2$; (b) $N = 5$; and (c) $N = 20$. Note how the thin junctions have a figure-of-merit independent of the properties of the barrier when the barrier height is large enough, and how the thicker barriers have a significant change in their properties at the metal-insulator transition. *Adapted with permission from* [Freericks, Nikolić and Miller (2001)] (original figure © 2001 the American Physical Society).

on the insulating side, but the data is difficult to generate for U^{FK} values larger than 6.

Since the Thouless energy played such a prominent role in describing the behavior of the transport in normal devices, we can ask the question about what role the Thouless energy plays in transport in superconductors. There is much guidance on this issue from quasiclassical calculations. There, it is found that the Thouless energy determines the figure-of-merit when the Thouless energy is the smallest energy scale, and the superconducting gap determines the figure-of-merit when it is the smallest energy scale [Dubos, *et al.* (2001)]. The quasiclassical approach predicts a universal behavior of the figure-of-merit versus the Thouless energy, which approaches the Kulik-Omelyanchuk limit when the superconducting gap is the smallest energy

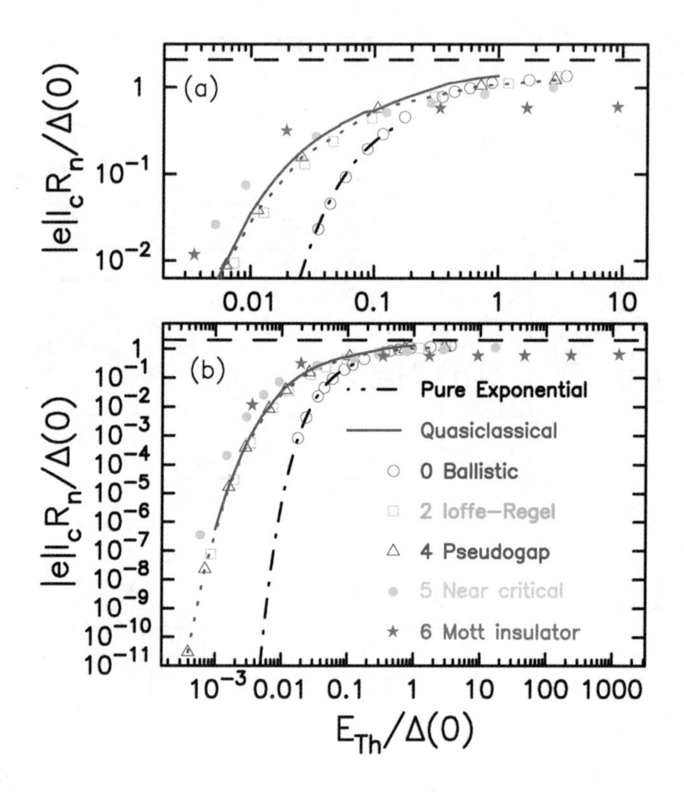

Fig. 5.10 Figure-of-merit at $T = 0.01$ versus the Thouless energy on a log-log plot. Panel (a) is a blow-up of the upper region of panel (b). Five numerical cases are presented: (i) a ballistic-metal barrier with $U^{\mathrm{FK}} = 0$ (red circles); (ii) a diffusive metal near the Ioffe-Regel limit with $U^{|rmFK} = 2$ (green squares); (iii) an anomalous ("pseudogap") metal with $U^{\mathrm{FK}} = 4$ (blue triangles); (iv) a near-critical Mott insulator with $U^{\mathrm{FK}} = 5$ (cyan filled circles); and (v) a small-gap Mott insulator with $U^{\mathrm{FK}} = 6$ (magenta asterisks). In addition, we show the quasiclassical prediction in yellow and an exponential curve in black. *Adapted with permission from* [Freericks, Tahvildar-Zadeh and Nikolić (2005)] (original figure ©IEEE).

and it is proportional to $10E_{\mathrm{Th}}$ when the Thouless energy is the smallest energy. The specific result is plotted in Fig. 5.10 in yellow for $T = 0.01$. On that same plot, we show numerical results for I_cR_n versus E_{Th} for a ballistic metal SNS junction (red circles), a strongly scattering metal near the Ioffe-Regel limit of minimal metallic conductivity [Ioffe and Regel (1960)] (green squares), an anomalous metal (blue triangles), a near-critical Mott insulator (cyan filled circles), and a small-gap Mott insulator (magenta asterisks). Note how both diffusive metals fall right on top of the quasiclassical curve. The ballistic metal case falls on top of an exponential curve (since the resistance is independent of the thickness, the figure-of-merit decays in the same way that the critical current decays). The insulating phases are close to the diffusive metal ones, but do show some deviations: (i) they fall lower than the quasiclassical result when the Thouless energy is large; (ii) they are higher when the Thouless energy is small; (iii) the crossover point lies near the region where $10E_{\rm Th} = \Delta$; and (iv) it appears that they may illustrate new universal behavior for the insulating case at small $E_{\rm Th}$, but this cannot be confirmed from the available data. Hence, we find a second important use for the generalized Thouless energy derived in Chapter 5. It appears to be able to determine the properties of the figure-of-merit of a Josephson junction. We hope that these results motivate experimental groups to try to analyze their data within the generalized Thouless energy concept.

Since the critical current of a Josephson junction always decreases exponentially with the thickness of the barrier, but the resistance increases more slowly (it is constant for a ballistic metal, increases linearly for a diffusive metal and a thick enough Mott insulator, and increases exponentially for a thin Mott insulator), we expect the figure-of-merit to always ultimately decrease as the barrier is made thicker. Since the Thouless energy decreases with increasing barrier thickness as well, as can be read off of Fig. 5.10, it is useful to replot the relationship between the figure-of-merit and the Thouless energy to make this dependence on the thickness more apparent. We do this in Fig. 5.11 [Tahvildar-Zadeh, Freericks and Nikolić (2006)]. The horizontal axis may appear to be a strange set of units, but if we recall for a diffusive conductor that the Thouless energy behaves like $E_{\rm Th} \approx \hbar\mathcal{D}/L^2 = 2\pi T\xi_T^2/L^2$ when expressed in terms of the thermal diffusion length, then we see $\sqrt{T/E_{\rm Th}} = L/\sqrt{2\pi}\xi_T$. Once again, we see different behavior for the ballistic junctions and the other junctions, but the quasiclassical prediction, and the two diffusive metal SNS junctions lie on top of each other. The insulating barriers have a similar behavior, and a similar shape to that of the diffusive junctions, but their quantitative behavior is different. Once again, we do see the possibility that there is a universal insulating curve, but it is difficult to verify that conjecture from the data that we have.

We interpret these results in the following way. First, it is clear that the Thouless energy is a valuable way to summarize transport data for Josephson junctions in addition to junctions in the normal state. Second, these results illustrate one reason why the quasiclassical approach works so well. When fitting quasiclassical results to data, one often chooses an appropriate diffusion constant by fitting resistance data and using an Einstein

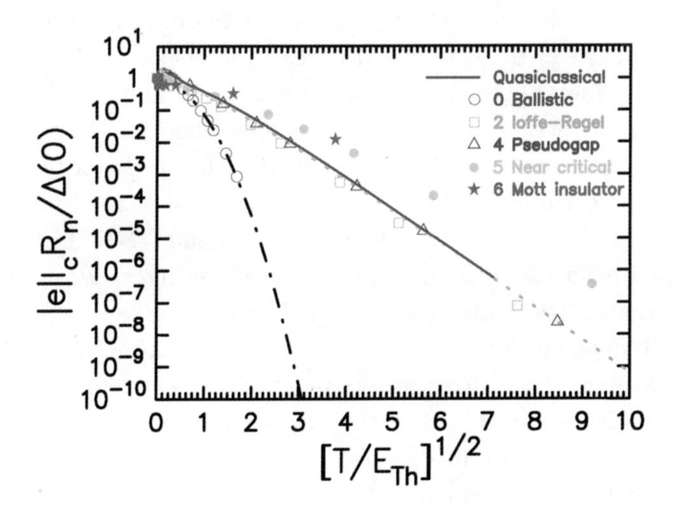

Fig. 5.11 Similar plot of the figure-of-merit at $T = 0.01$ versus the square-root of the ratio of the temperature to the Thouless energy. This horizontal axis is proportional to L/ξ_T for diffusive conductors, and it displays the results in a reasonable way for all of the considered cases. *Adapted with permission from* [Tahvildar-Zadeh, Freericks and Nikolić (2006)].

relation. Since the results shown here reproduce the quasiclassical behavior for all diffusive cases, even if the mean-free-path is much less than a lattice spacing, it shows that the quasiclassical approach may be valid beyond the regime where the mean-free-path is large compared to the Fermi wavelength. Furthermore, since the insulating barriers have similar behavior, albeit with slightly different shapes, the quasiclassical approach is a good first approximation to them as well.

5.5 Effects of Temperature

In externally shunted tunnel junctions, the resistance of the junction (corresponding to the intrinsic resistance through the barrier and the resistance through the shunt resistor) does not vary much with temperature over the operating range of the circuit (typically a few degrees Kelvin at a temperature below 10 K). Hence the majority of the temperature dependence of properties of the Josephson junction occur due to variations of the critical current with T. If we consider barriers tuned close to the metal-insulator transition, particularly junctions where the temperature is similar in mag-

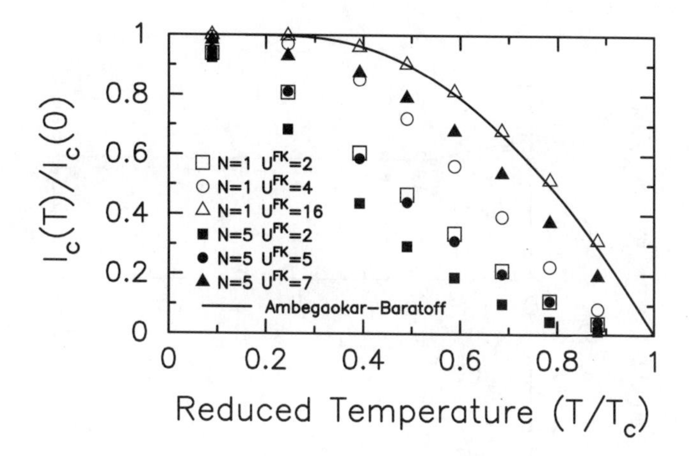

Fig. 5.12 Reduced critical current versus reduced temperature for a variety of thin ($N = 1$, open symbols) and moderately thick ($N = 5$, closed symbols) Josephson junctions with varying scattering in the barrier. The solid line is the prediction of [Ambegaokar and Baratoff (1963)]. All cases are at half filling with $U^{\mathrm{H}} = -2$ in the leads, and vanishing in the barrier. The barrier has $w_1 = 0.5$ and U^{FK} values as shown in the caption. *Reprinted with permission from* [Freericks, Nikolić and Miller (2003)] (©2003 American Institute of Physics).

nitude to the Thouless energy, then we expect to see stronger temperature dependence of the superconducting properties. In addition, it turns out that the critical current in metallic barriers also has significantly stronger T dependence than seen in SIS junctions.

These different properties are summarized in Fig. 5.12 [Freericks, Nikolić and Miller (2003)]. The solid line is the Ambegaokar-Baratoff prediction [Ambegaokar and Baratoff (1963)]. Open symbols are for thin barrier ($N = 1$) and solid symbols are for moderately thick barriers ($N = 5$). If we start by examining a thin insulating barrier ($N = 1$, $U^{\mathrm{FK}} = 16$, open triangles), we see that it tracks well with the Ambegaokar-Baratoff result. It is essentially flat at low temperature (because the gap hardly changes with T there), and it approaches T_c linearly. As the interaction strength is reduced, and we move to an anomalous metal ($U^{\mathrm{FK}} = 4$, open circles) and then to a strongly scattering metal ($U^{\mathrm{FK}} = 2$, open squares), the critical current has a stronger temperature dependence and it drops more rapidly as T increases. While the curves appear to approach $T = 0$ linearly, they actually are expected to saturate at the lowest temperatures. Interestingly, at higher temperatures, the curvature changes sign. Furthermore, there is a wide range of temperatures where the slope is similar in size to the slope of

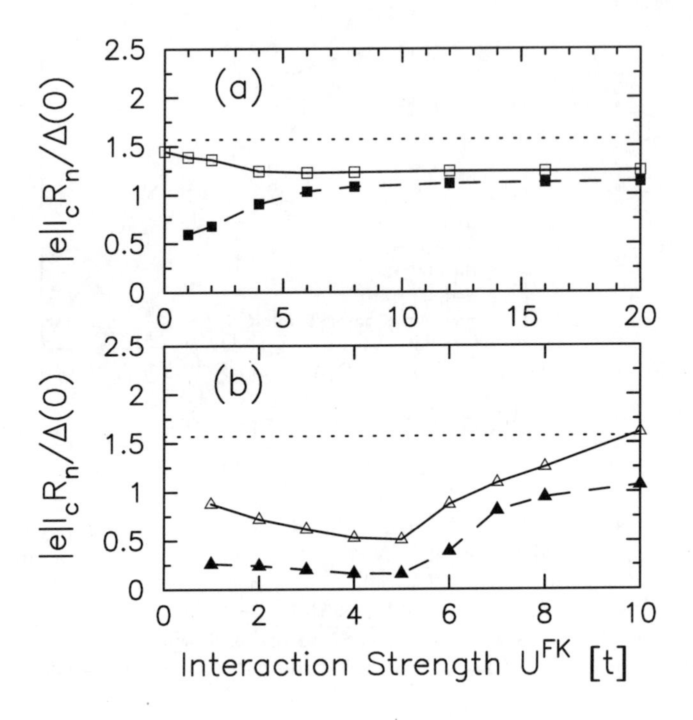

Fig. 5.13 Plot of the figure-of-merit versus U^{FK} for two different temperatures $T \approx T_c/11$ (open) and $T \approx T_c/2$ (solid). The top panel is for a single-plane barrier and the bottom panel is for a $N = 5$ barrier. Note how the SNS junctions have a higher figure-of-merit at low T for the $N = 1$ junctions, but this switches as T is increased. The situation is somewhat similar in the $N = 5$ junctions, but here we have clear optimization in the Mott-insulating phase. *Adapted with permission from* [Freericks, Nikolić and Miller (2003)] (original figure ©2003 American Institute of Physics).

the SIS junction (for $0.5 < T/T_c < 0.9$). This means that the temperature dependence will be similar to that of the SIS junction in this regime. But if the I_cR_n product is too small, the SIS junction will still be preferred, because it will switch faster.

When we move to the moderately thick junctions ($N = 5$, solid curves), we see some interesting behavior too. The metallic junction has even stronger T dependence, dropping rapidly as T increases ($U^{\mathrm{FK}} = 2$, solid squares). The character does not change much as the scattering increases within the metallic phase (compare to the $U^{\mathrm{FK}} = 5$, solid circles data), but then rapidly increases as we go into the small-gap Mott insulating regime $U^{\mathrm{FK}} = 7$ (solid triangles). This last set of data is intriguing because it has essentially the same slope as the SIS junction curve, but it is only reduced

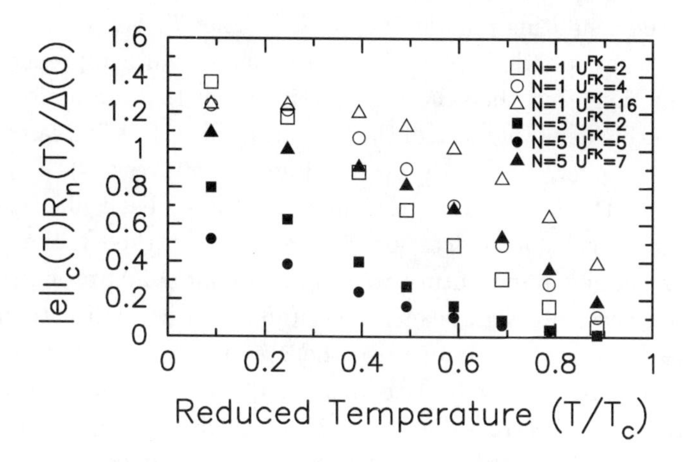

Fig. 5.14 Plot of the figure-of-merit versus reduced temperature for the same parameters as in Fig. 5.12. The thin metallic junctions are the best at low temperature, and the thin insulating junctions are best at high temperature. The moderately thick junction shares a similar shape as the thin tunnel junctions, but is reduced somewhat in size for like Ts.

by about 10–20%. If the resistance is large enough, we can expect to see an enhancement of the I_cR_n product, but the main interest is that these kinds of junctions may have less pinholes, and better junction uniformity across a chip, because they have thicker barriers.

Moving on to the figure-of-merit, we show a plot of $|e|I_cR_n/\Delta(0)$ versus U^{FK} for the $N=1$ and $N=5$ junctions in Fig. 5.13. The open symbols are $T\approx T_c/11$, and the solid symbols are $T\approx T_c/2$. In the thin junction case (top panel), the figure-of-merit is optimized at low temperature for metallic junctions, but their T-dependence is so strong, the figure-of-merit is significantly suppressed even at $T\approx T_c/2$. One can clearly see a much reduced temperature dependence in the insulating regime. Turning to the moderately thick junction ($N=5$, lower panel), we see a local maximum for the metallic barriers, but the figure-of-merit is optimized for more insulating junctions. The temperature dependence is much larger here, but is almost uniform as a function of U^{FK}. We see optimization just on the insulating side of the metal-insulator transition.

Our final results are summarized in Fig. 5.14. Here we plot the figure-of-merit versus reduced temperature for the same cases as in Fig. 5.12. These results summarize what we have been discussing so far. We have optimization for the thin junctions ($N=1$): the metallic junction is best at low T and the insulating junction is best for higher T. When we look at the moderately thick junction ($N = 5$), we see a significant suppression for the metallic junctions, but then there is a rapid rise as the interaction strength increases on the insulating side of the transition (if U^{FK} is increased further, we expect it to overtake the thin insulating junction results).

In general, the temperature dependence of Josephson junctions is not examined too much for devices, because one has, in theory, the option to run at low enough temperature to achieve high-quality properties. But refrigeration costs are significant, and running at as high a temperature as possible is desired. As we raise the operating temperature, we find that the switching speed will be reduced, and the circuit will become more susceptible to timing errors due to temperature fluctuations over the chip. Finding the balance between all of these different concerns is a complex engineering and optimization problem.

5.6 Density of States and Andreev Bound States

Our formalism development focused on imaginary-axis properties in the superconducting state, because they allow us to study much of the behavior of the junctions. There are, however, a number of interesting dynamical properties of Josephson junctions, which require a real-axis treatment within the ordered phase. A detailed description of such a formalism is beyond the scope of this book, and interested readers need to develop the formalism for themselves; the original literature [Miller and Freericks (2001); Nikolić, Freericks and Miller (2002b); Freericks, Nikolić and Miller (2002)] discusses these calculations, but does not provide complete formulas. The modifications of the formalism in going from the imaginary to the real axis are not too complicated, but the denominators can now become singular, which adds some significant complications for how the numerics needs to be carried out. Here we will briefly describe bulk calculations and then summarize a number of physical ideas that can be inferred from calculated results and be presented in figures.

The analytic continuation of the formulae for the Green's functions from the imaginary to the real axis is straightforward except for the 22 components of the Nambu-Gor'kov matrices, since they are the negative of the complex conjugate of the function evaluated at negative frequency. Hence a factor like $i\omega_n - \mu - |U|\langle n\rangle/2 + \lambda(i\omega_n)$ will be analytically continued to $\omega - \mu - |U|\langle n\rangle/2 + \lambda^*(-\omega)$. Using such a technique allows one to determine the impurity solution for the Green's function on the real axis. To complete the DMFT loop, we need to first define two functions

$$Z^{\mathrm{FK}}(\omega) = 1 - \frac{\Sigma^{\mathrm{FK}}(\omega + i0^+) - \Sigma^{\mathrm{FK}*}(-\omega + i0^+)}{2\omega},$$
$$\chi^{\mathrm{FK}}(\omega) = \frac{\Sigma^{\mathrm{FK}}(\omega + i0^+) + \Sigma^{\mathrm{FK}*}(-\omega + i0^+)}{2}. \tag{5.3}$$

Then the Hilbert transform for the Nambu-Gor'kov form of the Green's function becomes

$$\mathbb{G}(\omega) = \int d\epsilon \rho(\epsilon) \tag{5.4}$$
$$\begin{pmatrix} \omega Z^{\mathrm{FK}}(\omega) - \mu - \frac{1}{2}|U|\langle n\rangle + \chi^{\mathrm{FK}}(\omega) + \epsilon & \Delta + \Phi^{\mathrm{FK}}(\omega) \\ \Delta^* + \Phi^{\mathrm{FK}*}(\omega) & \omega Z^{\mathrm{FK}}(\omega) + \mu + \frac{1}{2}|U|\langle n\rangle - \chi^{\mathrm{FK}}(\omega) - \epsilon \end{pmatrix}$$
$$\times \frac{1}{\omega^2 Z^{\mathrm{FK}2}(\omega) - [\mu + \frac{1}{2}|U|\langle n\rangle - \chi^{\mathrm{FK}}(\omega) - \epsilon]^2 - |\Delta + \Phi^{\mathrm{FK}}(\omega)|^2}.$$

Once the DMFT algorithm has converged, then the DOS is found from the 11 component of the Green's function matrix. The DOS is plotted in Fig. 5.15 for the Hubbard-Falicov-Kimball model with $U^{\mathrm{H}} = -2$ and U^{FK} ranging from 0 to 3 and $w_1 = 0.5$. Note the main effect of the scattering is to reduce the gap, shifting the peak down in energy (the frequency is renormalized by the gap for the $U^{\mathrm{FK}} = 0$ case).

There is a wide range of interesting behavior to consider when examining the DOS within a Josephson junction. On the one hand, in the bulk superconductor, the presence of the superconducting energy gap, pushes states away from the Fermi energy, and creates a pile-up of states at an energy equal to the gap energy (both above and below the chemical potential). As $T \to 0$, this pile-up of DOS actually diverges! On the other hand, if the barrier is a normal metal, then there is significant DOS near the Fermi energy because it is a metal. In a quasiclassical calculation, this nonzero DOS in the barrier survives at low energy, but is predicted to go to zero linearly as the chemical potential is approached. Andreev [Andreev (1964)] discussed this physical situation first. The states that lie inside the normal-metal barrier cannot move into the superconductor, because there are no low-energy states present, so they must be localized within the barrier region. These localized, or bound states can actually carry current through the Josephson junction, as depicted in Fig. 1.12. The idea is that a

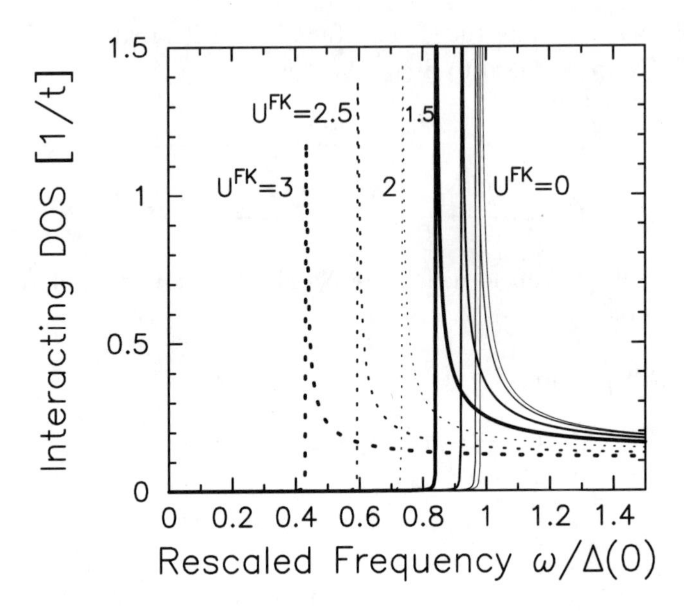

Fig. 5.15 Bulk superconducting DOS for the attractive Hubbard model with $U^{\mathrm{H}} = -2$ and the Falicov-Kimball interaction $U^{\mathrm{FK}} = 0$, 0.5, 1.0, 1.5, 2.0, 2.5, and 3.0. We renormalize the frequency with respect to the $U^{\mathrm{FK}} = 0$ gap at $T = 0$; the calculations here are done at $T = 0.01$. Note how the main effect of the scattering is to reduce the gap, but maintain a similar shape to the DOS. These results should be compared to those that follow for Josephson junctions with similar interactions in the barrier.

superconducting pair of electrons is incident from the left. This pair of electrons meets a hole in the normal-metal barrier, leaving an electron behind that moves to the right. That electron travels to the barrier-superconductor interface on the right, and a superconducting pair and a retroreflected hole emerge; the hole travels to the left in the metal, and the superconducting pair travels to the right in the right superconducting lead. The net effect is that a superconducting pair has been carried across the barrier from the left to the right, and there is a localized electron-hole state that remains in the barrier. A similar process can carry current in the opposite direction, and one can view the lack of supercurrent when there is no phase difference across the junction as corresponding to an equal occupation of the left and right current-carrying states. In a one-dimensional system, the Andreev bound state has a well-defined energy, but in a three-dimensional system, because of the possible nonzero value for the transverse ($\mathbf{k}_x$ and $\mathbf{k}_y$) momenta, these Andreev bound states appear as finite-width peaks in the DOS. Because the electrons involved in the Andreev bound states have energies close to the chemical potential, we expect these states to be located at an energy below the bulk superconducting gap.

When scattering is added to the barrier, we can characterize the level of scattering with the Thouless energy for the diffusive metal barrier (plus the contribution from the contact resistance, of course). When such scattering is included, quasiclassical calculations say there will be a so-called "hard" minigap in the DOS, where the DOS vanishes over a region proportional to E_{Th} about the chemical potential [Golubov and Kupriyanov (1989); Golubov, Wilhelm and Zaikin (1997); Zhou, *et al.* (1998); Pilgram, Belzig, and Bruder (2000)]. Since there is no gap when there is no scattering, and the Thouless energy decreases as the thickness of the barrier increases, the gap region is expected to first grow, and then decrease as scattering is turned on. Since a ballistic-metal-barrier junction has a nonzero Thouless energy, this analysis cannot be consistent with the Thouless energy solely determining the minigap, but most quasiclassical approaches ignore the contact resistance contribution to the Thouless energy, so from their perspective, the Thouless energy vanishes for a ballistic-metal-barrier junction, and the analysis is consistent.

There are additional sources of scattering that are not incorporated into these quasiclassical approaches. One important source is the so-called Δ/μ scattering [Hurd and Wendin (1994)], which says that as electrons move through the junction, if we take into account Fermi-surface effects, then the superconducting amplitude is modified on the order of Δ/μ by these Fermi-surface effects (μ is the distance from either the upper or lower band edge to the chemical potential). What is less known is what effect this will have on the DOS. Another important source, if there is scattering that is described by a self-energy, is the life-time effects associated with the many-body excitations. Generically, life-time broadening effects cause sharp features in the DOS to be washed out, and can lead to the disappearance of gaps within the spectrum.

We show a contour plot of the local interacting DOS for a $N = 10$ SNS Josephson junction ($U^{\mathrm{H}} = -2$, no scattering in the barrier, $T = 0.01$) in Fig. 5.16 [Miller and Freericks (2001)]. We show half of the junction, starting with the barrier planes (numbered from 1–5, starting from the center of the barrier), and continuing into the leads (numbered 6–35). The energy axis is plotted in units of the bulk superconducting gap at $T = 0$, so that the energy field of view consists of all of the subgap states in the device. The first thing to note is that the DOS does not go linearly to zero at $\omega = 0$, as predicted by quasiclassical theories for ballistic-metal

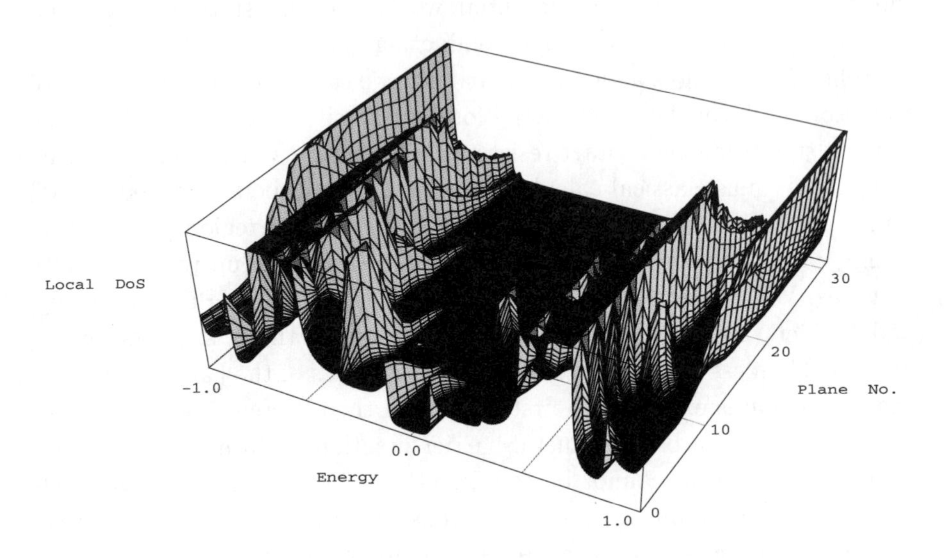

Fig. 5.16 Three-dimensional plot of the low-energy DOS of a $N = 10$ SNS Josephson junction with $U^{\mathrm{H}} = -2$ in the leads, and no scattering in the barrier. The plane numbers start in the center of the barrier and move outward; hence the barrier extends from 1 to 5, and the superconductor from 6 to 35. The horizontal (energy) axis is plotted in units of $\Delta(0) = 0.198$, the bulk superconducting gap at $T = 0$. *Reprinted with permission from* [Miller and Freericks (2001)].

barriers [Saint-James (1964); McMillan (1968)]. Instead, there is a "hard" minigap that has formed. In addition, we see peaks in the subgap states that are localized in the vicinity of the barrier, but they clearly extend for some distance into the superconductor. Furthermore, we can see an alternating parity to the spatial profile of the Andreev bound state peaks: they either are peaked at the central barrier plane, or have a node there, and this behavior alternates. Finally, as the Andreev states approach the gap energy, they extend further and further into the superconducting leads.

Since the minigap is not supposed to form in a SNS barrier that has no additional scattering, and because there are no life-time effects, the most likely explanation for the origin of the minigap is the finite Δ/μ scattering. We confirm that this notion is correct in Fig. 5.17. The two curves are for the same thickness barrier, but with different values of the attractive Hubbard interaction. As the interaction is reduced, both the gap and the minigap shrink, whereas, we would have expected the minigap to be

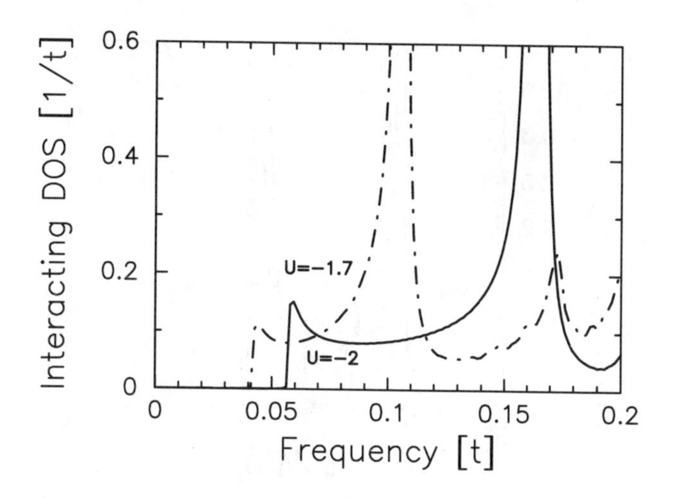

Fig. 5.17 DOS at the central plane of a $N = 5$ barrier Josephson junction with $U^{\mathrm{H}} = -2$ and $U^{\mathrm{H}} = -1.7$. Note how the minigap gets smaller in each case, which supports the notion that it arises from Δ/μ scattering, while the Thouless energy is unchanged. *Adapted with permission from* [Nikolić, Freericks and Miller (2002b)] (original figure © 2002 the American Physical Society).

unchanged if its size was governed by the Thouless energy. Instead, the shrinking of the size of the minigap points to its origin lying in the Δ/μ scattering effect, since Δ is reduced when $U^{\mathrm{H}} = -1.7$. Note how different the inhomogeneous results are from the homogeneous (bulk) results shown in Fig. 5.15.

Nevertheless, one can wonder whether the Thouless energy plays a role in the size of the minigap, and for an SNS junction we need to include the Thouless energy that arises from the contact resistance of the junction. Comparing the data for $U^{\mathrm{FK}} = 0.1$ in Fig. 5.18 for the $N = 5$ and $N = 10$ data, we would expect that the minigap would be a factor of two smaller for the $N = 10$ case, if the minigap was proportional to the Thouless energy (since the Thouless energy is two times smaller), but the minigap is almost a factor of three smaller. Furthermore, as scattering is introduced, by increasing U^{FK} further, the minigap shrinks rapidly, and then disappears. These results are hard to understand from any model that says it is the Thouless energy alone that determines the size of the minigap. Indeed, it is likely that there is an alternative explanation to those results. The data depicted in Fig. 5.18 have a number of interesting features and trends in them, though, that are at least consistent with the notion of the Thouless energy playing a role in the behavior. For example, as the thickness of the

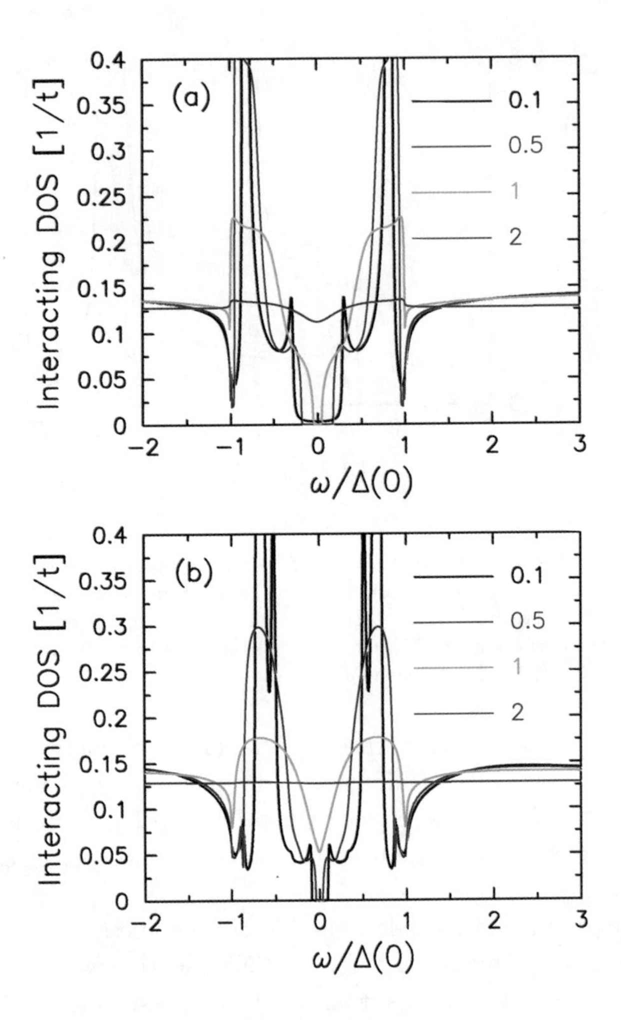

Fig. 5.18 Local DOS at the central plane of the barrier of an (a) $N = 5$ and (b) $N = 10$ Josephson junction with varying U^{FK} (the value is included in the label of the legend). Note how the minigaps for corresponding cases are smaller for the thicker barrier, and how the DOS loses all low energy structure once U^{FK} becomes large enough, even though it is well before the metal-insulator transition. The horizontal axis is plotted in units of $\Delta(0)$. *Adapted with permission from* [Nikolić, Freericks and Miller (2002b)] (original figure © 2002 the American Physical Society) and from [Freericks, Nikolić and Miller (2002)] (original figure ©World Scientific Publishing Co. Pte. Ltd. Singapore).

barrier increases, the minigaps decrease, and by the time we reach $N = 20$, no more minigap can be observed in the data. At this point, we see the expected quasiclassical prediction that the DOS vanishes linearly at $\omega = 0$ (not shown here). As scattering increases in the system, we never see a situation where the minigap increases due to increased scattering, instead, it seems to rapidly decrease and then disappear. It disappears sooner for thicker barriers, as we probably expected it would. The best idea for the source of the disappearance of the minigap is a life-time effect that arises from the scattering off the Falicov-Kimball defects. This scattering grows like $[U^{\rm FK}]^2$, and it can rapidly cause the minigap feature to be washed away.

We study the issue of how the DOS heals to the bulk superconducting DOS as we move away from the barrier in Fig. 5.19. The behavior of the DOS in this figure is quite complex. One might have assumed that the DOS would go back to its bulk value on a length scale on the order of the healing length for the pair-field amplitude, that was analyzed earlier in this chapter. But the healing length is around 5.1 lattice spacings, and we can still see significant variation of the DOS (especially at energies close to the gap energy) when we are even 30 planes away from the interface. While the

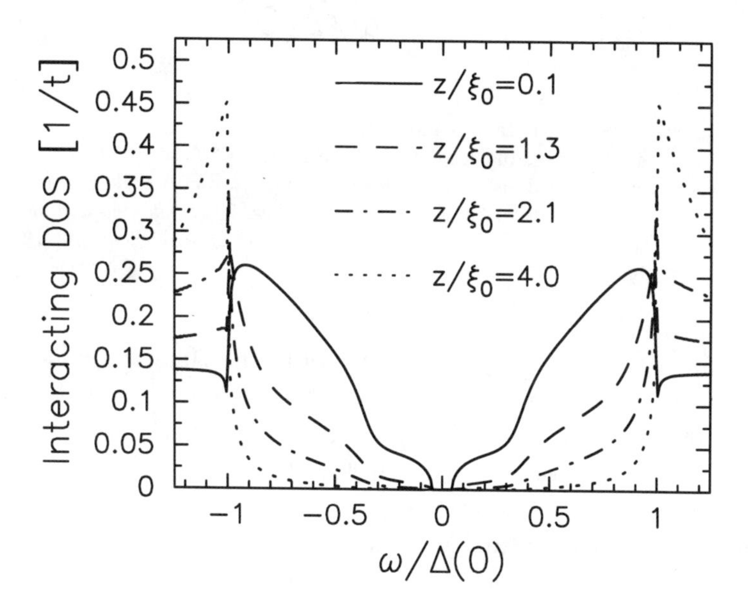

Fig. 5.19 Local DOS as a function of position within the superconducting leads showing how the full gap is eventually restored as we move further away from the barrier, but the healing of the DOS requires much longer length scales than the healing length for the pair-field amplitude in the superconductor, especially for energies close to the gap energy. The barrier has $U^{\rm FK} = 1$, $N = 5$, and $\xi_0 \approx 5.1a$; the distance z is measured from the SN interface. *Adapted with permission from* [Nikolić, Freericks and Miller (2002b)] (original figure © 2002 the American Physical Society).

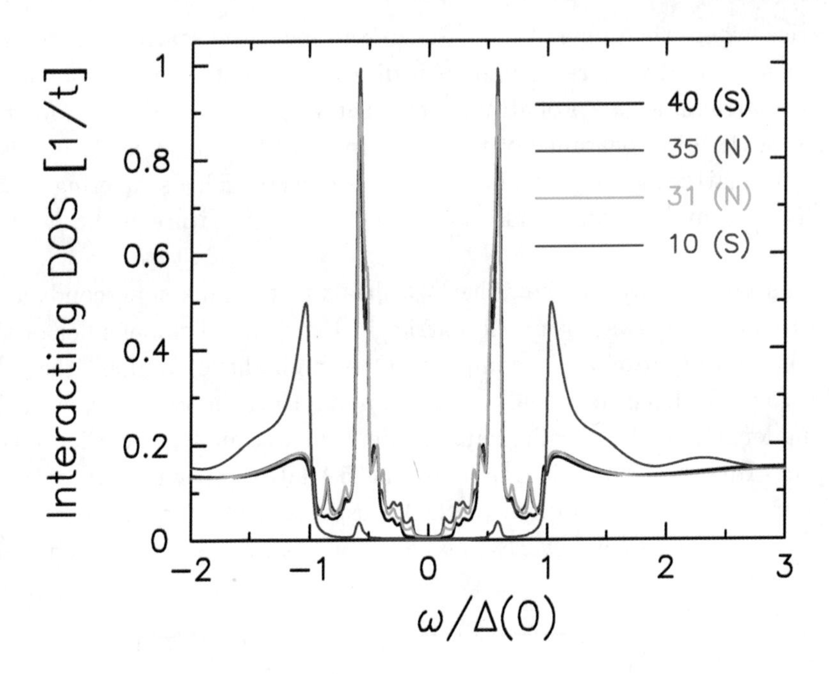

Fig. 5.20 Local DOS on different planes for the SNSNS Josephson junction shown in Fig. 5.3. Plane 40 is the center of the central S layers. Plane 35 is at the first NS interface away from the center. Plane 31 is at the first SN interface away from the center. Plane 10 is twenty planes into the superconductor from the last interface. *Adapted with permission from* [Freericks, Nikolić and Miller (2002)] (original figure ©World Scientific Publishing Co. Pte. Ltd., Singapore).

length scales for these two processes need not be the same, it would be useful to better understand why the decay length is so long. Anomalously long proximity effects for single-particle properties are usually assumed to occur due to low-energy excitations, with energies far below the gap energy, that result in superconducting correlations that stretch for longer length scales than the conventional proximity effect. This is the reason why we expect the minigap to be present when the Thouless energy is smaller than the superconducting gap. But the situation here is occurring at higher energy scales due to the long decay lengths of the Andreev bound states that sit close to the superconducting gap edge. Exactly why this occurs is not clear, and remains unexplained.

Interesting behavior also occurs in more complicated Josephson-like structures. Here we examine more closely the SNSNS junction that we first encountered in Fig. 5.3. Even though the $N = 20$ junction has no minigap,

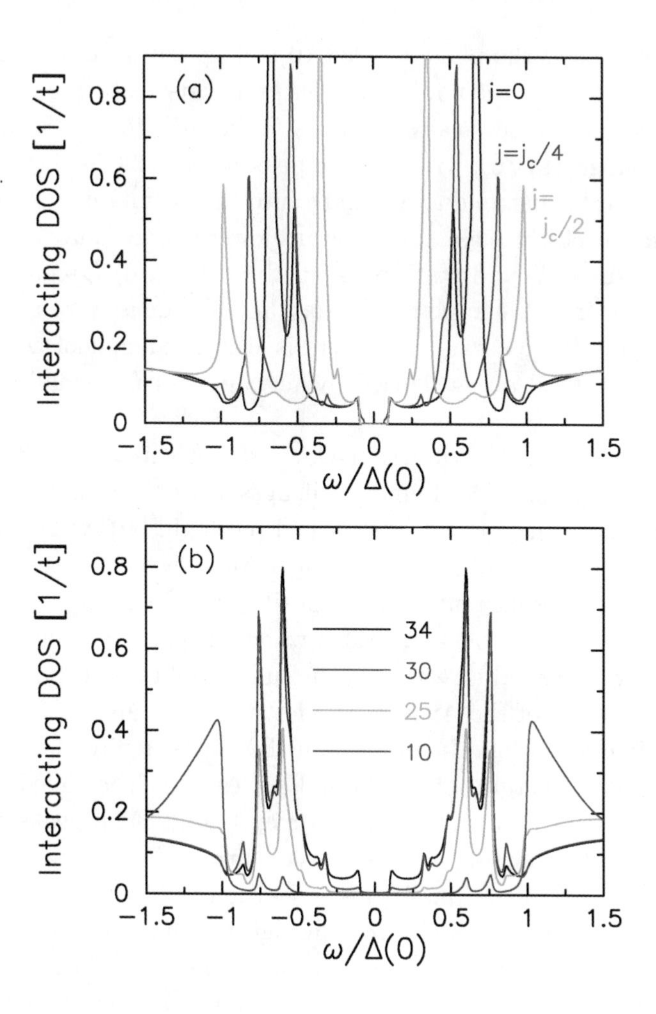

Fig. 5.21 (a) Local DOS on the central plane of a $N = 10$ Josephson junction with no scattering in the barrier, as a function of the current in the device. We show no current (black curve), $j = j_c/4$ (red curve) and $j = j_c/2$ (green curve). Note how the Andreev-bound-state peaks split when current flows due to the Doppler shift. (b) Local DOS when $j = j_c/4$ for various planes in the same Josephson junction. *Adapted with permission from* [Freericks, Nikolić and Miller (2002)] (original figure ©World Scientific Publishing Co. Pte. Ltd., Singapore).

the minigap returns here because we now have four superconductor-normal-metal interfaces, which enhance the Δ/μ scattering. The plane labeled 40 is in the center of the superconducting planes in the middle of the SNSNS junction, the plane labeled 35 is near the middle NS interface (within the normal metal), the plane labeled 31 is near the first SN interface (within the normal metal), and the plane labeled 10 is deep within the superconducting lead. One can see the minigap structure has developed, but the central superconducting region (black curve) is far from being BCS-like and has significant subgap states, with a large peak at $\omega \approx 0.6\Delta(0)$. As we move from the central superconducting plane into the normal metal (red curve), we see some spectral weight shifts toward lower energy, and the emergence of a new Andreev bound state peak at about $0.9\Delta(0)$ (green curve). As we move deep into the superconductor the BCS density of states begins to be recovered (blue curve), but there is still a remnant of the large Andreev bound state at about $0.6\Delta(0)$, which appears as a small peak in the figure.

Finally, we examine how the DOS varies when a current is flowing through the junction. We found challenges with the convergence of the calculations when the current was pushed too high, but we still find interesting behavior for small currents. Since Andreev bound states come in pairs, each carrying current in an opposite direction, we expect to see a modification of the the peak locations when current is flowing because we now have a well defined direction in the junction (the direction of the current flow), and one of the Andreev bound states is transporting current in the same direction, while the other is in the opposite direction. This leads to a shift in the energies of the bound states, which is called a Doppler shift. We expect to see the single Andreev bound state peaks to split in two, and have the splitting increase as the current flow increases. This can be seen in Fig. 5.21 (a). Finally, we show the results for the DOS when a fixed current density $j = j_c/8$ flows through the junction, as a function of position within the junction. There are more peaks present here, because the Doppler shift splits all of the Andreev bound state peaks, but the minigap is still present, and we can see remnants of these structures leak far into the superconductor before they disappear [Fig 5.21 (b)].

Chapter 6

Thermal Transport

We examine the phenomenon of electronic charge reconstruction for systems that have barriers that lie close to a metal-insulator transition. We will consider a Mott-insulating barrier (the focus here is on spinless systems, adding spin is simple, and will have minor modifications to the results), where the entire system is held at half-filling. This system is interesting, because, in the bulk, both constituents of the device have no thermoelectric effects, because both the Seebeck and Peltier coefficients vanish due to particle-hole symmetry. But if we attach these two thermoelectrically inert materials together, and there is a Fermi level mismatch between them, then the electronic charge reconstruction breaks the particle-hole symmetry and allows them to have thermoelectric effects. What is different in the Mott-insulating case from a metallic case is that in the bulk Mott insulator, the DOS has sharp structure near the chemical potential, which is likely to lead to a large thermoelectric effect. From an applications perspective, we are using electronic charge reconstruction as an engineering tool, to try to enhance the thermoelectric response of the full device over and above that of its constituent parts. In this sense, the geometry of the multilayered nanostructure, and the way that the Mott insulator is deconfined between the metallic leads plays a crucial role in determining the thermoelectric response.

6.1 Electronic Charge Reconstruction Near a Metal-Insulator Transition

As before, we will consider a system made of ballistic metal leads and a strongly correlated electron barrier. We fix $e_{\mathrm{Schot}} = 0.4$, which corresponds to a screening length of a few lattice spacings, for both materials. When there is a mismatch of Fermi levels, the electronic charge reconstruction will provide an inhomogeneous doping of the Mott-insulating barrier.

Since the electronic charge reconstruction generically makes the barrier metallic (although it can have strong scattering), we perform the calculations for the DOS and the real-axis self-energies by simply adding a small

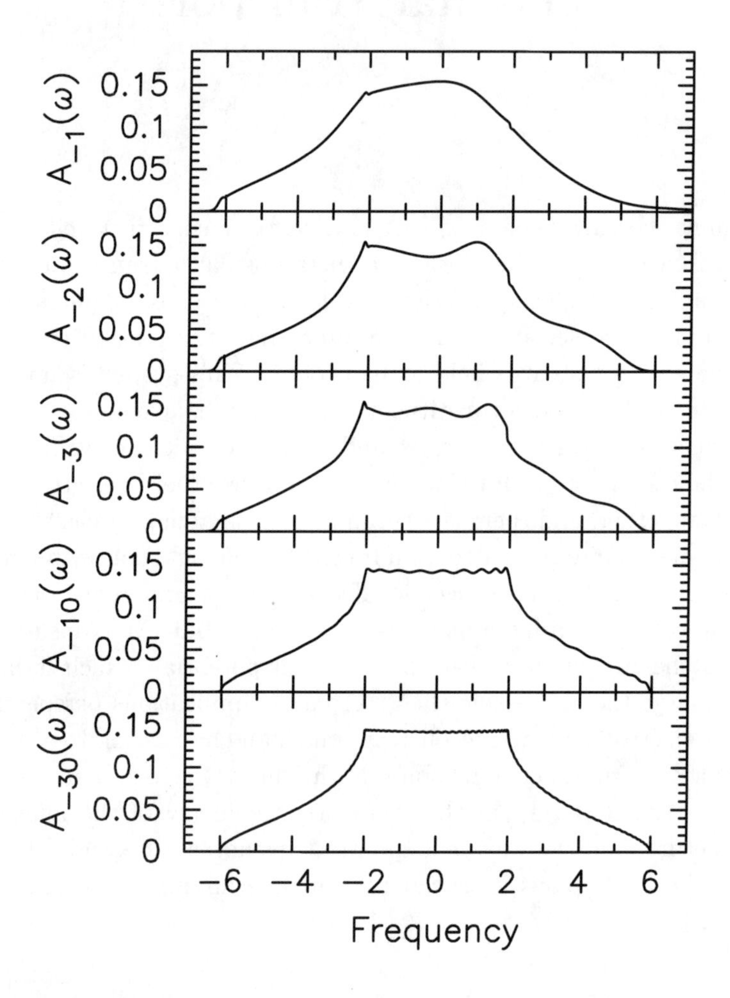

Fig. 6.1 DOS for different planes in the metallic lead of a multilayered heterostructure at half filling with ballistic metal leads and a barrier described by the Falicov-Kimball model at half filling with $U^{\mathrm{FK}} = 6$. The center of the band of the barrier is shifted up in energy by $2t$ ($\Delta E_F = -2$). The top panel is the first metal plane to the left of the interface, then we follow with the second, third, tenth, and thirtieth planes. Note the asymmetry in the DOS that enters due to the charge reconstruction and how it is reduced the farther we move from the interface.

imaginary part ($0.001i$) to the frequency, so that we can avoid any singularities that involve principal-value integrations. We choose at least 10,000 quadrature points, so that there is sufficient accuracy to properly describe any of the near singular behavior. This allows for a quick means to calculate the DOS and the transport of these systems. If, on the other hand, the barrier is insulating, and the chemical potential lies close to the location of the pole in the bulk material, then the calculations need to be handled more carefully.

The lead is a ballistic metal and the ($N = 20$) barrier is a Mott insulator described by the spinless Falicov-Kimball model (with $w_1 = 0.5$ and $U^{\mathrm{FK}} = 6$). The center of the band of the barrier is shifted to create a chemical potential mismatch which then leads to an electronic charge reconstruction. We calculate the reconstruction for each temperature that we examine, but the change with temperature is minimal at low temperature (an example of the charge rearrangement can be found in Fig. 3.12 for $\Delta E_F = -1$). The shift of the band energies of the barrier, and the electronic charge reconstruction break the particle-hole symmetry of the system. We illustrate this with a plot of the DOS for a number of different planes in the leads in Fig. 6.1. Note how the asymmetry is quite strong near the interface, but fades away as we move further away. These results should be compared to Fig. 3.3, which has no shift of the barrier band energies (and has $N = 5$, but the width of the barrier has little impact on the lead DOS).

Next we examine the DOS in the barrier for the same device in Fig. 6.2. Here the asymmetry is not as marked as it was in the leads, but the shift of the center of the band for the central plane is dramatically less than 2 (it decreases the further we are from the interface), because it is partially reduced by the magnitude of the Coulomb potential energy at the given plane of the barrier; on the central plane, the magnitude of the Coulomb potential is close to 1.6, which explains why the shift of the band center is only about 0.4 instead of 2.

We plot a false-color contour plot of the low temperature DOS of the multilayered nanostructure in Fig. 6.3. One can clearly see how the DOS is shifted as we increase the chemical potential mismatch of the barrier, and include the resulting electronic charge reconstruction. Note how there is an interesting bending of the bands, how the shift of the barrier is reduced by the Coulomb potential from the electronic charge reconstruction, and how the DOS develops extra oscillations near the band edges of the leads. Indeed, the change of the Coulomb potential with distance in the barrier

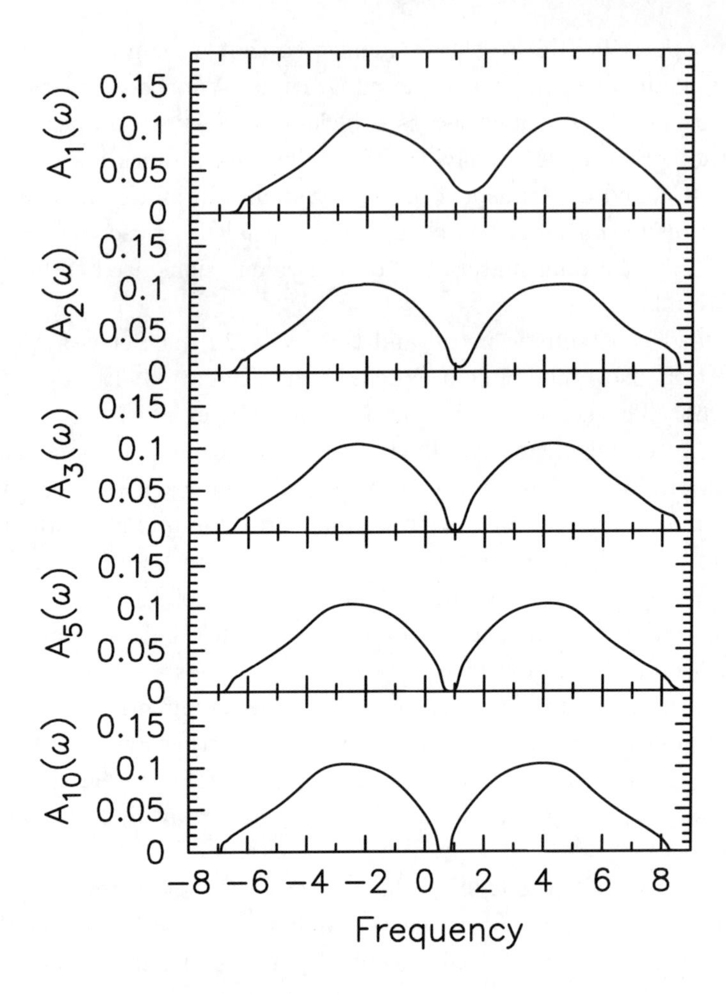

Fig. 6.2 DOS for different planes in the barrier of a multilayered heterostructure at half filling with ballistic metal leads and a $N = 20$ barrier described by the Falicov-Kimball model at half filling with $U^{\mathrm{FK}} = 6$. The center of the band of the barrier is shifted up in energy by $2t$ ($\Delta E_F = -2$). The top panel is the first metal plane to the right of the interface, then we follow with the second, third, fifth, and tenth planes. Note how the band of the barrier has shifted by less than two units because of the electronic charge reconstruction (this is because the electrical potential is approximately -1.6 at the central planes of the barrier).

explains the curvature of the blue region in the center of the barrier, because the magnitude of the potential shift decreases as we approach the interface, so the center of the DOS shifts to the right as we approach the interface from within the barrier.

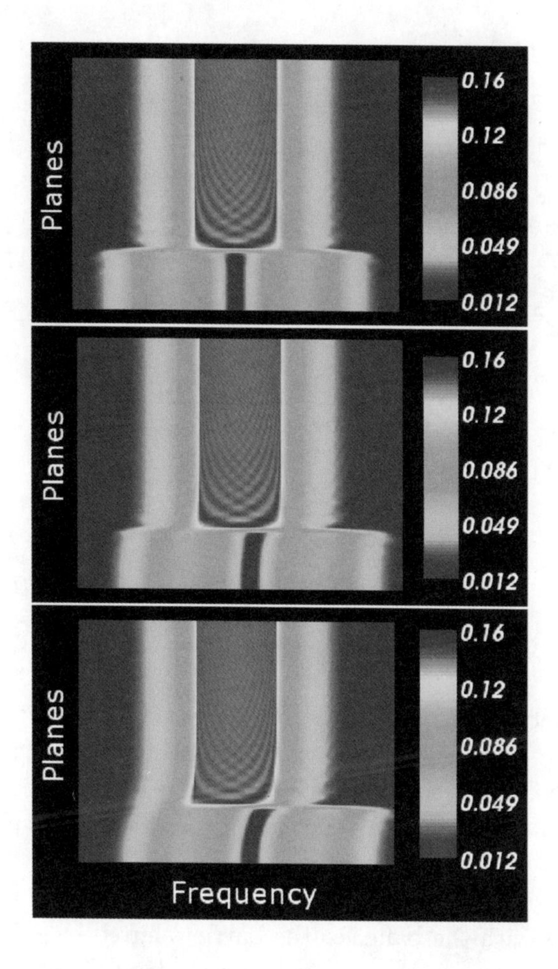

Fig. 6.3 False color plot of the $T = 0.01$ DOS for a $N = 20$ metal-barrier-metal junction with different shifts of the bands (top, $\Delta E_F = 0$; middle, -2; and bottom, -4). The barrier is described by the half-filled Mott-insulating Falicov-Kimball model at $U^{\mathrm{FK}} = 6$. One can clearly see how the DOS is modified by the combination of the shift of the band energies and the resultant charge reconstruction.

6.2 Thermal Transport Through a Barrier Near the Metal-Insulator Transition

Transport can also be calculated in these systems using the formalism developed in Chapter 3. As with the DOS, we add a small imaginary part to the frequency (or equivalently to the self-energy) to smooth out any potential poles that require principal-value integration (we choose the same size shift of $\delta = 0.001i$). Using the different L_{ij} matrices, we can determine the

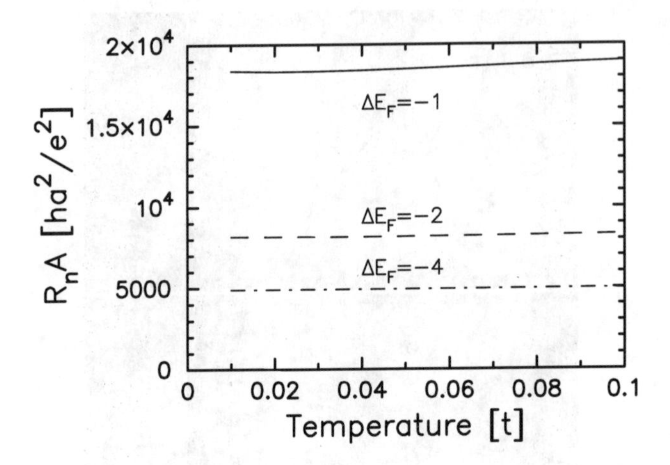

Fig. 6.4 Electronic charge resistance-area product in a $N = 20$ plane device described by the $U = 6$ Falicov-Kimball model at half filling with a shift of the band given by $\Delta E_F = -1$ (solid line), -2 (dashed line) and -4 (chain-dashed line). The temperature runs from 0.01 to 0.1 and the resistance-area units are ha^2/e^2.

charge and thermal resistance, the Seebeck and Peltier coefficients of the device, and the efficiency for use as a thermal cooler or as a power generator. Here we will focus solely on the resistances, and on the thermoelectric effects, but will not discuss calculations of the efficiency of parts of a cooler or generator.

We start with the results for the $U = 6$ Falicov-Kimball model barrier at half filling, but with the band for the barrier shifted by ΔE_F as described above. The resistance in the case with no electronic charge reconstruction saturates at larger than 10^{12} at low temperature. When the barrier-plane bands are shifted, we see that the resistance is sharply reduced, because the system has become metallic, but they have little temperature dependence. As the chemical potential is shifted farther away from the chemical potential of the leads, the system becomes more metallic, because the reduction of scattering due to the reduction of the correlations is more important than the increase in scattering that arises from the increased electronic charge reconstruction.

The thermal resistance-area product is plotted in Fig. 6.5 for the same device. Note how the thermal resistance diverges as C/T for $T \to 0$. This is because the thermal conductivity of each constituent goes to zero linearly in T, so the thermal resistance should diverge. As with the charge resistance, we see that the farther we shift the barrier band from the metallic lead

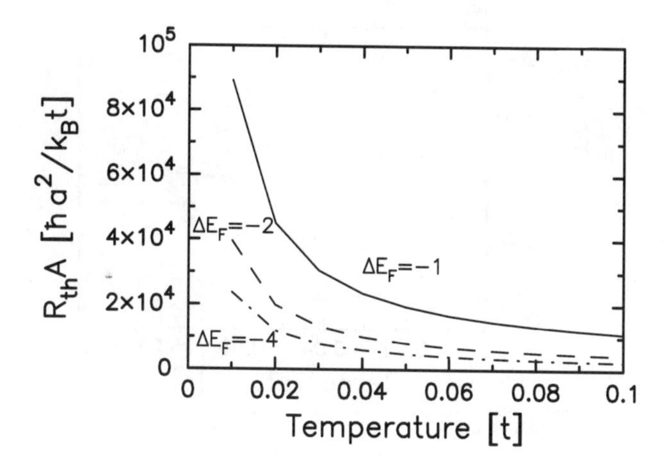

Fig. 6.5 Electronic thermal resistance-area product in a $N = 20$ plane device described by the $U = 6$ Falicov-Kimball model at half filling with a shift of the band given by $\Delta E_F = -1$ (solid line), -2 (dashed line) and -4 (chain-dashed line). The temperature runs from 0.01 to 0.1 and the thermal-resistance-area units are $\hbar a^2/k_Bt$.

band, the smaller the thermal resistance becomes. This is also expected because the system is becoming more metallic.

We can define a device Lorenz number by taking the ratio of the resistances and the temperature via $\mathcal{L} = e^2R_n/R_{th}k_B^2T$. This is the analog of the bulk Lorenz number, but is now well defined for a device as well. We expect that the Lorenz number should approach the metallic value of $\pi^2/3$ given by the Wiedemann-Franz law, unless the device does not reproduce the law even if its constituents do. As can be seen in Fig. 6.6, we do produce the correct $T \to 0$ limit and the more metallic the device is, the more constant the Lorenz number is with T. The Lorenz number does change as T increases, and it decreases in magnitude, which is a good sign for thermoelectric calculations, since the Lorenz number enters in the denominator of the figure of merit (at least in the bulk).

Having completed an examination of the charge and thermal resistances of the device, we now move onto the thermoelectric effects. In the bulk, the Peltier effect and the Seebeck effect are closely related due to Onsager's reciprocal relations. But in a nanostructure, we saw when we derived the inhomogeneous analogs of the Peltier and Seebeck effects that they are expected to be different, even though the transport coefficients L_{12} and L_{21} do satisfy Onsager reciprocity microscopically. This arose because the Seebeck effect was weighted by a matrix related to the thermal resistance, while

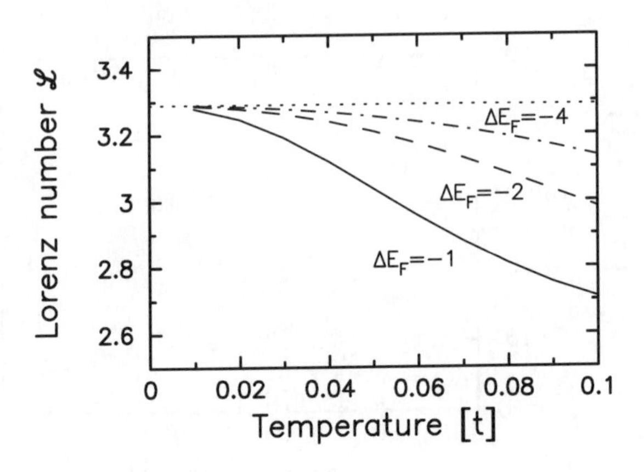

Fig. 6.6 Device Lorenz number $e^2R_n/R_{th}k_B^2T$ plotted for the same device as in the preceeding two figures. Note how it approaches the universal metal result of $\pi^2/3$ at low temperature even though it is made out of heterogeneous multilayers. This implies that the changes to the charge resistance and the thermal resistance are identical due to the inhomogeneous multilayered structure, which is what one would naively guess by adding the respective resistances in series, if each plane satisfies the Wiedemann-Franz law. This naive picture seems to hold for the full quantum system.

the Peltier effect was not. Furthermore, there are many different "Peltier effects" that can be examined because the heat current is not conserved for a Peltier-effect experiment, so we now detail the different ones that can be considered. First, one can examine the ratio of the heat current carried through the device to the charge current carried through the device. Because the heat current vanishes as we move deep into the metallic leads (remember the leads are particle-hole symmetric so they have no net Peltier effect), we find the local Peltier coefficient decreases the farther away from the interface that we move. We define one Peltier coefficient to include the average heat transport through the entire barrier plus the first plane in the lead to the left and to the right. The second Peltier coefficient we can examine is to determine the total change in the heat current as we move from the left to the right of the device. Unfortunately, this change turns out to be identically zero, probably because the device has mirror symmetry about the center of the barrier, but we have not proved this result analytically. Because of the vanishing of this difference, we instead examine the ratio of the local heat current at the central plane of the barrier to the charge current. These are the two different Peltier coefficients that we will calculate for our devices. The Seebeck effect is much more straightforward, as it results from an unambiguous experiment. From a numerical standpoint it

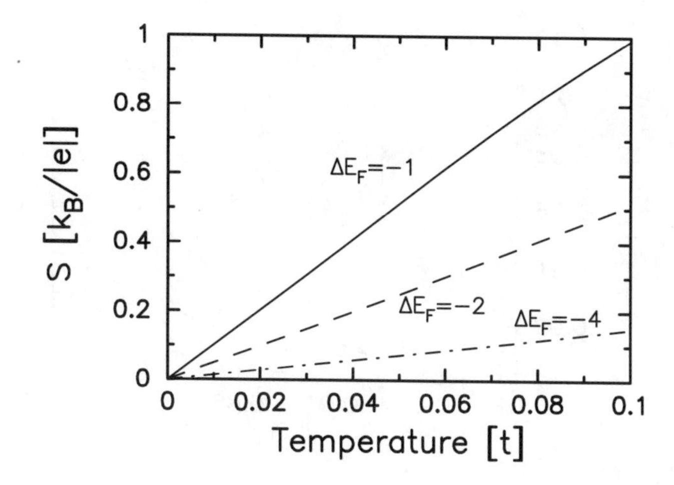

Fig. 6.7 Theoretical prediction for the Seebeck effect on the same device as in the preceeding figures. The Seebeck coefficient continues to increase in magnitude over this region of temperature in a linear fashion. It is not large enough to likely be important for thermoelectric applications for these parameters, but it is amazing that one can get such a large Seebeck effect when constructing a heterostructure out of two materials which both have exactly zero Seebeck coefficient.

is also well-defined, because the Seebeck coefficient approaches a constant as we expand the matrices to incorporate the entire barrier, just like the resistance and thermal-resistance calculations.

We start by showing the Seebeck-effect results, because they are well defined. We plot the numerical results for the Seebeck coefficient of the device, measured by determining the ratio of the voltage difference to the temperature difference of the junction in an open circuit. These results are shown in Fig. 6.7. Interestingly, the Seebeck effect is largest for the metallic system that is closest to the insulating phase. Since it must vanish when $\Delta E_F = 0$, it must eventually drop as we approach the insulator, and hence there should be a region where the Seebeck effect has a maximal response. Similar to what we saw in the bulk case, this large Seebeck effect is arising from a "fine-tuning" of parameters.

Next we examine the two Peltier effects discussed above in Fig. 6.8. The thick curves look at the average heat current transmitted through the device, averaging over the barrier planes plus the first metallic plane on either side, and the thin curves are the local Peltier coefficient at the central plane of the barrier. If we recall that the relation between the Peltier coefficient and the Seebeck coefficient in the bulk is $\Pi = ST$, we

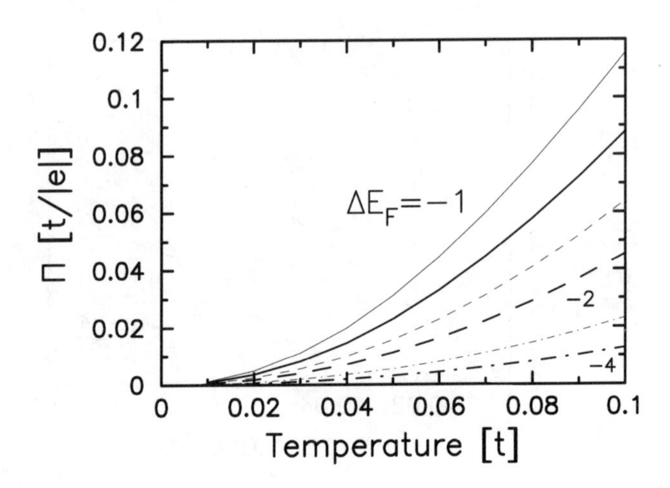

Fig. 6.8 Theoretical prediction for the Peltier effect on the same device as in the preceeding figures. The Peltier coefficient satisfies $\Pi = ST$ in the bulk, and appears to be close to that here. The thick curves are for the average Peltier effect for the heat current carried through the device (averaging over the planes of the barrier and the first metallic plane on either side) and the thin curves are for the local Peltier effect at the central plane of the barrier.

see that the magnitude of either Peltier coefficient is similar to the Seebeck coefficient, but both are quantitatively different. It appears that the heat current in the center of the barrier is somewhat larger than the average heat current for this device. Such a result is to be expected since the heat current vanishes far from the barrier.

Note that both the Seebeck and Peltier effects are positive, indicating hole-like transport, which is to be expected due to the way we shifted the band of the barrier—the barrier was shifted into the lower Hubbard band, so the carriers are holes.

In summary, we have shown some remarkable thermal-transport effects for inhomogeneously doped Mott insulators. Starting with materials that have exactly zero thermoelectric effect, we find that if there is a chemical potential mismatch between the two materials, then if we place them into a multilayered nanostructure, the electronic charge reconstruction will inhomogeneously dope the Mott insulator into a metal, and create a large thermoelectric response due to the breaking of particle-hole symmetry. This effect is an interesting one that can be employed to try to enhance thermoelectric effects by using multilayered heterostructures. Since the heterostructure has many boundaries between the different materials, it should act to reduce the phonon contribution of the thermal conductivity as well. Hence this approach may have promise for use in creating new thermoelectric devices from materials that may not have a large intrinsic Seebeck coefficient. These conclusions are in the same spirit as the work of Rontani and Sham [Rontani and Sham (2001)], but here the focus is on Mott insulators rather than excitonic insulators. It will be interesting to consider other cases as well, such as what happens when a doped Mott insulator is used as a barrier and the electronic charge reconstruction modifies the electron filling to make it insulating.

Chapter 7

Future Directions

While we spent some time in this book developing a formalism for arbitrary impurity solvers, all of the numerical results for multilayered nanostructures came from solutions of the Falicov-Kimball model. There is no inherent reason why this has to be so, the field is sufficiently new, however, that more sophisticated solvers like the numerical renormalization group, have not yet been applied to these inhomogeneous DMFT problems. While significant work can be done with a working inhomogeneous DMFT-NRG code, we will not discuss this straightforward generalization further, but instead will concentrate on three new directions where theoretical work can go in the future: (i) the field of spintronics; (ii) real-materials modeling; and (iii) nonequilibrium and nonlinear field effects.

7.1 Spintronics Devices

The field of spintronics was born with the proposal by Datta and Das for a spin-based transistor [Datta and Das (1990)]. The basic idea of spintronics arises from the pursuit to find alternative ways to create digital electronics circuits with electrons. Currently, most electronics devices involve manipulating the electronic charge in the device. For example, a *pn* junction is essentially a one-way switch, which lets current flow in one direction, but not in another. The reason why so many devices rely on manipulating the charge is that it is easy to do so by simply applying an electric field to the device (as in a conventional transistor). But the electron also has a spin degree of freedom. Usually this spin degree of freedom is not exploited for any kind of digital electronics calculations, although it is heavily used in magnetic-field sensors, which are employed in hard-drive storage, but also are being used more frequently for sensors of position (by attaching a magnet to an object, and detecting the field as the object is moved close to, or far away from the sensor). The use in hard-drive technology is based on the giant magnetoresistance effect, where the resistance of a multilayered metal structure (made of magnetic and nonmagnetic metals), changes depending on the direction of an external magnetic field. A normal metal is sandwiched between two ferromagnetic metals which can have their magnetizations either aligned or anti-aligned with each other. Current is passed through the nonmagnetic layers, parallel to the ferromagnetic planes. If the two ferromagnetic layers are aligned in the same way, then as an electron moves through the nonmagnetic layer, and bounces off each ferromagnetic layer, it will be in a high resistance state if the magnetizations are antiparallel, because each electron will have spin-flip scattering at one interface, and it will be in a low resistance state if they are parallel, because one spin species will conduct more easily than the other (since it has no spin-flip scattering), and will shunt the current flow. Hence, a simple measurement of the resistance can determine whether the two magnetizations are parallel or antiparallel. If one of the magnetizations can be pinned to always point in one direction, and the other is free to move in response to an external field, then the device can be employed as a magnetic-field sensor. Indeed, this is how all read heads on hard disk drives currently operate.

This use of spin-dependent properties for magnetic-field sensing is a very useful tool, but it is not a new form of computation. Recently there has been a significant effort to try to find spin-based analogs of the charge transistor, or to find ways to use spins for alternative forms of computation, such as quantum computing. Since it is difficult to shield magnetic fields, it will be much more useful if the spin dynamics can be controlled by electric fields rather than magnetic fields. One promising way that this can occur is via something called the Rashba effect [Rashba (1960)], which involves a coupling, via the relativistic spin-orbit interaction, between an electric field and the spins of the electrons in materials that don't have inversion symmetry. The hope was that this effect could be used to generate spin-polarized currents, especially in semiconductors. Other ideas are based on half-metallic ferromagnets, which are not like conventional ferromagnets (such as iron or chromium) that have both spin species present at the Fermi energy, with the total number of up spin and down spin electrons being different. In conventional ferromagnets, the DOS at the Fermi energy can be dominated by either the majority or the minority spin species, because the magnetization arises from all the electrons not just those at the Fermi energy and the band structures for the two different spins are different.

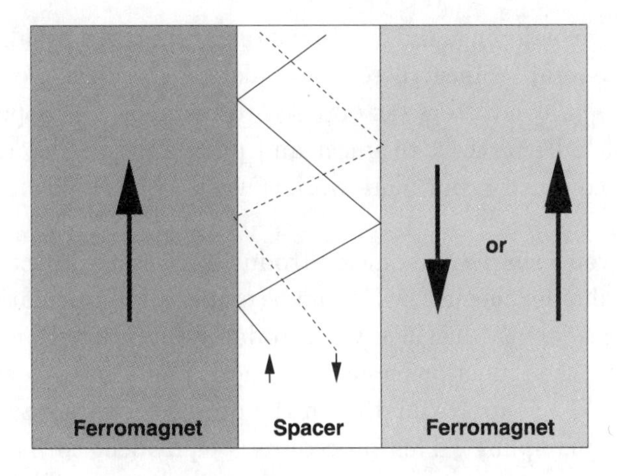

Fig. 7.1 Schematic picture of the GMR spin valve. The structure has a pinned ferromagnet on the left, a nonmagnetic spacer metal in the middle, and a free ferromagnet on the right. Current is passed through the device parallel to the planes. Depending on the configuration of the free ferromagnet, the electrons will see more or less spin-flip scattering. Each spin feels spin-flip scattering at every other reflection off the interfaces when the magnets are anti-aligned. The majority spin feels no spin-flip scattering, while the minority spin feels spin-flip scattering at every reflection, when the magnets are aligned. The difference in resistance, which is typically on the order of 1-10%, can be used to sense an external magnetic field (which causes the free ferromagnet to change direction).

Instead, in a half-metallic ferromagnet, we can theoretically have a situation where the DOS of the minority carriers is exactly zero at the Fermi energy, and the minority spin band has a gap in it. Chromium dioxide is one candidate for such a half-metallic ferromagnet, but to date no such material has been completely verified to exist (other candidates include so-called half-Heusler alloys). Using such a half-metallic ferromagnetic, one can create a spin polarized current at low enough T, which can be used as either a spin filter, or as a spin injector. Unfortunately, when one tries to inject such current into semiconductors, it has turned out to be quite difficult to maintain the spin polarization for long periods of time (long enough to perform significant computation), and this has proved to be one of the biggest challenges in the field. In addition, the issue of finding a way to amplify the spin current, in the same way that a transistor can amplify the charge current, is needed, but has not yet been discovered.

Other promising paths for spintronics-based work include examining dilute magnetic semiconductors, which usually involve II-VI semiconductors with small concentrations of Mn atoms, that create ferromagnetic materials over some range of doping, and over some range of T (but usually well below room temperature). Since these materials are so much more similar to conventional semiconductors (as opposed to metals), it is hoped that they could provide a better way to inject and maintain spin-polarized current within a semiconductor, but more work is needed before this idea can come to fruition.

Having given a simple and quick introduction to the subject, we now will discuss how inhomogeneous DMFT can be employed in calculations relevant for spintronic systems. Just like we introduced an ordered phase when we discussed Josephson junctions, a ferromagnetic phase can be described by a simple Hartree-Fock approximation in the presence of a nonzero ferromagnetic exchange coupling. This procedure is equivalent to Slater's picture of itinerant ferromagnetism. Next, one can add different barrier layers, or construct other multilayered systems, by adding materials near the Mott metal-insulator transition, or semiconductors, or nonmagnetic metals, and then the transport (for current perpendicular to the planes) can be calculated using the Kubo approach in real space. In this way, a wide range of different spintronic systems can be examined with the DMFT approach. If different materials are placed next to each other, the electronic charge and spin reconstruction at the interface will need to be determined—both the charge and the spin can be screened at the interface.

Since spin-orbit coupling plays such an important role in these systems, it is likely that one would need to modify the formalism to take into account the spin-orbit interaction (and other interactions like the Rashba coupling). This complicates the formalism, because we need to treat the electronic Green's functions as 2×2 matrices, with off-diagonal terms that represent the spin-flip processes. Treatment of such problems is more complicated, but the formalism should be able to be developed in analogy to how we developed the formalism for the superconducting state, which also involved a 2×2 matrix formalism. Doing so is beyond the scope of this book.

Finally, it is likely true that the character of the orbitals, be they s, p, d, or f-orbitals, may be important in determining the magnetic properties, since degeneracy, and Hund's rules lie at the heart of ferromagnetic phenomena. This could require the theories for spintronics to involve more complex multiband models that can take into account the change in character of the orbitals from one material to another. A general description for how to do this follows in the next section.

7.2 Multiband Models for Real Materials

Another new direction to proceed in is to consider multiband models of real materials and add electron correlations as needed to the formalism. This is within the spirit of the DFT+DMFT approach described in Chapter 1, but it needs to be generalized to the inhomogeneous problem. One strategy for proceeding is to first work out a DFT calculation for all materials that will be put into the multilayered nanostructure. Next, a tight-binding analysis is performed, to try to fit the DFT band structure to a tight-binding bandstructure; it is important to keep just nearest-neighbor tight-binding parameters in the z-direction (otherwise the quantum-zipper algorithm cannot be employed). The tight-binding fitting procedure is not completely determined. Choices will need to be made to try to fit the bandwidths, or the shape of the Fermi surface, or the Fermi velocities, with the tight-binding procedure, since all will not be able to be fit when the tight-binding approach is truncated to a small number of neighbors. How to choose to best make the fit will depend on the physical properties that need to be modeled (for transport, probably getting the Fermi surface shape and the Fermi velocities correct is more important than the bandwidths). There are two possible results that can come out of such an analysis. Either the tight-binding bandstructure is separable [in the form $\epsilon(\mathbf{k}_x, \mathbf{k}_y) + \bar{\epsilon}(\mathbf{k}_z)$] or it is not. If it is separable, then one can replace the summation over the planar momentum by an integral over the respective two-dimensional DOS. If it is not separable, this means that the hopping integral in the z-direction depends on either $\mathbf{k}_x$, $\mathbf{k}_y$, or both. Then the quantum zipper algorithm can still be used, but now one will need to perform a two-dimensional integral over $\mathbf{k}_x$ and $\mathbf{k}_y$ separately; this can make the algorithm require significantly more computer resources than if one can use just a single integral over the two-dimensional DOS. The issue may even be more subtle. The two-dimensional DOS has a logarithmically divergent van Hove singularity in it. If we work directly with the DOS, then we can properly handle the singularity by performing a change of variables in the quadrature. If, instead, we are summing over the two-dimensional planar Brillouin zone, then the singularity might either be masked, or it might require many integration points to properly produce it. This can further complicate the numerical implementation of the algorithm.

Next, a DFT calculation should be performed of the interface between the two materials that are being grown on top of each other. When the interface is fully relaxed, and the charge redistribution from the electronic charge reconstruction has taken place, one can try to extract the modification of the hopping parameters for atoms close to the interface, and one can extract the charge screening length, and the mismatch of the chemical potentials. All of these quantities will be important inputs into the DMFT transport code.

The other issue that needs to be discussed is the issue of what models will be used, and what impurity solvers will be employed. If one wants to investigate properties of disordered charge scatterers on transport (which is challenging to implement within a DFT perspective), one can include such effects by using an appropriate Falicov-Kimball-like model which will properly handle annealed disorder. If one wants to investigate other types of interactions, like those found in the Hubbard, Kondo, or periodic Anderson models, then it is unlikely that an NRG approach can be used, because the NRG cannot easily handle multiband situations, since one needs to maintain too many states to be practical at each iteration, and the numerical accuracy cannot be maintained. Instead, a combination of quantum Monte Carlo plus numerical analytic continuation would need to be employed, which has its own share of numerical problems, that can make such calculations very time-consuming, and possibly beyond the scope of today's computers.

To illustrate how such a calculation might proceed, we discuss the problem of making c-axis Josephson junctions out of MgB_2 superconductors. MgB_2 was discovered in 2001 as a medium temperature s-wave superconductor with a transition temperature that satisfies $T_c \approx 39$ K [Nagamatsu, *et al.* (2001)]. The material forms a layered hexagonal lattice structure of alternating Mg and B planes (with twice as many B atoms in a plane as the Mg atoms). The B atom states are the states close to the Fermi energy, so the electrons sit in s and p hybridized states. These electrons either move within the planes, or between the planes, and the hopping is much larger within the planes. Similarly, the superconductivity has two different gaps: a large gap that is on the planar states and a small gap that involves electrons moving between planes. One might be able to view the small gap as a kind of a "bulk" proximity effect, where the planar electrons have a strong enough pairing to become superconducting at a high temperature, and they make the longitudinal electrons become superconducting via an "internal" proximity effect. Since the two gaps have a large difference in size (more than a factor of two), one important question is which gap will govern the Josephson properties—the large gap or the small one? As disorder is added to the system, the superconductivity is reduced, and the

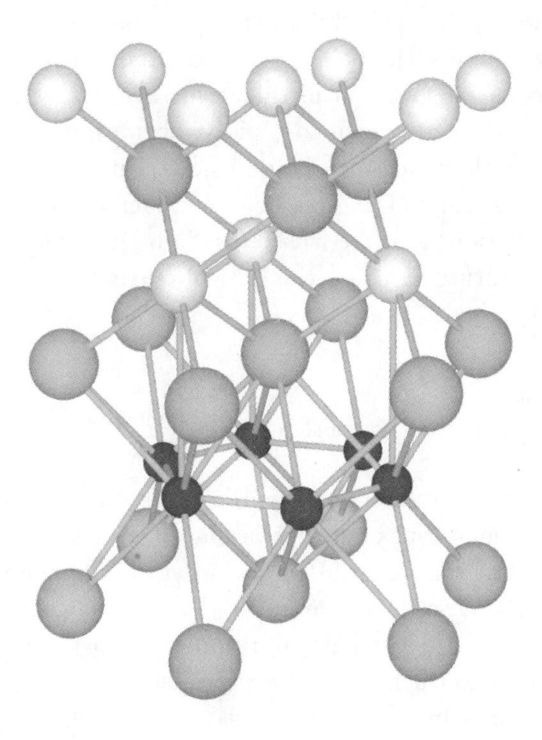

Fig. 7.2 Atomic positions of the magnesium (black), boron (dark gray), and oxygen (light gray) atoms at a MgB_2–MgO interface. The magnesium and boron atoms lie on hexagonal lattices, and the MgO, which has a face-centered-cubic structure, is oriented in the (111) direction, so it also stacks hexagonal planes on top. *Figure reprinted with permission from* [Schnell (unpublished)].

two gaps come closer in magnitude to each other, and once the T_c drops below about 25 K, the gap is uniform throughout. If the small gap dominates the supercurrent transport in a Josephson junction, then is it better to disorder the system a bit to have a uniform-sized gap, or is the purest system the best superconductor to use for the leads? Finally, there is the question of what barrier will yield optimal performance. Aluminum oxide is a candidate that is an obvious one to try, because it works so well for so many different systems, but it may be dangerous here, because aluminum impurities cause a substantial reduction to the T_c of MgB_2, and introducing Al will likely cause the interfacial region to have many Al defects. Magnesium oxide is an interesting candidate, because it is well lattice matched to MgB_2 when grown in the (111) direction (see Fig. 7.2). Unfortunately, the bandstructure in the (111) direction is not separable, so calculations on such a device will necessarily be complicated. Another potentially interesting barrier to consider is boron nitride, which requires no oxygen during growth; oxygen has been known to rapidly form a native oxide barrier on MgB_2 during growth, which may be detrimental if one is trying to grow a different barrier. Finally, TiB_2 has been considered for a SNS type junction, but such junctions are rarely used for digital electronics circuits unless run at low temperature, because the critical current decreases too rapidly with temperature.

7.3 Nonequilibrium Properties

Another interesting new direction involves studying nonequilibrium properties. When a a material or device is placed in a large electric field (if one volt is placed over a material on the order of one nanometer thick, an electric field of 10^9 V/m is applied over the material, and this is a huge field) one needs to go beyond linear response theory to determine what is happening. Hence, when one studies properties of multilayered nanostructures, it becomes important to consider nonlinear response effects of the large electric fields that will be applied to the systems. There is no simple way to generalize the Kubo approach to consider nonlinear effects, instead one needs to examine the nonequilibrium version of the many-body problem, whose formalism was worked out independently by Kadanoff and Baym [Kadanoff and Baym (1962)] and by Keldysh [Keldysh (1964)].

Fortunately, these approaches provide an exact means to calculate nonequilibrium and nonlinear response effects, because the exact nonequilibrium Green's functions are known for the noninteracting system, and there is a complex formalism that can add the many-body interactions to solve the full problem. The numerics associated with trying to carry this out is quite challenging, and only now do we have sufficient computer power to be able to solve such problems. The basic idea that underlies the formalism is that when a system is driven by an external field, we do not know what state it evolves to at long times after the field has been turned on. In the equilibrium case, we always know we have an equilibrium distribution, but in the nonequilibrium case, the system usually evolves to some form of steady state, which can have current flowing, and is impossible to determine *a priori*. As a result, the theory becomes well defined if we imagine evolving time forward to some maximal time, and then evolving it backward again, back to the initial time we started from. Since the initial system is in an equilibrium distribution, we can properly determine the equations of motion for the Green's functions.

We now have two different Green's functions that need to be determined—one is the familiar retarded Green's function, which provides information about the different quantum states that are available in the system. The other is the so-called lesser Green's function, which tells us how the electrons occupy those states (in equilibrium, the electronic states are occupied according to a Fermi-Dirac distribution, so we need only consider the properties of the states themselves, since the occupation is already known). These two Green's functions are coupled together, and furthermore, we need to work with two-time Green's functions, because we no longer have time-translation invariance (after all, we turn on an external field at some time and see how the system evolves). A moment's reflection shows that this is indeed a very complex situation, because we will be analyzing both the steady-state contributions and the transient contributions to the response of the system, and quite complex behavior can arise.

Nevertheless, DMFT is well situated to be able to solve these nonequilibrium problems. The same arguments that led to the conclusion that the self-energy was local for the equilibrium problem also hold for the nonequilibrium problem, so we can assume the self-energy has no momentum dependence. Recall that the impurity problem involves a single site in the presence of a time-dependent field. This field can depend on two times without causing any serious problems to the formalism for solving the problem, although it can no longer be diagonalized by making a Fourier transformation to frequency space. Instead, the continuous matrix operator in time must be diagonalized, which is carried out by first discretizing the time axis with a fixed time step, and then performing matrix operations with the resulting discrete matrices. If the limit where the step size goes to zero can be taken, then the result should yield the continuous operator solution. The difficult part of the DMFT loop is actually the step where we sum over all momenta. In equilibrium, the summation can be replaced by a single integral over the noninteracting DOS. In the nonequilibrium case, the summation involves at least a double integral over two energies and the integrand is now a matrix of two time variables instead of just a scalar function.

Fortunately there are a number of important sum rules that the Green's functions satisfy, and using these sum rules, one can quantify the accuracy of the calculations, and verify whether scaling approaches are properly scaling to the zero step-size limit [Turkowski and Freericks (2006)]. To date there has been much progress on solving the bulk nonequilibrium response with numerical algorithms that require huge amounts of computer time. The results are promising, and show how the Bloch oscillations are attenuated as scattering increases, and how their character changes completely as the system moves through the Mott transition from a metal to an insulator. Generalizing this approach from the bulk to an inhomogeneous nanostructure is a problem for the future. Solving this problem will allow one to determine the voltage profile through the device as a function of the current flowing, and will allow one to calculate the current-voltage characteristic. It is likely that one will need to take into account dielectric effects from the ionic cores of the material, because those cores have bound electrons that cannot contribute to the current, but they can be polarized, and contribute to the dielectric function, thereby modifying the capacitance of the device. The capacitance must play a role in determining the shape of the current-voltage characteristic. This a rich problem that should be able to provide a multitude of useful results.

7.4 Summary

This book ends with a short summary section. The material presented here should enable readers who master the exercises to begin exploring research problems of transport in strongly correlated multilayered nanostructures. In this sense, we have provided a set of tools that can make one efficient in performing calculations and analyzing new systems. As hinted at in the previous three sections, there is much new work that can be done, and we hope that this book facilitates entry into this research field. Dynamical mean-field theory is still rapidly growing and developing; hopefully the readers will aid in its future evolution.

Appendix A

Problems

These series of problems accompany the text and are designed to help readers master the material. Those who successfully complete these problems will have a set of tools that will allow them to carry out calculations for strongly correlated multilayered nanostructures at the research level.

A.1 Jellium model

Using the result of the four-fermion expectation value in Eq. (2.25) gives an expression for the expectation value of the potential energy

$$\langle \psi^0_{\rm gs}|\hat{V}|\psi^0_{\rm gs}\rangle = -\frac{1}{2}2\frac{V^2}{(2\pi)^6}\int_{k<k_f} d^3k \int_{k'<k_f} d^3k' \frac{4\pi e^2}{V|\mathbf{k}-\mathbf{k}'|^2}; \tag{A.1}$$

the factor of 2 is from spin and the factor of $V^2/(2\pi)^6$ comes from the momentum integral measure. Perform the integration over k' first, using the direction of $\mathbf{k}$ as the z-axis (*i.e.*, $\mathbf{k}\cdot\mathbf{k}' = kk'\cos\theta$, or $|\mathbf{k}-\mathbf{k}'|^2 = k^2 - 2kk'\cos\theta + k'^2$). Show that the integral of $1/|\mathbf{k}-\mathbf{k}'|^2$ over k' yields

$$\frac{2\pi}{k}\int_0^{k_F} dk' k' \ln\left|\frac{k+k'}{k-k'}\right| = 2\pi\left\{\frac{k_f^2-k^2}{2k}\ln\left|\frac{k_f+k}{k_f-k}\right| + k_f\right\}. \tag{A.2}$$

Then perform the integral over k to find

$$\langle \psi^0_{\rm gs}|\hat{V}|\psi^0_{\rm gs}\rangle = e^2\frac{k_f^4}{4\pi^3}V. \tag{A.3}$$

Combining this with the average kinetic energy gives

$$E_{\rm jellium\ HF} = \left[\frac{\hbar^2}{2m}\frac{k_F^5}{5\pi^2} - e^2\frac{k_f^4}{4\pi^3}\right]V. \tag{A.4}$$

Use the fact that $\rho_e = k_F^3/3\pi^2$, and the definition of a Rydberg (Ry$= e^2/2a_0$ with $a_0 = \hbar^2/me^2$ the Bohr radius) to show that

$$\frac{E_{\text{jellium HF}}}{N} = \left[\frac{3}{5}(k_f a_0)^2 - \frac{3}{2\pi}k_f a_0\right] \text{Ry}. \tag{A.5}$$

Finally, use the definition of r_s to get

$$\frac{E_{\text{jellium HF}}}{N} = \left[\frac{3}{5}\left(\frac{9\pi}{4}\right)^{\frac{2}{3}} \frac{1}{r_s^2} - \frac{3}{2\pi}\left(\frac{9\pi}{4}\right)^{\frac{1}{3}} \frac{1}{r_s}\right] \text{Ry}, \tag{A.6}$$

and plug in the numbers to verify Eq. (2.27).

A.2 Density of states for the hypercubic lattice in 1, 2, 3, and ∞ dimensions

(a) Calculate the density of states on a hypercubic lattice with nearest-neighbor hopping $t = t^*/2\sqrt{d}$ using t^* as the energy unit and taking the limit $d \to \infty$. Use the following ideas (or your own techniques):

$$\rho_{d\to\infty}(\epsilon) = \lim_{d\to\infty} \prod_{i=1}^{d} \left(\int_{-\pi/a}^{\pi/a} \frac{dk_i}{2\pi}\right) \delta(\epsilon - \epsilon_k) \,, \tag{A.7}$$

with the bandstructure

$$\epsilon_k = -2t\sum_{i=1}^{d} \cos(k_i a) = -\frac{t^*}{\sqrt{d}}\sum_{i=1}^{d} \cos(k_i a) \,. \tag{A.8}$$

Write the delta function as

$$\delta(\epsilon - \epsilon_k) = \frac{1}{2\pi t^*}\int_{-\infty}^{\infty} d\lambda\, e^{-i\lambda(\epsilon-\epsilon_k)/t^*} \,, \tag{A.9}$$

and substitute into the above formula. Now the integral over each k_i can be carried out. Note that you can expand in small quantities (like $1/d$) and that

$$\begin{aligned}\lim_{d\to\infty}\left(1+\frac{\alpha}{d}+\frac{\beta}{d^2}\right) &= \lim_{d\to\infty}\left(\exp\left[\frac{\alpha}{d}+\frac{\beta-\frac{1}{2}\alpha^2}{d^2}+...\right]\right)^d ,\\ &\approx \lim_{d\to\infty}\exp\left[\alpha+\frac{\beta-\frac{1}{2}\alpha^2}{d}\right] \approx \exp[\alpha] \,. \end{aligned} \tag{A.10}$$

Ans: $\rho_\infty(\epsilon) = \exp(-\epsilon^2/t^{*2})/\sqrt{\pi}t^*\Omega$, with $\Omega = \lim_{d\to\infty} a^d$ the unit cell volume.

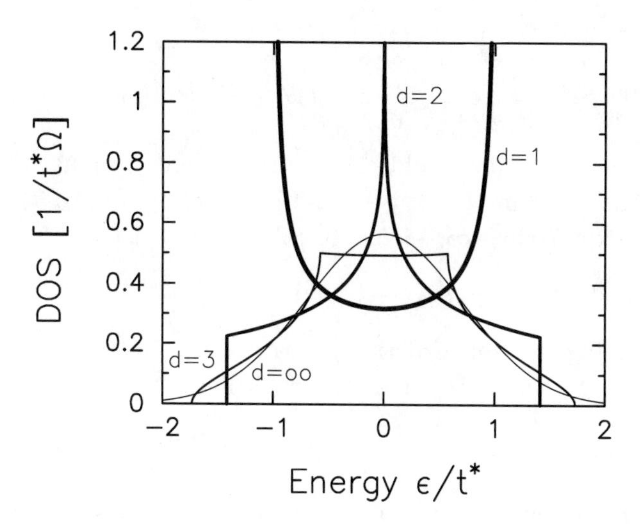

Fig. A.1 Scaled DOS for d = 1, 2, 3, and ∞. Note how the DOS looks more like the infinite-dimensional result as d increases.

(b) Using units where the lattice constant satisfies $a = 1$, and where the hopping satisfies $t = t^*/2\sqrt{d}$ for each dimension d, plot the density of states for $d = 1$, $d = 2$, $d = 3$, and $d = \infty$ on the same plot (see [Economou (1983)] for assistance in calculating the DOS for different d). Comment on the behavior as a function of d.

Ans: the DOS are plotted in Fig. A.1.

(c) Using the general formula for the density of states of a coordination Z Bethe lattice [as derived in Eq. (2.71)], take the limit where the coordination number approaches infinity $Z \to \infty$, but the hopping satisfies $t = t^*/\sqrt{Z}$. Describe the similarities and differences of the infinite-coordination-number Bethe lattice DOS with the hypercubic lattice DOS.

Ans: $\rho_{\rm Bethe}(\epsilon) = \sqrt{4t^{*2} - \epsilon^2}/2\pi t^*$; the Bethe lattice has a finite bandwidth while the hypercubic lattice has an infinite bandwidth.

A.3 Noninteracting electron in a time-dependent potential

Begin with the impurity Hamiltonian

$$\mathcal{H} = -\mu c^\dagger c, \tag{A.11}$$

for a spinless electron ($\{c^\dagger, c\}_+ = 1$). Let

$$S(U) = \exp\left[-i\int dt\,\theta(t)\theta(t_0 - t)Uc^\dagger(t)c(t)\right], \tag{A.12}$$

be a time-dependent phase added as a time-ordered product to the definition of the Green's function (take $t_0 > 0$) with $c(t) = \exp(i\mathcal{H}t)c\exp(-i\mathcal{H}t)$ and $c^\dagger(t) = \exp(i\mathcal{H}t)c^\dagger\exp(-i\mathcal{H}t)$. In other words, define

$$G(t,t') = -i\langle \mathcal{T}_t c(t)c^\dagger(t')S(U)\rangle, \tag{A.13}$$

with $\mathcal{T}_t$ denoting the time-ordering operator.
(a) Calculate the generalized partition function

$$\mathcal{Z}(U) = \mathrm{Tr}\left\{\mathcal{T}_t e^{-i\int dt\,\theta(t)\theta(t_0-t)Uc^\dagger(t)c(t)}e^{-\beta\mathcal{H}}\right\}, \tag{A.14}$$

directly from the anticommutation relations of the c's and $c^\dagger$'s and the time-ordering operator. *Hint*: there are only two Fermionic states in the trace: the vacuum state $|0\rangle$ which is annihilated by c and the one-electron state $|1\rangle = c^\dagger|0\rangle$ which is annihilated by $c^\dagger$.
Ans: $\mathcal{Z} = 1 + \exp(\beta\mu - it_0U)$.
(b) Evaluate all twelve cases for the Green's function $G(t,t')$ defined above for $t < t' < 0$, $t < 0 < t' < t_0$, $t < 0 < t_0 < t'$, etc. Once again, use only the operator definitions and the explicit form of the Green's function to perform your calculations; be sure to write out the time ordering explicitly.
Ans: for $0 < t' < t < t_0$ we have $G(t,t') = -i\exp[i(\mu - U)(t-t')]/[1 + \exp(\beta\mu - it_0U)]$; the results are similar for other cases.

A.4 Relation between imaginary-time summations and real-axis integrals

Consider the average kinetic energy

$$K = \langle\hat{T}\rangle = \sum_{\mathbf{k}}\epsilon_{\mathbf{k}}\langle c^\dagger_{\mathbf{k}}c_{\mathbf{k}}\rangle, \tag{A.15}$$

expressed as the expectation value of the momentum creation and annihilation operators (with ϵ_k the bandstructure).
(a) Using the definitions of the Matsubara Green's functions, show that the average kinetic energy satisfies

$$K = VT\sum_n\int d\epsilon\,\epsilon\rho_\infty(\epsilon)G(\epsilon, i\omega_n), \tag{A.16}$$

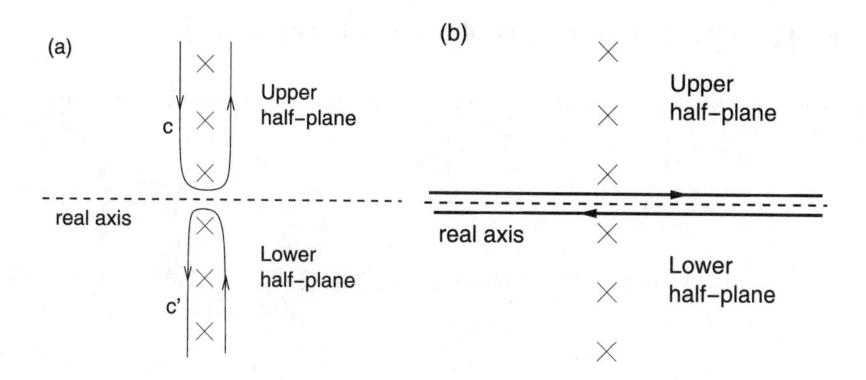

Fig. A.2 Contours used for relating Matsubara summations to integrals over the real axis. (a) Contours c and c' that give the Matsubara summation. (b) Contours that the contours in panel (a) can be deformed into that yield the integral over the real axis.

within dynamical mean-field theory with ρ_∞ the noninteracting DOS for the given lattice.

(b) Using the fact that the Fermi-Dirac distribution $f(\omega) = 1/[1+\exp(\beta\omega)]$ has simple poles at the Fermionic Matsubara frequencies with residue $-T$, show that the Matsubara frequency summation can be written as a contour integral over the (two-piece) contour in Fig. A.2 (a):

$$K = -\frac{V}{2\pi i}\int_c dz \int d\epsilon\, \epsilon\rho_\infty(\epsilon)G^R(\epsilon,\omega) - \frac{V}{2\pi i}\int_{c'} dz \int d\epsilon\, \epsilon\rho_\infty(\epsilon)G^A(\epsilon,\omega)\,, \tag{A.17}$$

where R and A refer to the retarded and advanced Green's functions. Since the integrands are respectively analytic in the upper and lower half planes, the contours can be deformed to the final result [see Fig. A.2 (b)]

$$K = -\frac{V}{\pi}\int_{-\infty}^{\infty} d\omega \int d\epsilon\, \epsilon\rho_\infty(\epsilon)\mathrm{Im}G(\epsilon,\omega)\,. \tag{A.18}$$

(c) Using the fact that the self-energy is local, perform the integral over ϵ to yield

$$K = -\frac{V}{\pi}\int_{-\infty}^{\infty} d\omega\, \mathrm{Im}\left\{[\omega+\mu-\Sigma(\omega)]G(\omega)\right\}\,. \tag{A.19}$$

Hence the average kinetic energy can be determined either by a summation over Matsubara frequencies, or by a one-dimensional integral on the real axis. Note that the same trick used on the real axis in part (c), to replace the integral over ϵ by local quantities, can be used in the imaginary-axis formula to get rid of the integration over ϵ.

A.5 The Green's functions of a local Fermi liquid

In a local Fermi-liquid system, one can write the retarded Green's function as

$$G^R(k,\omega) = \frac{Z}{\omega - \frac{\epsilon_k - \mu}{Z} + i\gamma} + G_{\text{inc}}(k,\omega)\,, \tag{A.20}$$

with $0 \le Z \le 1$ and γ proportional to T^2 ($\gamma = CT^2$, $C > 0$, for small ω). Suppose we are at a low enough temperature that we can neglect the incoherent piece of the Green's function. Determine the local density of states near the Fermi level for finite, but small T (you can express your answer in terms of the noninteracting local Green's function). Take the limit of $T \to 0$. At $\omega = 0$, you should find that the DOS for the interacting system is *the same* as the DOS for the noninteracting system. This is a result of Luttinger's theorem for a local self energy.
Ans: $A(\omega) = -\text{Im}G_{\text{non}}^{\text{local}}(\frac{\omega}{Z} + \mu + i\frac{CT}{Z})/\pi$.

A.6 Rigid-band approximation to the Falicov-Kimball model

Suppose we have a local DOS for spin-1/2 electrons that satisfies

$$A_\sigma(\omega) = \sum_k A_\sigma(k,\omega) = (1 - w_1)\rho_{\text{Bethe}}(\omega) + w_1\rho_{\text{Bethe}}(\omega - U)\,, \tag{A.21}$$

with $w_1(T) = 1 + \exp(-\beta E_f)$. Choose the total filling of the electrons to satisfy $\rho_e(T) = 1 - w_1(T)$, $U = t^*$, and $E_f = 0.1t^*$.
(a) Plot the local DOS $A(\omega)$ (per spin; we drop the spin index since the results are independent of σ in the paramagnetic phase), with the chemical potential located at $\omega = 0$, for $T = 0$, 0.1, and 1. Note that you must solve numerically for $\mu(T)$ and you must use Eq. (A.24) to determine the conduction electron filling. *Hint*: you will need a good one dimensional root solver [Brent (1973)] to determine the chemical potential.

(b) Write a computer program to calculate

$$G_{\text{loc}}(i\omega_n) = \int d\omega \frac{A(\omega)}{i\omega_n - \omega}\,, \tag{A.22}$$

for the Fermionic Matsubara frequencies $i\omega_n = i\pi T(2n+1)$ using the local DOS determined in part (a). Once you have generated $G_{\text{loc}}(i\omega_n)$, use those

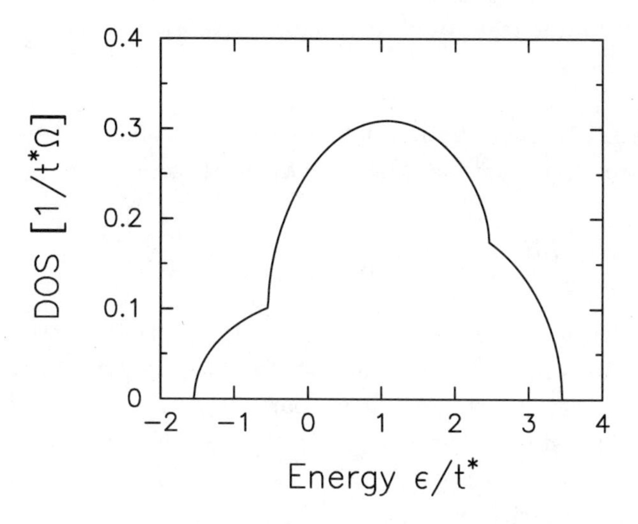

Fig. A.3 DOS for the rigid-band approximation when $T = 0.1$.

values to calculate the filling of the electrons from the Matsubara- frequency formula (the factor of 2 is from spin)

$$\rho_e = 2T \sum_n G_{\rm loc}(i\omega_n)\,, \tag{A.23}$$

and compare those results to the alternate real-axis formula

$$\rho_e = 2 \int d\omega A(\omega) f(\omega) = 2 \int d\omega \frac{A(\omega)}{1 + e^{\beta\omega}}\,, \tag{A.24}$$

with $\omega = 0$ located at the chemical potential. Perform this analysis for $T = 0.01$, 0.1, and 1. *Hint*: The sum over Matsubara frequencies must be truncated, but the sum must be handled carefully, since the terms go like $1/i\omega_n$ for large n. So use the fact that

$$T \sum_n \frac{1}{i\omega_n + x} = \frac{1}{1 + e^{-\beta x}} - \frac{1}{2}\,, \tag{A.25}$$

judiciously in evaluating your sum (you need to choose an appropriate value for x too). Try to prove the identity in Eq. (A.25). Be sure to estimate the error in your Matsubara summation, and explain how you chose the Matsubara frequency cutoff (which will depend on the temperature).
Ans: you should find you get four digits of accuracy to the electron filling when $T = 0.1$ and you use about 7500 positive Matsubara frequencies.

A.7 Comparing the spectral formula to the Hilbert transform

This problem is a continuation of the previous problem (A.6). Using the two equivalent forms for the retarded Green's function

$$G^R(\omega) = \int d\omega' \frac{A(\omega')}{\omega - \omega' + i\delta} = \int d\epsilon \frac{\rho_{\mathrm{Bethe}}(\epsilon)}{\omega + \mu - \Sigma(\omega) - \epsilon}, \tag{A.26}$$

and the explicit forms for $A(\omega)$ and $\rho_{\mathrm{Bethe}}(\epsilon)$, perform the integrals analytically, and solve your resulting equations for $\Sigma(\omega)$ in terms of w_1, μ, ω, etc. Plot both the real and the imaginary parts of $\Sigma(\omega)$ for $T = 0.0$, 0.01, 0.1, and 1. Is this system a Fermi liquid? Choose the same parameters as in Problem A.6. Note that the only way to evaluate the second integral above is to first making a trigonometric substitution that removes the square root, then introduce a complex variable $z = \exp(i\theta)$, and express the integral as a line integral around the unit circle, and finally, evaluate the integral by the calculus of residues and the residue theorem. You should find two poles inside the unit circle, and the residues need to be determined carefully. As a check, it is worthwhile to write a computer code to numerically perform the integration for some test choices of the self-energy (just pick some values in the complex plane), and compare the numerical integration against the analytical result from the residue theorem. *Hint*: you will need to develop a strategy to choose the signs of the square roots that arise in this calculation. When there is an imaginary part to G or Σ, choose the signs so the imaginary parts are negative; when G or Σ are real, choose them by continuity.
Ans: $\Sigma(\omega) = \omega + \mu - G^R(\omega) - G^R(\omega)^{-1}$; you will need to solve for an explicit expression for g^R to substitute into the above form.

A.8 Imaginary-time Green's functions

Consider a homogeneous translationally invariant interacting system. Assume that $\langle c^\dagger c\rangle = n$. Show that $\lim_{\tau\to 0^+} G(\tau) = n-1$, $\lim_{\tau\to 0^-} G(\tau) = n$, $\lim_{\tau\to\beta^-} G(\tau) = -n$, and $\lim_{\tau\to\beta^+} G(\tau) = 1-n$ for the Green's function $G(\tau) = -\langle T_\tau c(\tau)c^\dagger(0)\rangle$. Show the time-translation invariance property, that $G(\tau,\tau') = G(\tau-\tau')$, and the antiperiodicity property $G(\tau) = -G(\tau+\beta)$ for $-\beta < \tau < 0$. Show your work explicitly and perform your calculations directly from the definitions of the Green's functions in imaginary time. *Hint*: you will need to use the invariance of the trace $\mathrm{Tr}AB = \mathrm{Tr}BA$; a and B represent different quantum operators.

A.9 Partition function for a spinless electron in a general time-dependent field

In this problem we will verify that the partition function of a single-site impurity in a general time-translation-invariant field can be expressed as the following infinite product:

$$\mathcal{Z}_{\mathrm{imp}}(\lambda, \mu) = 2e^{\beta\mu/2} \prod_{n=-\infty}^{\infty} \frac{i\omega_n + \mu - \lambda_n}{i\omega_n}, \tag{A.27}$$

by a brute-force evaluation.
(a) By taking derivatives of both sides of Eq. (A.27), show that the partition function satisfies the differential equation

$$-\frac{1}{\mathcal{Z}_{\mathrm{imp}}(\lambda, \mu)} \frac{\partial \mathcal{Z}_{\mathrm{imp}}(\lambda, \mu)}{\partial \lambda_n} = \frac{1}{i\omega_n + \mu - \lambda_n}. \tag{A.28}$$

(b) Next show that when the field vanishes

$$\mathcal{Z}_{\mathrm{imp}}(\lambda = 0, \mu) = 2e^{\beta\mu/2} \prod_{n=-\infty}^{\infty} \frac{i\omega_n + \mu}{i\omega_n} = 1 + e^{\beta\mu}. \tag{A.29}$$

(c) Finally show that

$$\mathrm{Tr}[e^{\beta\mu c^\dagger c}] = 1 + e^{\beta\mu}, \tag{A.30}$$

by directly evaluating the trace over the impurity electron states.

These three parts show that the partition function, expressed as the infinite product in Eq. (A.27), satisfies the correct differential equation, and has the right boundary condition at $\lambda = 0$, hence it is the partition function.

A.10 Mapping the impurity in a field to an impurity coupled to a chain in the NRG approach

In this problem we will derive expressions for the chain parameters given a specific $\Delta(\omega)$. In his original work [Wilson (1975)], Ken Wilson found an analytic expression for the λ_n values when $\Delta(\omega) = \Delta_0\Theta(\omega + E)\Theta(-\omega +$

E), a constant running from $-E \leq \omega \leq E$. This result was generalized to arbitrary power laws in [Bulla, Pruschke and Hewson (1997)]. It is important to note that these coefficients typically decrease like $\lambda_n \propto \Lambda^{-n/2}$. This is what makes the truncation to a finite chain work. We will first derive some general formulas for the first three terms in the chain, then evaluate them for the constant case and for a quadratic "pseudogap" case.

(a) Use the notation

$$\langle \omega^m \rangle = \sum_n \left[(\gamma_n^+)^2 (\xi_n^+)^m + (\gamma_n^-)^2 (\xi_n^-)^m \right], \tag{A.31}$$

with the γ and ξ coefficients defined in Eqs. (2.106) and (2.107). Note that if $\Delta(\omega)$ is an even function, then $\xi_n^+ = -\xi_n^-$, which we shall assume to hold in this problem.

Start with the state $\tilde{c}_0^\dagger|0\rangle$, and operate on it with the full chain Hamiltonian on the left. Show that $\langle 0|c\mathcal{H}_{\text{chain}}\tilde{c}_0^\dagger|0\rangle = V$, and that $\epsilon_0 = \langle 0|\tilde{c}_0\mathcal{H}_{\text{chain}}\tilde{c}_0^\dagger|0\rangle = 0$. The new operator $\tilde{c}_1$ is defined via

$$\lambda_1 \tilde{c}_1^\dagger|0\rangle = \mathcal{H}_{\text{chain}}\tilde{c}_0^\dagger|0\rangle - Vc^\dagger|0\rangle. \tag{A.32}$$

Using the fact that $\{\tilde{c}_1^\dagger, \tilde{c}_1\} = 1$, show that $\lambda_1 = \sqrt{\langle\omega^2\rangle/\langle 1\rangle}$.

Now we continue for the next state. Operate on $\tilde{c}_1^\dagger|0\rangle$ with the Hamiltonian. Show that $\epsilon_1 = 0$, and then using

$$\lambda_2 \tilde{c}_2^\dagger|0\rangle = \mathcal{H}_{\text{chain}}\tilde{c}_1^\dagger|0\rangle - \lambda_1\tilde{c}_0^\dagger|0\rangle, \tag{A.33}$$

determine $\tilde{c}_2^\dagger$ and λ_2. Repeat for ϵ_2, $\tilde{c}_3^\dagger$ and λ_3.

Ans.: You should find $\lambda_2 = \sqrt{[\langle\omega^4\rangle\langle 1\rangle - \langle\omega^2\rangle^2]/\langle\omega^2\rangle\langle 1\rangle}$, $\epsilon_2 = 0$, and $\lambda_3 = \sqrt{[\langle\omega^6\rangle\langle\omega^2\rangle\langle 1\rangle - \langle\omega^4\rangle^2\langle 1\rangle]/[\langle\omega^4\rangle\langle\omega^2\rangle\langle 1\rangle - \langle\omega^2\rangle^3]}$.

(b) Using the form $\Delta(\omega) = \Delta\Theta(\omega + E)\Theta(-\omega + E)$, evaluate all γ_n's, ξ_n's, and the first three λ values. Note that you should evaluate the *summations* that you find for each $\langle\omega^m\rangle$ exactly.

Ans.:

$$\begin{aligned}
\lambda_1 &= \frac{1}{2}E\left(1+\frac{1}{\Lambda}\right)\sqrt{1-\frac{1}{\Lambda}}\bigg/\sqrt{1-\frac{1}{\Lambda^3}}, \\
\lambda_2 &= \frac{1}{2}E\left(1+\frac{1}{\Lambda}\right)\frac{1}{\sqrt{\Lambda}}\left(1-\frac{1}{\Lambda^2}\right)\bigg/\sqrt{\left(1-\frac{1}{\Lambda^3}\right)\left(1-\frac{1}{\Lambda^5}\right)}, \\
\lambda_3 &= \frac{1}{2}E\left(1+\frac{1}{\Lambda}\right)\frac{1}{\Lambda}\left(1-\frac{1}{\Lambda^3}\right)\bigg/\sqrt{\left(1-\frac{1}{\Lambda^5}\right)\left(1-\frac{1}{\Lambda^7}\right)}.
\end{aligned} \tag{A.34}$$

Note how the λ's decrease approximately as $1/\Lambda^{n/2}$ as n increases.
(c) Using the form $\Delta(\omega) = \Delta\omega^2\Theta(\omega+E)\Theta(-\omega+E)$, evaluate all γ_n's, ξ_n's, and the first three λ values.
(d) Note that the general case, of an arbitrary integer power law for $\Delta(\omega)$ has been solved analytically for all n. If you want to try an extended project, see whether you can reproduce the general result in [Bulla, Pruschke and Hewson (1997)]. To do this you need to find a general recursion relation for the λ coefficients, and then solve the recursion, similar to what is discussed in [Wilson (1975)]. Be cautioned that this last problem is equivalent to a significant research project.

A.11 Impurity Green's function for the chain Hamiltonian in the NRG approach

In this problem, you will show that the chain Hamiltonian with the interaction vanishing on the impurity site produces a discretized version of Eq. (2.105). The strategy is to use EOM techniques. First, define the following Green's functions:

$$\begin{aligned} G(\tau) &= -\langle \mathcal{T}_\tau c(\tau)c^\dagger(0)\rangle, \\ G_n^+(\tau) &= -\langle \mathcal{T}_\tau a_n(\tau)c^\dagger(0)\rangle, \\ G_n^-(\tau) &= -\langle \mathcal{T}_\tau b_n(\tau)c^\dagger(0)\rangle. \end{aligned} \tag{A.35}$$

Next write the Hamiltonian for the chain explicitly in terms of the a, b, and c operators

$$\begin{aligned} \tilde{\mathcal{H}}_{\rm chain} = \mathcal{H}_{\rm imp}^{\rm Hubb}(U=0) &+ \sum_{n\sigma}[\xi_n^+ a_{n\sigma}^\dagger a_{n\sigma} + \xi_n^- b_{n\sigma}^\dagger b_{n\sigma}] \\ &+ \sum_{n\sigma}(\gamma_n^+ c_\sigma^\dagger a_{n\sigma} + \gamma_n^- c_\sigma^\dagger b_{n\sigma} + \gamma_n^+ a_{n\sigma}^\dagger c_\sigma + \gamma_n^- b_{n\sigma}^\dagger c_\sigma). \end{aligned} \tag{A.36}$$

Taking derivatives with respect to imaginary time τ and then Fourier transforming to Matsubara frequencies, show that

$$\begin{aligned} (i\omega_m + \mu)G(i\omega_m) &= 1 + \sum_n [\gamma_n^+ G_n^+(i\omega_m) + \gamma_n^- G_n^-(i\omega_m)], \\ (i\omega_m - \xi_n^+)G_n^+(i\omega_m) &= \gamma_n^+ G(i\omega_m), \\ (i\omega_m - \xi_n^-)G_n^-(i\omega_m) &= \gamma_n^- G(i\omega_m). \end{aligned} \tag{A.37}$$

Solve the equations in Eq. (A.37) for G, analytically continue by substituting $i\omega_m \to \omega + i\delta$, and then extract $\lambda^R(\omega)$ from the form in Eq. (2.105). Finally, using $\Delta(\omega) = -\text{Im}\lambda^R(\omega)/\pi$, verify Eq. (2.110). Look at Fig. 2.7 to see an example of how this works.

A.12 Solving the NRG many-body Hamiltonian for the chain

(a) We start by examining the impurity Hamiltonian at the end of the chain (the box in Fig. 2.8). The Hamiltonian is given in Eq. (2.104). There are four states: (i) no electrons; (ii) a spin-up electron; (iii) a spin-down electron; and (iv) a doubly occupied site. Find the energies of each of these states and the quantum numbers for the number of electrons, the total spin, and the z-component of the spin.
(b) When we add the first site (labeled by 0), we add an additional four states corresponding to no, one, or two electrons on site 0. This creates a total of $4 \times 4 = 16$ many-body states. We can break those states up in terms of the following quantum numbers: (i) $N = 0$, $S = 0$, $S_z = 0$ (one state); (ii) $N = 1$, $S = 1/2$, $S_z = \pm 1/2$ (four states); (iii) $N = 2$, $S = 1$, $S_z = 0, \pm 1$ (three states); (iv) $N = 2$, $S = 0$, $S_z = 0$ (three states); (v) $N = 3$, $S = 1/2$, $S_z = \pm 1/2$ (four states); and (vi) $N = 4$, $S = 0$, $S_z = 0$ (one state). The (i), (iii) and (vi) cases are 1×1 matrices, the (ii) and (v) cases are 2×2 matrices, and the (iv) case is a 3×3 matrix. Construct all six matrices and find the ten distinct eigenvalues. You will need to determine an explicit representation for each of the 16 states in order to find the matrices. Be sure to use your knowledge of addition of angular momentum in completing this task and recall cases where there is spin degeneracy.
Ans.: for $\mu = 2$, $U = 4$, $V = 1.5$ and $\epsilon_0 = 0$, you should find the ten eigenvalues are ($N = 0$: 0; $N = 1$: -2.80278, 0.802776; $N = 2$: -4.16228, -2, 0, 2.16228; $N = 3$: -2.80278, 0.802776; and $N = 4$: 0). Note that these results have particle-hole symmetry due to our choice of parameters; *i.e.* the eigenvalues for $N = 0$ and $N = 4$ are the same, as are the eigenvalues for $N = 1$ and $N = 3$.
(c) Assuming you have a set of eigenfunctions and eigenvalues with definite quantum numbers N, S, and S_z at a given stage of the NRG. Describe how to construct an algorithm to compute the block-diagonal matrices and the quantum numbers for the many-body problem at the next iteration of

the NRG, where we add another site at the end of the chain. Why is it necessary to have the chain be one-dimensional in order to construct the new Hamiltonian without requiring additional information about the states (*i.e.*, just those three quantum numbers are all that is required).

A.13 Metal-insulator transition in the half-filled Falicov-Kimball model

Consider the spinless Falicov-Kimball model on the Bethe lattice. Choose the case of half filling for both particles: $w_1 = 1/2$ and $\rho_e = 1/2$. The DOS will split into upper and lower Hubbard bands at a critical value of the coupling strength U.
(a) Plot the DOS for appropriately chosen values of U to see the weakly correlated metal, the strongly correlated metal, the critical DOS at the MIT, and the strongly correlated insulator. Determine U_c for the MIT.
(b) Solve for the self-energy $\Sigma(\omega)$ for arbitrary U, and show explicitly that the self-energy develops a pole at U_c. Plot the strength of the pole (i.e., the residue) versus U. By extrapolating this curve to zero, you have another way of determining U_c. *Hint*: the residue of the pole can be found by examining $\lim_{\omega \to 0} \omega\Sigma(\omega)$; note that since we fix $w_1 = 1/2$, there is no need to calculate the atomic partition function in your analysis.
Ans: See Fig. A.4 for the size of the residue of the pole as a function of U.

A.14 Kramers-Kronig analysis for the Green's function, and the effect of the pole in the Mott insulator

Take the results of Problem A.13 for the values $U = 1$, $U = 2$, and $U = 3$. Using $\mathrm{Im}G(\omega)$, employ the Kramers-Kronig analysis to find $\mathrm{Re}G(\omega)$. Compare this to the $\mathrm{Re}G(\omega)$ that you can calculate directly. Do the same starting with $\mathrm{Re}G$ to find $\mathrm{Im}G$. Repeat for $\mathrm{Im}\Sigma$ and for $\mathrm{Re}\Sigma - \mu$. Note that you need to subtract the constant value that the real part of the self-energy approaches as $|\omega| \to \infty$, since the Kramers-Kronig relation assumes the function vanishes at large frequency.

In each case, plot the result of your Kramers-Kronig integration versus the result calculated directly from DMFT. Adjust the step size and frequency cutoffs until you can achieve good accuracy. What do you need to do when there is a pole in the self-energy? Comment on the overall accuracy of these calculations. Is it better to use the Kramers-Kronig relation

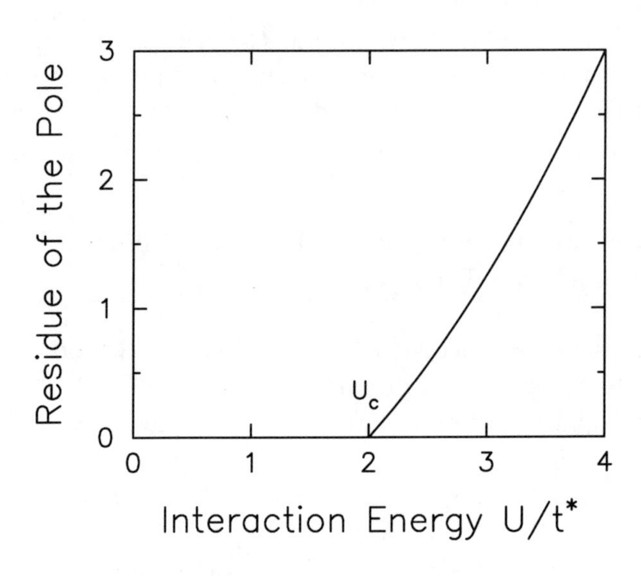

Fig. A.4 Residue of the pole in the self-energy for the Mott-insulating phase on the Bethe lattice at half filling.

for the real part or for the imaginary part, or does it make no difference? *Hint*: the imaginary part of the self-energy picks up a delta function contribution when there is a pole; you will find this difficult to calculate from the Kramers-Kronig transformation of the real part.

A.15 Metal-insulator transition on a simple cubic lattice

Use the dynamical mean-field approximation to calculate the Green's function of the simple cubic lattice in the local approximation. Rather than use the exact three-dimensional DOS in the Hilbert transform, use the approximate result of Uhrig [Uhrig (1996)], which is derived in parts (a–c).

Make an ansatz that the three-dimensional simple cubic DOS can be approximated by (taking $t = 1$)

$$\rho_{\rm ap}(\epsilon) = \frac{1}{\pi}\left[M(\epsilon)\sqrt{36-\epsilon^2} + N(\epsilon)\sqrt{4-(\epsilon-4)^2} + N(-\epsilon)\sqrt{4-(\epsilon+4)^2}\right], \tag{A.38}$$

where the first term is present for $-6 \le \epsilon \le 6$, the second for $2 \le \epsilon \le 6$, and the third for $-6 \le \epsilon \le -2$. The functions M and N are polynomials in ϵ: $M(\epsilon) = m_0 + m_2\epsilon^2 + m_4\epsilon_4 + ...$ is even and $N(\epsilon) = n_0 + n_1\epsilon + n_2\epsilon^2 + ...$

has no restrictions. In this form, we have an even DOS which has the right van Hove singularities in the three-dimensional band.

The strategy we will take is to calculate the moments of the approximate DOS

$$A_k = \int_{-6}^{6} d\epsilon\, \epsilon^k \rho_{\rm ap}(\epsilon)\,, \tag{A.39}$$

and of the exact DOS

$$E_k = \int_{-6}^{6} d\epsilon\, \epsilon^k \rho_{\rm 3d}(\epsilon)\,, \tag{A.40}$$

and then force the coefficients m_i and n_i to produce the exact moments for some number of k-values. Since both DOS are even functions, all odd moments vanish.

(a) For the approximate moments, note that

$$\begin{aligned} A_k = & \frac{1}{\pi}\int_{-6}^{6} d\epsilon\, \epsilon^k (m_0 + m_2\epsilon^2 + m_4\epsilon^4 + ...)\sqrt{36-\epsilon^2} \\ & + \frac{2}{\pi}\int_{2}^{6} d\epsilon\, \epsilon^k (n_0 + n_1\epsilon + n_2\epsilon^2 + ...)\sqrt{4-(\epsilon-4)^2}\,. \end{aligned} \tag{A.41}$$

Restrict yourself to include only m_0 and m_2 in the polynomial $M(\epsilon)$ and n_0, n_1, and n_2 in the polynomial $N(\epsilon)$. In terms of m_0, m_2, n_0, n_1, and n_2, evaluate the moments A_0, A_2, A_4, A_6, and A_8 exactly (use a symbolic manipulation program to carry out the algebra).

Ans: $A_0 = 18m_0 + 162m_2 + 4n_0 + 16n_1 + 68n_2$; $A_2 = 162m_0 + 2916m_2 + 68n_0 + 304n_1 + 1416n_2$; $A_4 = 2916m_0 + 65610m_2 + 1416n_0 + 6816n_1 + 33684n_2$; $A_6 = 65610m_0 + 1653372m_2 + 33684n_0 + 170032n_1 + 873272n_2$; and $A_8 = 1653372m_0 + 44641044m_2 + 873272n_0 + 4549600n_1 + 23987752n_2$.

(b) Now calculate the exact moments E_k by the following procedure. Begin with the three-dimensional DOS

$$\rho_{\rm 3d}(\epsilon) = \frac{1}{(2\pi)^3}\int_{-\pi}^{\pi} dk_x \int_{-\pi}^{\pi} dk_y \int_{-\pi}^{\pi} dk_z \delta(\epsilon + 2t\cos k_x + 2t\cos k_y + 2t\cos k_z)\,, \tag{A.42}$$

and then substitute into the definition for E_k and integrate over ϵ to give

$$E_k = \frac{(-2t)^k}{(2\pi)^3}\int_{-\pi}^{\pi} dk_x \int_{-\pi}^{\pi} dk_y \int_{-\pi}^{\pi} dk_z (\cos k_x + \cos k_y + \cos k_z)^k\,. \tag{A.43}$$

Evaluate the first 5 even moments E_0, E_2, E_4, E_6, and E_8.
Ans: $E_0 = 1$; $E_2 = 6$; $E_4 = 90$; $E_6 = 1860$; and $E_8 = 44730$.
(c) Solve the set of five equations $A_k = E_k$ for the coefficients m_0, m_2, n_0, n_1, and n_2 to determine the polynomials $M(\epsilon)$ and $N(\epsilon)$ for the approximate three-dimensional DOS. Plot the approximate DOS versus the exact DOS; determine how large the absolute error is for the approximation (the exact DOS was found in Prob. A.2).
Ans: $M(\epsilon) = \frac{13033}{174528} + \epsilon^2 \frac{8675}{6283008}$ and $N(\epsilon) = -\frac{1389}{12928} - \epsilon \frac{51}{25856} - \epsilon^2 \frac{81}{51712}$. Figure A.5 plots the difference of the exact and the approximate DOS. Note that the maximal error is about 0.2% near $|\omega| = 2$.
(d) Using the approximate form for the three-dimensional DOS, write a computer program to calculate the DOS for $w_1 = 1/2$ and $\rho_e = 1/2$ for the spinless Falicov-Kimball model on the simple-cubic lattice. Plot the DOS for $U = 0 - 8$ in steps of 1. Note that U_c for the MIT lies at about $U_c \approx 4.92$. *Hint*: the Hilbert transform can be evaluated analytically in three separate pieces, since the Hilbert transform of powers multiplied by a square-root function can be integrated exactly. Note that this requires a proper choice of the signs for the different square-root factors. Using this analytic form for the Hilbert transform will speed up the computer program significantly.
Ans: See the plot of the $U^{\mathrm{FK}} = 6$ results in the figure.

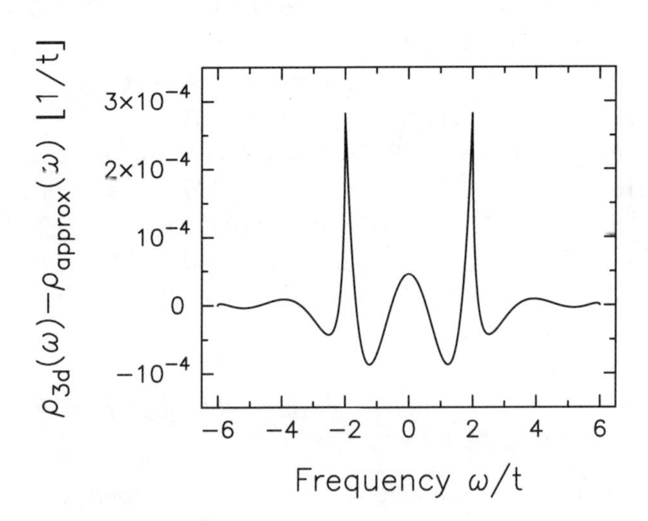

Fig. A.5 Difference between the approximate DOS and the exact DOS in three dimensions. Since the DOS is about 0.14, the absolute error is less than approximately 0.2%.

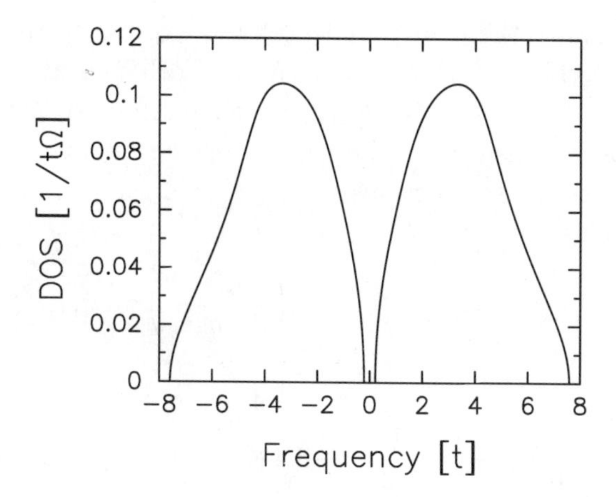

Fig. A.6 DOS for the half-filled Falicov-Kimball model on a simple cubic lattice with $U^{\mathrm{FK}} = 6$, which is a small-gap Mott insulator.

A.16 DC conductivity for the simple cubic lattice

(a) Evaluate the transport DOS for the simple-cubic lattice

$$\rho_{\mathrm{trans}}(\epsilon) = \frac{1}{(2\pi)^3} \int_{-\pi}^{\pi} dk_x \int_{-\pi}^{\pi} dk_y \int_{-\pi}^{\pi} dk_z \sin^2 k_x \delta(\epsilon - \epsilon_k) , \qquad \text{(A.44)}$$

by using the approximate DOS from Problem A.15, and deriving a differential equation for the transport DOS in terms of the approximate DOS. Begin by noting that $(t = 1)$

$$\frac{d}{d\epsilon}\delta(\epsilon - \epsilon_k) = -\frac{dk_x}{d\epsilon_k}\frac{d}{dk_x}\delta(\epsilon - \epsilon_k) = -\frac{1}{2\sin k_x}\frac{d}{dk_x}\delta(\epsilon - \epsilon_k) . \qquad \text{(A.45)}$$

Use this result to prove that $d\rho_{\mathrm{trans}}(\epsilon)/d\epsilon = -\epsilon\rho_{\mathrm{3d}}(\epsilon)/12$.

(b) Taking the approximate DOS for Problem A.15, solve the differential equation derived in part (a) for the transport DOS, and use that to calculate the *dc*-conductivity on a simple-cubic lattice. Evaluate $\rho_{dc} = 1/\sigma_{dc}$ as a function of temperature for $0 < T < 2$ and $U = 2$, 3, 4, 5, 6. Comment on the behavior of the resistivity. *Hint:* the transport DOS cannot be expressed as a polynomial multiplied by the approximate 3-d DOS, so it is best to solve the differential equation to determine the transport DOS analytically, and then perform the integration over both the energy variable and the frequency variable numerically. Your work can be simplified by defining a function $G_{\mathrm{tr}}(z) = \int \rho_{\mathrm{tr}}(\epsilon)/(z - \epsilon)$ (in terms of the

Hilbert transform of the transport DOS) and expressing the integrand for the transport coefficients in terms of $G_{\rm tr}$ and other local functions. Note that $dG_{\rm tr}(z)/dz = -[1 - zG(z)]/12$, which relates the derivative of the transport Hilbert transform to the local Green's function, will be useful in deriving the *dc* response functions.
Ans: You should find

$$\sigma_{\rm dc} = \sigma_{0,3d} \int_{-\infty}^{\infty} d\omega \left(-\frac{df(\omega)}{d\omega}\right) \tau(\omega), \tag{A.46}$$

with

$$\tau(\omega) = \frac{1}{2\pi^2} \left[\frac{{\rm Im}G_{\rm tr}(\omega)}{{\rm Im}\Sigma(\omega)} + \frac{1}{12} - \frac{1}{12}{\rm Re}\{[\omega + \mu - \Sigma(\omega)]G(\omega)\}\right]. \tag{A.47}$$

The plot shows the results for $U^{\rm FK} = 6$.

A.17 Jonson-Mahan theorem

In this problem, we will verify the Jonson-Mahan theorem, as formulated in [Mahan (1998)]. For simplicity, we will work on the spinless case, but the generalization to include spin is straightforward. The first step is to work with a general form for the heat-current operator. Since the heat-current operator depends on the potential energy, we will consider a general form for the local potential energy

$$\hat{V} = \sum_i V_i c_i^\dagger c_i + \sum_i \left[\bar{V}_i f_i^\dagger c_i + \bar{V}_i^* c_i^\dagger f_i\right] \tag{A.48}$$

where $f_i^\dagger$ (f_i) are localized electron creation (annihilation) operators at site i for a hybridization piece of the Hamiltonian as in the PAM. The form in Eq. (A.48) even holds for the Hubbard model if the coefficients V_i can depend on the opposite spin operators (which they can).

(1) Verify that a direct computation of the heat-current operator yields

$$\begin{aligned}\mathbf{j}_Q = &\sum_{\mathbf{k}} \mathbf{v}_{\mathbf{k}}(\epsilon_{\mathbf{k}} - \mu)c_{\mathbf{k}}^\dagger c_{\mathbf{k}} + \frac{1}{2}\sum_{\mathbf{k}\mathbf{k}'}(\mathbf{v}_{\mathbf{k}} + \mathbf{v}_{\mathbf{k}'})V(\mathbf{k}-\mathbf{k}')c_{\mathbf{k}}^\dagger c_{\mathbf{k}'} \\ &+ \frac{1}{\sqrt{\Lambda}}\sum_{i\mathbf{k}} \mathbf{v}_{\mathbf{k}}\left(e^{i\mathbf{k}\cdot\mathbf{R}_i}\bar{V}_i f_i^\dagger c_{\mathbf{k}} + e^{-i\mathbf{k}\cdot\mathbf{R}_i}\bar{V}_i^* c_{\mathbf{k}}^\dagger f_i\right),\end{aligned} \tag{A.49}$$

where $V(\mathbf{k}-\mathbf{k}') = \sum_i \exp[-i(\mathbf{k}-\mathbf{k}')\cdot\mathbf{R}_i]V_i/\Lambda$. Note that this general form agrees with what we derived for the Falicov-Kimball model, as it must.

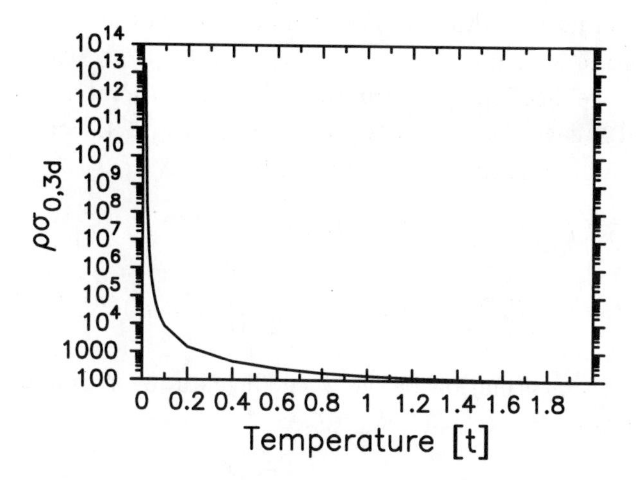

Fig. A.7 Resistivity for the half-filled Falicov-Kimball model on a simple cubic lattice with $U^{\rm FK} = 6$, which is a small-gap Mott insulator.

(2) Using the equations of motion for the imaginary time creation and annihilation operators, show that

$$\begin{aligned}\partial_\tau c^\dagger_{\mathbf{k}}(\tau) &= [\mathcal{H} - \mu\mathcal{N}, c^\dagger_{\mathbf{k}}(\tau)]\\ &= (\epsilon_{\mathbf{k}} - \mu)c^\dagger_{\mathbf{k}}(\tau) + e^{\tau(\mathcal{H}-\mu\mathcal{N})}[V, c^\dagger_{\mathbf{k}}]e^{-\tau(\mathcal{H}-\mu\mathcal{N})},\\ \partial_\tau c_{\mathbf{k}}(\tau) &= [\mathcal{H} - \mu\mathcal{N}, c_{\mathbf{k}}(\tau)]\\ &= -(\epsilon_{\mathbf{k}} - \mu)c_{\mathbf{k}}(\tau) + e^{\tau(\mathcal{H}-\mu\mathcal{N})}[V, c_{\mathbf{k}}]e^{-\tau(\mathcal{H}-\mu\mathcal{N})}.\end{aligned} \tag{A.50}$$

(3) Now we are ready to proceed with the proof. Start with the definition of two correlation functions

$$\begin{aligned}F_{ab}(\tau,\tau') &= \sum_{\mathbf{k}} \mathbf{v}_{\mathbf{k}a}\langle \mathcal{T}_\tau c^\dagger_{\mathbf{k}}(\tau)c_{\mathbf{k}}(\tau')\mathbf{j}_b(0)\rangle,\\ S_{ab}(\tau,\tau') &= \frac{1}{2}(\partial_\tau - \partial_{\tau'})F(\tau,\tau').\end{aligned} \tag{A.51}$$

Note that a and b refer to coordinate axes of the corresponding vectors. It is easy to show that $F_{ab}(\tau,\tau^-) = L_{11}(\tau)_{ab} = \langle \mathcal{T}_\tau \mathbf{j}_a(\tau)\mathbf{j}_b(0)\rangle$. Using the equations of motion derived above, show that $S_{ab}(\tau,\tau^-) = L_{21}(\tau)_{ab} = \langle \mathcal{T}_\tau \mathbf{j}_{Qa}(\tau)\mathbf{j}_b(0)\rangle$.

Hint: You will need to evaluate the commutators of the creation and annihilation operators with the potential.

(4) Now we need to perform the appropriate analytic continuation. Since F and S each depend on two time variables, their Fourier transform involves a double transformation. Show that

$$F_{ab}(\tau,\tau') = T^2 \sum_{nm} F_{ab}(i\omega_n, i\omega_m) e^{(i\omega_n\tau - i\omega_m\tau')},$$
$$S_{ab}(\tau,\tau') = T^2 \sum_{nm} F_{ab}(i\omega_n, i\omega_m) \frac{1}{2}(i\omega_n + i\omega_m) e^{(i\omega_n\tau - i\omega_m\tau')}. \quad (A.52)$$

The Fourier transformation of the L coefficients involves a single Fourier transformation:

$$L_{11}(i\nu_l)_{ab} = \int_0^\beta d\tau e^{i\nu_l\tau} F_{ab}(\tau,\tau^-) = T \sum_n F_{ab}(i\omega_n, i\omega_n + i\nu_l),$$
$$L_{21}(i\nu_l)_{ab} = \int_0^\beta d\tau e^{i\nu_l\tau} S_{ab}(\tau,\tau^-) = T \sum_n F_{ab}(i\omega_n, i\omega_n + i\nu_l)(i\omega_n + \frac{1}{2} i\nu_l). \quad (A.53)$$

The last part of each equation above needs to derived explicitly.

(5) The final step is to complete the analytic continuation as described in Chapter 2. It may appear somewhat trickier than before, because the F function depends on two frequencies, not one, but if you replace the F function by its actual value (knowing what the L_{11} analytic continuation is, you will see that it involves the product of two Green's functions at different frequencies). The analytic continuation can now be performed as before by taking into account the analytic properties of the Green's functions. Show that the Jonson-Mahan theorem holds for the relation between the L_{11} and L_{21} coefficients. Since $L_{21} = L_{12}$ we are done.

It turns out that the Jonson-Mahan theorem always holds when the potential is a sum of local terms (as in the general potential we took for this problem).

A.18 Charge and thermal conductivity for the Falicov-Kimball model

Taking the results of Problems A.15 and A.16, calculate the Lorenz number

$$\mathcal{L} = \frac{\kappa_e}{\sigma_{dc} T}, \quad (A.54)$$

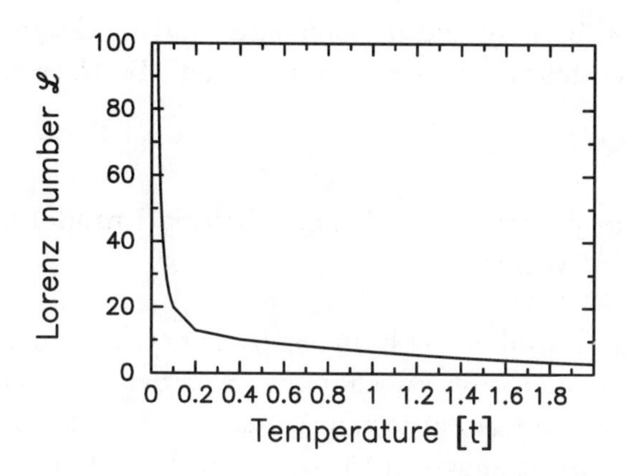

Fig. A.8 Lorenz number for the half-filled Falicov-Kimball model on a simple cubic lattice with $U^{\mathrm{FK}} = 6$, which is a small-gap Mott insulator.

for $0 < T < 2$ and $U = 2$, 3, 4, 5, 6 ($\rho_e = 1/2$ and $w_1 = 1/2$). Recall that in a Fermi liquid, the Lorenz number approaches a constant $\pi^2/3$. How do these results compare to those in a conventional metal? Note that you can fix the chemical potential at $\mu = U/2$ for this calculation for all T.
Ans: The plot shows the results for $U^{\mathrm{FK}} = 6$.

A.19 The particle-hole asymmetric metal-insulator transition

(a) Compute the Green's function of the spinless Falicov-Kimball model on a simple-cubic lattice with $\rho_e = 1 - w_1$, $w_1 = 0.1$, 0.2, 0.3, 0.4, and 0.5. Find U_c for the critical opening of a gap in the DOS for each filling. Plot the critical DOS for each case.
Ans: $U_c \approx 4.95$ for $w_1 = 0.3$.
(b) How much larger does U have to be than U_c in order for there to be a pole in the self energy [consider the same fillings as in (a)]?
Ans: $U_{\mathrm{pole}} \approx 5.35$ for $w_1 = 0.3$.

A.20 Non Fermi-liquid behavior of the Falicov-Kimball model

Compute the self energy of the spinless Falicov-Kimball model on a simple-cubic lattice with $\rho_e = 1 - w_1$, $w_1 = 0.1$, 0.2, 0.3, 0.4, and 0.5.

Take $U = 1, 2, 3, 4, 5$, and 6. Compare $\Sigma(\omega)$ to the results of Fermi-liquid theory. Is this system ever a Fermi liquid? Do the results depend on temperature?

A.21 Thermopower of the Falicov-Kimball model and the figure-of-merit

(a) In the case of half-filling for both particles of the Falicov-Kimball model, show that the thermopower vanishes due to particle-hole symmetry.
(b) Calculate the thermoelectric figure-of-merit ZT for the spinless Falicov-Kimball model on the simple cubic lattice for $U = 2, 3, 4, 5$, and 6, and $w_1 = 0.1, 0.2, 0.3, 0.4$, as a function of T for $0.001 < T < 2$. Set the electron filling to $\rho_e = 1 - w_1$. Plot ZT versus T for all w_1 values at a given value of U on one plot. *Hint*: one must be careful to determine the chemical potential properly for low temperature in the insulating phases, as the simple numerical root-finding technique will not normally produce a chemical potential in the gap. The $T \to 0$ limit of the chemical potential must lie at the center of the gap, and using the constraint where the number of "holes" equals the number of "electrons" at finite temperature will allow the chemical potential to be found for all T. This procedure requires a real-axis code to be carried out. Note that you may still run into precision problems in determining the thermopower, due to the need to take the ratio of two small numbers, each of which is becoming inaccurate at low temperatures for $U = 5$ and 6.
Ans: The plot shows the results for $U = 3$ and $w_1 = 0.1, 0.2, 0.3$, and 0.4.

A.22 $U \to \infty$ Green's functions

Consider the spinless Falicov-Kimball model on the infinite-coordination Bethe lattice. As $U \to \infty$, we have

$$G(\omega) = \frac{1 - w_1}{G_0^{-1}(\omega)} + \frac{w_1}{G_0^{-1}(\omega) - U} \to (1 - w_1)G_0(\omega), \tag{A.55}$$

if $\mu \ll U$. Employ the DMFT algorithm to compute an *analytic expression* for the DOS and for the self-energy when $U \to \infty$ and $0 \le w_1 \le 1$ and $0 \le \rho_e + w_1 < 1$. You can express your result in terms of ω, w_1, μ, and U.
Ans: $A(\omega) = \sqrt{4(1 - w_1) - (\omega + \mu)^2}/2\pi$ which has the same form as the noninteracting DOS, except it is band-narrowed, and has weight $1 - w_1$.

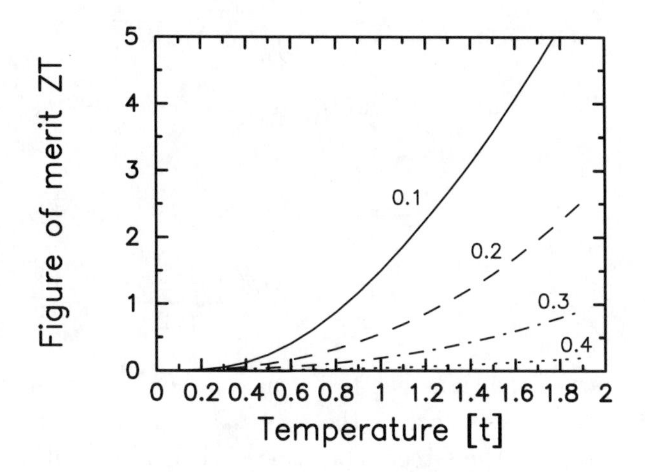

Fig. A.9 Figure-of-merit for the Falicov-Kimball model on a simple cubic lattice with $U^{\mathrm{FK}} = 3$ and various fillings. As expected, the figure-of-merit grows as T increases, and as the system becomes more particle-hole asymmetric; it vanishes at half-filling.

A.23 Determining $G_{\alpha\beta}$ from the quantum zipper algorithm

(a) Starting from the expression for the quantum zipper algorithm given in Eq. (3.5), examine the cases $\gamma = \alpha$, $\gamma = \alpha - 1$, $\gamma = \alpha - 2$, ..., $\gamma = \alpha - n$ to show that

$$G_{\alpha\alpha-n}(\mathbf{k}^{\|}, z) = -\frac{G_{\alpha\alpha-n+1}(\mathbf{k}^{\|}, z)t_{\alpha-n+1\alpha-n}}{L_{\alpha-n}(\mathbf{k}^{\|}, z)}, \tag{A.56}$$

where z denotes a complex variable in the upper half plane and $L_{\alpha-n}(\mathbf{k}^{\|}, z)$ satisfies the recursion relation in Eq. (3.9). Note that you need to prove that L satisfies the appropriate recursion relation.

(b) Similarly, show that

$$G_{\alpha\alpha+n}(\mathbf{k}^{\|}, z) = -\frac{G_{\alpha\alpha+n-1}(\mathbf{k}^{\|}, z)t_{\alpha+n-1\alpha+n}}{R_{\alpha+n}(\mathbf{k}^{\|}, z)}, \tag{A.57}$$

with $R_{\alpha+n}(\mathbf{k}^{\|}, z)$ satisfying the recursion relation in Eq. (3.12).

(c) Show that for a bulk (no α dependence to t) homogeneous system of a noninteracting metal, we have

$$L(\epsilon,\omega) = R(\epsilon,\omega) = \frac{\omega+\mu-\epsilon}{2} \pm \frac{1}{2}\sqrt{(\omega+\mu-\epsilon)^2-4}\,, \tag{A.58}$$

for the left and right functions, and that the off-diagonal Green's functions become

$$G_{\alpha\alpha+n}(\epsilon,\omega) = \pm\frac{1}{\sqrt{(\omega+\mu-\epsilon)^2-4}} \times \left[-\frac{\omega+\mu-\epsilon}{2} \pm \frac{\sqrt{(\omega+\mu-\epsilon)^2-4}}{2}\right]^n, \tag{A.59}$$

where the choice of the signs has to be made according to the analyticity requirements. Show how to choose these signs, and, in particular, determine the imaginary part of the Green's functions, when ω lies within the band. Note that the imaginary part of the off-diagonal Green's functions need not have a definite sign as the local Green's function has. Note further that we have chosen $\epsilon = \epsilon_\alpha$ for each α, since we are in the bulk.

A.24 The stability of the left and right recursion relations of the quantum zipper algorithm

In this problem, we analyze the recursion relations for the left and right functions in Eqs. (3.9) and (3.12).
(a) Show that if a given L or R function has a positive imaginary part, then the next L or R function in the recursion relation also has a positive imaginary part. This shows that the recursion relation is stable when there is an imaginary part.
(b) Show that when $L_{-\infty}$ or R_∞ is real, then the physical root is stable in the recursive iterations. In other words, show that the large magnitude root is the physical root, and show (either by an eigenvalue analysis, or by numerical testing) that the large root is stable in the iterations.

A.25 Efficient numerical evaluation of integrals via changes of variables

It is instructive to understand how computers perform numerical quadrature and how to prepare integrands for efficient evaluation. When using packaged routines, adaptive integration schemes work the best. But sometimes, it is easier to write your own integration code, or to prepare the

integrand for use with a package, if you know the potential "singular" behavior in your integrand.
(a) Using a midpoint rectangular integration scheme, evaluate the integral

$$\int_0^1 dx\,\sqrt{x}\,, \tag{A.60}$$

whose value is equal to 2/3. How many points are needed to achieve an accuracy of two decimal points, three decimal points, four decimal points, and five decimal points. How many points are needed for eight digits of accuracy? What is the best accuracy you can achieve with a midpoint integration routine?
Ans: 0.6695 for $N=7$, 0.66699 for $N=31$, 0.666699 for $N=147$, and 0.66666999 for $N=688$. One needs $N\approx 175,000$ for eight digits of accuracy!
(b) The reason why the integration was so difficult was because the function $\sqrt{x}$ has an infinite slope at $x=0$. We can remove this "singularity" by making a change of variables $x=y^2$. Repeat part (a) for the transformed integral. Discuss why preparing the integrand for numerical quadrature is useful. Note that the substitution $x=y^2$ is easy to perform in a computer program. All quadrature routines have the form $I=\sum_i weight(i)f(i)$, where the weight function changes as we change variables. If we set up a grid with respect to y instead of x, we simply set $x=y^2$ for each y point, and substitute that into our function $f(x)$ to calculate that piece of the integrand. We need to also modify the weight function, which will now include the new term $2y\Delta y$ for each point in the y grid. When the integrand is a complicated function, using this strategy can save significant coding time. Is it easier to achieve high accuracy for the transformed integral?
(c) The function $\ln x$ has an integrable singularity in it. Compare a midpoint integration routine for $\ln x$ from 0 to 1 with that of a transformed integrand, when we pick $x=y^\alpha$, for some exponent α. What choice of α is the smallest integer such that the derivative of the integrand is finite at $x=0$?
(d) In the evaluation of the Green's functions for a multilayered nanostructure, we need to perform integrals, which behave like

$$\int_{-4}^{4} d\epsilon \frac{1}{\sqrt{(a-\epsilon)^2-b^2}}\rho_{2d}(\epsilon)h(\epsilon)\,; \tag{A.61}$$

this integrand has $\ln\epsilon$ behavior near $\epsilon=0$ due to the logarithmic divergence of the two-dimensional DOS at the center of the band, and it has

an inverse square-root singularity when $\epsilon = a \pm b$ [the function $h(\epsilon)$ has no singularities]. Devise a scheme to divide the region of integration into pieces that contain no singularities, and hence can be integrated directly, and those that have isolated singularities that can be removed by an appropriate change of variable. Determine a possible change of variables for each of the different regions. Note that your results will depend on what the values of a and b are, since there is no singularity if $a \pm b < -4$ or $a \pm b > 4$. *Hint*: a square-root singularity of the form given above is best removed by either a trigonometric or a hyperbolic trigonometric substitution like $\epsilon = a + b\cos\theta$ or $\epsilon = a \pm b\cosh\theta$.
(e) If the integrand contains a principal-value integration, of the form

$$\int d\epsilon \, \frac{h(\epsilon)}{\epsilon - a + i\delta}, \tag{A.62}$$

where $h(\epsilon)$ has no singularities but is nonzero in a region around $\epsilon = a$, and $\delta = 0^+$. This integrand has a true nonintegrable singularity, unless we evaluate the integrand with a special procedure called a principal-value integration. The strategy is to create an integration grid that is uniformly spaced, and which goes through the point $\epsilon = a$. Then, because the integrand changes sign at $\epsilon = a$, we can "cancel" the singularity by evaluating the integrand on a symmetric grid above and below a. We have to add an imaginary part coming from the delta function associated with the limit $\delta \to 0^+$, which equals $-i\pi\delta(\epsilon - a)$. Write a computer code to perform a principal-value integration when $a = 1$ and $h(\epsilon) = \epsilon\cos(\pi\epsilon/4)$. Take the integration limits to run from $0 \le x \le 2$. Note that we do not include the point $\epsilon = a$ in the numerical integration when we evaluate in a principal-value way.
Ans: $0.19988556 - 2.22144147i$.

A.26 Equilibrium solutions with charge reconstruction

In this problem, we will show that an equilibrium solution for the charge reconstruction in a nanostructure carries no charge current.

An examination of the local charge current operator shows that if $G_{\alpha\alpha+1} = G_{\alpha+1\alpha}$, then the expectation value of $j_\alpha^{\text{long}} = 0$. The quantum zipper algorithm shows that

$$G_{\alpha\alpha+1}(\mathbf{k}^{\|},z)=\frac{G_{\alpha\alpha}(\mathbf{k}^{\|},z)t_{\alpha\alpha+1}}{R_{\alpha+1}(\mathbf{k}^{\|},z)},$$
$$G_{\alpha+1\alpha}(\mathbf{k}^{\|},z)=\frac{G_{\alpha+1\alpha+1}(\mathbf{k}^{\|},z)t_{\alpha+1\alpha}}{L_{\alpha}(\mathbf{k}^{\|},z)}. \tag{A.63}$$

Use the recursion relations for the Green's functions and for the R and L functions to express the result for the Green's functions in terms of $R_{\alpha+1}$ and L_α. Note that the equations derived in the text need to be modified to include the electrostatic potentials from the charge reconstruction. Show that

$$G_{\alpha\alpha+1}(\mathbf{k}^{\|},z)=G_{\alpha+1\alpha}(\mathbf{k}^{\|},z)=\frac{1}{L_{\alpha}(\mathbf{k}^{\|},z)R_{\alpha+1}(\mathbf{k}^{\|},z)-t_{\alpha\alpha+1}t_{\alpha+1\alpha}}, \tag{A.64}$$

which shows that the expectation value of the local current operator vanishes even in the presence of the electronic charge reconstruction. This is what we expect, because the equilibrium state will not have any currents flowing, even though there are nonzero electric fields in the system. A similar result should hold for the local heat current, but such a derivation is significantly more complicated.

A.27 Local charge and heat current operators for a nanostructure

We work out a concrete example for deriving local current operators in a nanostructure described by the Falicov-Kimball model. Start with the Hamiltonian in real space

$$\mathcal{H}^{\mathrm{FK}}-\mu\mathcal{N}=-\sum_{\alpha ij}t^{\|}_{\alpha ij}c^{\dagger}_{\alpha i}c_{\alpha j}-\sum_{\alpha i}t_{\alpha\alpha+1}(c^{\dagger}_{\alpha+1i}c_{\alpha i}+c^{\dagger}_{\alpha i}c_{\alpha+1i})$$
$$+\sum_{\alpha i}(U_{\alpha}w_{\alpha i}-\mu+V_{\alpha}-\Delta E_{F\alpha})c^{\dagger}_{\alpha i}c_{\alpha i}, \tag{A.65}$$

where α refers to the plane, i and j are spatial coordinates on the plane, $t^{\|}$ is the intraplane hopping, t is the interplane hopping, and U is the Falicov-Kimball interaction. We let $w_{\alpha i}=f^{\dagger}_{\alpha i}f_{\alpha i}$ be the number operator for the localized electrons, which can be thought of as a classical variable taking the values of 0 or 1. The parameter $\Delta E_{F\alpha}$ is the shift of the band zero for planes whose band zero does not match with the band zero of the leads. V_α is the static potential from the electronic charge reconstruction, as described in Sec. 3.6.

Using the procedure outlined in the text for calculating the longitudinal charge-current operator from the commutator of the polarization operator (in the z-direction) with the Hamiltonian, determine the total longitudinal charge-current operator in real space. Show that we can define the local longitudinal charge-current operator by

$$\mathbf{j}_{\alpha}^{\text{long}} = iat_{\alpha\alpha+1}\sum_{i}(c_{\alpha+1i}^{\dagger}c_{\alpha i} - c_{\alpha i}^{\dagger}c_{\alpha+1i}), \tag{A.66}$$

where the charge current involves multiplying the number-current operator above by the electric charge $-|e|$. Verify that the total current operator satisfies $\mathbf{j}^{\text{long}} = \sum_{\alpha}\mathbf{j}_{\alpha}^{\text{long}}$. *Note*: a symmetrized version of the current operator is $\mathbf{j}_{\alpha}^{\text{long,sym}} = (\mathbf{j}_{\alpha-1}^{\text{long}} + \mathbf{j}_{\alpha}^{\text{long}})/2$, which averages the current operators to the left and to the right of plane α.

Continuing, use the procedure outlined in the text for calculating the longitudinal heat-current operator from the commutator of the heat polarization operator (in the z-direction) with the Hamiltonian, determine the total longitudinal heat-current operator in real space. Show that we can define the local longitudinal heat-current operator by

$$\begin{aligned}
\mathbf{j}_{\alpha}^{\text{Q,long}} &= iat_{\alpha\alpha+1}\Bigg\{ -\sum_{ij}\frac{1}{2}(t_{\alpha ij}^{\|} + t_{\alpha+1ij}^{\|})(c_{\alpha+1i}^{\dagger}c_{\alpha j} - c_{\alpha i}^{\dagger}c_{\alpha+ij}) \\
&- \frac{1}{2}t_{\alpha+1\alpha+2}\sum_{i}(c_{\alpha+2i}^{\dagger}c_{\alpha i} - c_{\alpha i}^{\dagger}c_{\alpha+2i}) \\
&- \frac{1}{2}t_{\alpha-1\alpha}\sum_{i}(c_{\alpha+1i}^{\dagger}c_{\alpha i} - c_{\alpha i}^{\dagger}c_{\alpha+1i}) \\
&+ \frac{1}{2}\sum_{i}(U_{\alpha}w_{\alpha i} + U_{\alpha+1i}w_{\alpha+1i})(c_{\alpha+1i}^{\dagger}c_{\alpha i} - c_{\alpha i}^{\dagger}c_{\alpha+1i}) \\
&+ \left[-\mu + \frac{1}{2}(V_{\alpha} + V_{\alpha+1}) + \frac{1}{2}(\Delta E_{F\alpha} - \Delta E_{F\alpha+1})\right] \\
&\times (c_{\alpha+1i}^{\dagger}c_{\alpha i} - c_{\alpha i}^{\dagger}c_{\alpha+1i})\Bigg\},
\end{aligned} \tag{A.67}$$

with $\mathbf{j}^{\text{Q,long}} = \sum_{\alpha}\mathbf{j}_{\alpha}^{\text{Q,long}}$. We also can define the symmetrized heat current operator as $\mathbf{j}_{\alpha}^{\text{Q,long,sym}} = (\mathbf{j}_{\alpha-1}^{\text{Q,long}} + \mathbf{j}_{\alpha}^{\text{Q,long}})/2$.

Note that the strategy we have adopted here is to first calculate the total current operators, and then to extract "reasonable" choices for the local current operators from the expressions for the total current. This procedure is not unique, and the choice made for the heat-current operator in Eq. (A.67) is precisely the choice needed to generalize the Jonson-Mahan theorem to the local correlation functions. This makes it the most reasonable choice to make.

A.28 Operator identity for the Jonson-Mahan theorem

By explicit computation, verify that

$$\lim_{\tau'\to\tau}\frac{1}{2}\left(\frac{\partial}{\partial\tau}-\frac{\partial}{\partial\tau'}\right)iat_{\alpha\alpha+1}\sum_{i\in\text{plane}}\left[c^{\dagger}_{\alpha+1i}(\tau)c_{\alpha i}(\tau')-c^{\dagger}_{\alpha i}(\tau)c_{\alpha+1i}(\tau')\right]$$
$$=j^{\mathrm{Q,long}}_{\alpha}(\tau). \tag{A.68}$$

A.29 BCS gap equation

In this problem we will find the T_c and superconducting gap for the attractive Hubbard model on a simple-cubic lattice using the BCS (Hartree-Fock) approximation. The gap equation appears in Eq. (3.100). The strategy for solving for the gap as a function of temperature requires us to first find the transition temperature, and then the gap as a function of T.

(a) *Solving for the transition temperature.* To solve for the transition temperature we need to use a one-dimensional root finder and find the temperature where Eq. (3.100) is satisfied with $\Delta = 0$. This is easy to do if we pick two temperatures that bracket the root, and use a one-dimensional root finder like Brent's false-position method [Brent (1973)]. The key is that we need to choose an appropriate number of Matsubara frequencies, and perform the summation over them, after evaluating the integrals over the DOS. These integrals appear formidable if one is to do them exactly. But we can evaluate them approximately, yet highly accurately, by using the methods developed for the Green's functions on a simple-cubic lattice. Verify that

$$\int d\epsilon\rho(\epsilon)\frac{1}{\omega_n^2+(\mu+\frac{1}{2}|U|\langle n\rangle-\epsilon)^2+|\Delta|^2}$$
$$=\int d\epsilon\rho(\epsilon)\left[\frac{1}{i\sqrt{\omega_n^2+|\Delta|^2}+\mu+\frac{1}{2}|U|\langle n\rangle-\epsilon}\right. \tag{A.69}$$

$$
\left. - \frac{1}{-i\sqrt{\omega_n^2 + |\Delta|^2} + \mu + \frac{1}{2}|U|\langle n\rangle - \epsilon}\right] \frac{1}{-2i\sqrt{\omega_n^2 + |\Delta|^2}}
$$

$$
= \left. \frac{G(z) - G^*(z)}{-2i\sqrt{\omega_n^2 + |\Delta|^2}} \right|_{z=i\sqrt{\omega_n^2+|\Delta|^2}+\mu+\frac{1}{2}|U|\langle n\rangle}
$$

This allows you to quickly calculate the terms for each Matsubara frequency by using the results of Prob. A.15 for the Green's functions of the simple-cubic lattice. Write a computer program to calculate the transition temperature (*i.e.*, set $\Delta = 0$) for enough values of U that you can reproduce Fig. 3.17 (a). Note you should use an energy cutoff for your Matsubara frequency summations (something like $|\omega_n| \leq 100$) to provide a consistent truncation to the summations as you vary the temperature.
Ans. you should find $T_c = 0.03463$ for $|U| = 1.5$ and $T_c = 0.6472$ for $|U| = 4$.

(b) *The superconducting gap.* Choose $|U| = 2$, 3 and 4. For each case, solve the gap equation to find $\Delta(T)$ as a function of T. Plot your results as $\Delta(T)/\Delta(T \to 0)$ versus T/T_c. You should be able to verify the results in Fig. 3.17(b). Note that you will need to find a way to comfortably determine the gap as $T \to 0$. Explain how you do this.

A.30 Equations of motion needed for the Nambu-Gor'kov formalism

Starting from the definitions for F and $G_\downarrow$ given by

$$
F_{ij}(\tau) = -\langle \mathcal{T}_\tau c_{i\uparrow}(\tau) c_{j\downarrow}(0)\rangle, \quad -G_{ji\downarrow}(-\tau) = -\langle \mathcal{T}_\tau c^\dagger_{i\downarrow}(\tau) c_{j\downarrow}(0)\rangle, \tag{A.70}
$$

take derivatives with respect to τ and use Wick's theorem to approximately evaluate the four-fermion operator average, and derive Eqs. (3.102) and (3.103).

A.31 Spin one-half atom in a time-dependent normal and anomalous dynamical mean field

Derive Eq. (3.114) by first computing the equation of motion in imaginary time by taking the derivative of each Green's function with respect to τ, and then Fourier transform to the Matsubara frequencies.

Invert Eq. (3.114) to find the four Green's functions,

$$G_\uparrow(i\omega_n) = \frac{i\omega_n - \mu + \lambda_\downarrow^*(i\omega_n)}{[i\omega_n + \mu - \lambda_\uparrow(i\omega_n)][i\omega_n - \mu + \lambda_\downarrow^*(i\omega_n)] - \bar{\alpha}(i\omega_n)\alpha(i\omega_n)}$$
$$F(i\omega_n) = \frac{\bar{\alpha}(i\omega_n)}{[i\omega_n + \mu - \lambda_\uparrow(i\omega_n)][i\omega_n - \mu + \lambda_\downarrow^*(i\omega_n)] - \bar{\alpha}(i\omega_n)\alpha(i\omega_n)}$$
$$\bar{F}(i\omega_n) = \frac{\alpha(i\omega_n)}{[i\omega_n + \mu - \lambda_\uparrow(i\omega_n)][i\omega_n - \mu + \lambda_\downarrow^*(i\omega_n)] - \bar{\alpha}(i\omega_n)\alpha(i\omega_n)}.$$
$$-G_\downarrow^*(i\omega_n) = \frac{i\omega_n + \mu - \lambda_\uparrow(i\omega_n)}{[i\omega_n + \mu - \lambda_\uparrow(i\omega_n)][i\omega_n - \mu + \lambda_\downarrow^*(i\omega_n)] - \bar{\alpha}(i\omega_n)\alpha(i\omega_n)} \tag{A.71}$$

Verify that the partition function, given in Eq. (3.115), is consistent with the above equations, by calculating the derivatives of $\mathcal{Z}_{\rm imp}$ with respect to each of the dynamical mean fields. Finally, show that the partition function reduces to the noninteracting result when the dynamical mean fields vanish. This establishes the form for the partition function used in the text.

Note that the fact that $G_\uparrow = G_\downarrow$ and $\bar{F} = F^*$, tells us that the self-consistent solution to the DMFT equations will have $\lambda_\uparrow = \lambda_\downarrow$ and $\bar{\alpha} = \alpha^*$.

A.32 Hilbert transformation in the Nambu-Gor'kov formalism

In this problem we will show how to efficiently evaluate the Hilbert transformation in the superconducting state [given in Eq. (3.124)]. The first step is to explicitly calculate the matrix inverse yielding

$$\mathbb{G}(i\omega_n) = \int d\epsilon\rho(\epsilon)\frac{-(i\omega_n - i\mathrm{Im}\Sigma_n)\mathbb{I} + (\mu - \mathrm{Re}\Sigma_n - \epsilon)\tau_3 - \mathrm{Re}\Phi_n\tau_1 + \mathrm{Im}\Phi_n\tau_2}{(\omega_n - \mathrm{Im}\Sigma_n)^2 + (\mu - \mathrm{Re}\Sigma_n - \epsilon)^2 + |\Phi_n|^2}. \tag{A.72}$$

Next, we need to evaluate two integrals. Verify that

$$\int d\epsilon\rho(\epsilon)\frac{a - \epsilon}{(a - \epsilon)^2 + b^2} = \mathrm{Re}G(a + ib), \tag{A.73}$$

$$\int d\epsilon\rho(\epsilon)\frac{1}{(a-\epsilon)^2+b^2}=-\frac{\text{Im}G(a+ib)}{b},\tag{A.74}$$

with $G(z)=\int d\epsilon\rho(\epsilon)/(z-\epsilon)$ the ordinary Hilbert transform that we use to calculate the normal-state Green's function in the bulk. Finally, use those integrals to show that

$$\mathbb{G}(i\omega_n)=\text{Re}G(z)\tau_3+[(i\omega_n-i\text{Im}\Sigma_n)\mathbb{I}+\text{Re}\Phi_n\tau_1-\text{Im}\Phi_n\tau_2]\frac{\text{Im}G(z)}{b},\tag{A.75}$$

with $z=a+ib$. To complete the calculation, you need to recognize that $a=\mu-\text{Re}\Sigma_n$ and $b=\sqrt{(\omega_n-\text{Im}\Sigma_n)^2+|\Phi_n|^2}$. This then is an efficient starting point to modify normal-state codes for use in superconductivity problems.

A.33 Evaluating Hilbert transformation-like integrals needed for determining the bulk critical current on a simple-cubic lattice

In this problem we will show how to perform the integration over k_z by contour integration and the residue theorem. Our starting point is Eq. (3.138) which is the generalization of the Hilbert transformation when we have current flowing in the bulk.

(a) Show that the Green's function can be expressed as the following integral

$$\mathbb{G}(i\omega_n)=\int d\epsilon^{\|}\rho^{\text{2d}}(\epsilon^{\|})\int_0^{2\pi}\frac{dk}{2\pi}$$

$$\begin{bmatrix} ia-2t\sin\frac{Q}{2}\sin k-b-2t\cos\frac{Q}{2}\cos k & c \\ c^* & ia-2t\sin\frac{Q}{2}\sin k+b+2t\cos\frac{Q}{2}\cos k \end{bmatrix}$$

$$\times\frac{1}{(ia-2t\sin\frac{Q}{2}\sin k)^2-(b+2t\cos\frac{Q}{2}\cos k)^2-|c|^2},\tag{A.76}$$

with $a=\omega_n-\text{Im}\Sigma_n$, $b=\mu-\epsilon^{\|}-\text{Re}\Sigma_\text{n}$, and $c=\Phi_n$.

(b) Transform the integral over k to an integral over the unit circle in the complex plane by making the substitution $z=\exp[ik]$, to yield

$$\mathbb{G}(i\omega_n) = \int d\epsilon^{\|}\rho^{2\mathrm{d}}(\epsilon^{\|}) \oint \frac{dz}{2\pi i z}$$

$$\begin{bmatrix} iaz - it\sin\frac{Q}{2}(z^2-1) - bz - t\cos\frac{Q}{2}(z^2+1) & cz \\ c^*z & ia - it\sin\frac{Q}{2}(z^2-1) + bz + t\cos\frac{Q}{2}(z^2+1) \end{bmatrix}$$

$$\times \frac{1}{-(az - t\sin\frac{Q}{2}[z^2-1])^2 - (bz + t\cos\frac{Q}{2}[z^2+1])^2 - |c|^2z^2}. \tag{A.77}$$

The contour integral is taken around the unit circle in the counter-clockwise direction. Now show that the denominator of the integrand can be written as

$$-t^2\left[z^4 + 2\left(\frac{a}{t}\sin\frac{Q}{2} + \frac{b}{t}\cos\frac{Q}{2}\right)z^3 + \left(\frac{a^2+b^2+|c|^2}{t^2} + 2\cos Q\right)z^2 \right.$$
$$\left. + 2\left(-\frac{a}{t}\sin\frac{Q}{2} + \frac{b}{t}\cos\frac{Q}{2}\right)z + 1\right], \tag{A.78}$$

which will be used in part (c).

(c) Because the quartic polynomial in z has a 1 for its constant term, the product of all of the roots of the polynomial equals 1. Hence, if we assume that no roots lie on the unit circle, at least one root, yielding a pole, must lie inside the unit circle, and at least one root must lie outside. Using the residue theorem, we can evaluate the integrals under the assumption that the roots are all distinct, and none lie on the unit circle. Do this to find

$$\mathbb{G}(i\omega_n) = \int d\epsilon^{\|}\rho^{2\mathrm{d}}(\epsilon^{\|}) \sum_{r_j:|r_j|<1}$$

$$\begin{bmatrix} iar_j + it\sin\frac{Q}{2}(r_j^2-1) - br_j - t\cos\frac{Q}{2}(r_j^2+1) & cr_j \\ c^*r_j & iar_j + it\sin\frac{Q}{2}(r_j^2-1) + br_j + t\cos\frac{Q}{2}(r_j^2+1) \end{bmatrix}$$

$$\times \frac{1}{-t^2\prod_{k\neq j}(r_j - r_k)}. \tag{A.79}$$

Now one can perform the integration over the two-dimensional density of states remembering to change variables near $\epsilon^{\|} = 0$ to remove the singularity from the numerical quadrature. Note that it is a good practice to check the number of roots that lie inside the unit circle, as these should not change for any given calculation as $\epsilon^{\|}$ changes (otherwise roots have crossed the unit circle), and one should also check that none of the roots become multiple roots either. This method for evaluating the integral is much more efficient than a two-dimensional integration, because the root-finder for a quartic polynomial in z is very fast, and one has just a one-dimensional integral that remains, which is simple to evaluate if the singularity is properly removed.

(d) We also need the integral defined in Eq. (3.141) to find the current flowing through the superconductor. This integral is similar to the (1,1) component of the integral evaluated above, except we have an additional factor of

$$\cos k \cos \frac{Q}{2} - \sin k \sin \frac{Q}{2} + i \sin k \cos \frac{Q}{2} + i \cos k \sin \frac{Q}{2} \rightarrow$$
$$\frac{1}{2}\left(z + \frac{1}{z}\right) e^{i\frac{q}{2}} + \frac{1}{2}\left(z - \frac{1}{z}\right) e^{i\frac{q}{2}} = z e^{i\frac{q}{2}} \tag{A.80}$$

in the integrand, which is simple to evaluate in the same way as in part (c). Determine what the final formula for the integral is after evaluating the residues.

(e) Since the only change to the superconducting algorithm when current is flowing is the change in the generalized Hilbert transformation, use the result in part (c) to modify your superconducting bulk code to be able to determine the critical current. Run some cases to reproduce, for example, the curves given in Fig. 3.20.

A.34 The single-plane Mott-insulating barrier

In this problem, we will calculate the local DOS at the chemical potential for the central plane of a $N = 1$ strongly correlated nanostructure (see Fig. 3.7). Begin by neglecting the self-consistency in the noninteracting metallic leads (*i.e.*, set $L_\alpha = L_{-\infty}$ and $R_\alpha = R_\infty$ for all α except the plane with the Falicov-Kimball interaction) and choose the system to be at the symmetric half-filling point for both particles ($\rho_e = 1/2$ and $w_1 = 1/2$).
(a) Show that the local Green's function for the central plane ($\alpha = 0$) satisfies

$$G_{\alpha=0}(\omega) = \int d\epsilon\, \rho_{2d}(\epsilon) \frac{1}{2(\omega-\epsilon) - \Sigma_0(\omega) \pm \sqrt{(\omega-\epsilon)^2 - 4}}, \tag{A.81}$$

and determine a procedure to choose the sign of the square root.
(b) Assume $\Sigma_0(\omega)$ is large in magnitude for small ω (which occurs in a Mott insulator) and show that the Green's function can be expanded as

$$G_{\alpha=0}(\omega) \approx -\frac{1}{\Sigma_0(\omega)} - \frac{2\omega \pm s(\omega)}{\Sigma_0^2(\omega)} + \ldots, \tag{A.82}$$

with $s(\omega) = \int d\epsilon\, \rho_{2d}(\epsilon)\sqrt{(\omega-\epsilon)^2 - 4}$.
(c) Now plug the above formula for the Green's function into the DMFT algorithm to find the leading behavior of the self-energy; verify that the self-energy is large for small ω. Numerically evaluate the integral $s(\omega)$ for small ω to finally determine the DOS at the central plane

$$\rho_0(\omega = 0) = \frac{4.2}{\pi U^2}. \tag{A.83}$$

You can compare these numerical results with the full self-consistent solution: $U = 6$ $\rho_0(0) = 0.0378$; $U = 8$ $\rho_0(0) = 0.0211$; and $U = 12$ $\rho_0(0) = 0.0093$.

A.35 Green's functions of the particle-hole symmetric Falicov-Kimball model nanostructure

This problem is more of a project than a homework problem since it will require a significant amount of time to complete. It is a necessary prerequisite for the last problem.

We will examine a multilayered nanostructure consisting of perfect (ballistic) metal leads and a barrier described by the Falicov-Kimball model at half filling. If we choose the leads to also be at half filling, then the chemical potential is equal to 0 for all T, and there is no charge reconstruction. Take a frequency grid that runs from -10 to 10 with a step size of 0.01, and write a computer code to calculate the real-axis local Green's functions for the multilayered nanostructure. Take 30 self-consistent ballistic metal planes on each side of the barrier, and examine the cases with a barrier thickness of 1, 5, and 10 planes. Perform calculations for $U = 2$, $U = 4$, and $U = 6$. Note that if you want to parallelize the code, the calculations at each frequency grid are independent of each other and can be distributed to different nodes. Be sure to use the results of problems A.23–A.25 in developing your code. Be sure to write out the self-energies into a file if you plan to also complete Problem A.37. *Hint*: Use a quadrature grid of about 2000 points for the $U = 2$ and $U = 4$ calculations. You may need to increase the number of grid points for the thicker $U = 6$ calculations. You will need to come up with an appropriate convergence criterion for the Green's function at each frequency; it is better to iterate the equations at each frequency separately, because some frequencies require far fewer iterations than do others. You should also use particle-hole symmetry and calculate only the positive, or the negative frequencies, and determine the other ones via the symmetry. Note that the Green's function should look like a smooth function of frequency (on a logarithmic scale for $U = 6$), and the integral of $-1/\pi$ times the imaginary part should equal 1 for each plane.

A.36 Parallel implementation for the resistance calculation of a nanostructure

Show that the resistance calculation for a ballistic-metal–barrier–ballistic-metal nanostructure at half filling can be performed for all temperatures of interest once the local self-energy has been found. Describe how to construct a parallel algorithm for performing this calculation. Is it better to send the calculations for different temperatures to different nodes or for different frequencies to different nodes? Be sure to describe how you will calculate the Kubo response, namely how will you generate the off-diagonal Green's functions (see Problem A.23). *Hint*: think about the temperature dependence of the elements of the conductivity matrix (note that the chemical potential does not depend on T for half filling). Also think about how to check whether the conductivity matrix is invertible.

A.37 Resistance and Thouless energy of a nanostructure

This problem is a project that builds on the results of Problems A.35 and A.36.

Taking the results of your calculations for the Green's functions and self-energies in Problem A.35, create a program to calculate the resistance of a multilayered nanostructure for the same cases as discussed in Problem A.35. Take 100 temperature points from 0.01 to 1, and plot the resistance as a function of temperature. Next, using the results of Problem A.19

for $w_1 = 0.5$, calculate the Thouless energy and plot it as a function of temperature. Describe the differences in the temperature dependence of both the resistance and the Thouless energy for the three different cases. *Hint*: Be sure to use the same integration grids used to generate the self-energies when you calculate the off-diagonal Green's functions. Access to a parallel computer will greatly speed up your calculations.

Bibliography

Abrikosov, A. A. and Gor'kov, L. P. (1960). Theory of superconducting alloys with paramagnetic impurities, *Zh. Eksp. Teor. Fiz.* **39**, pp. 1781–1796 (in Russian); (1961). *Sov. Phys. JETP* **12**, pp. 1243–1253.

Ambegaokar, V. and Baratoff, A. (1963). Tunneling between superconductors, *Phys. Rev. Lett.* **11**, p. 104.

Anderson, P. W. (1959a). New approach to the theory of superexchange interactions, *Phys. Rev.* **115**, pp. 2–13.

Anderson, P. W. (1959b). Theory of dirty superconductors, *J. Phys. Chem. Solids* **11**, pp. 26–30.

Anderson, P. W. (1961). Localized magnetic states in metals, *Phys. Rev.* **124**, pp. 41–53.

Anderson, P. W. and Rowell, J. M. (1963). Probable observation of the Josephson superconducting tunneling effect, *Phys. Rev. Lett.* **10**, pp. 230–232.

Anderson, P. W. and Yuval, G. (1969). Exact results in the Kondo problem: Equivalence to a classical one-dimensional Coulomb gas, *Phys. Rev. Lett.* **23**, pp. 89–92.

Anderson, P. W., Yuval, G. and Hamann, D. R. (1970). Exact results in the Kondo problem. II. Scaling theory, qualitatively correct solution, and some new results on one-dimensional classical statistical models, *Phys. Rev. B* **1**, pp. 4464–4473.

Andreev, A. F. (1964). The thermal conductivity of the intermediate state in superconductors, *Zh. Eksp. Teor. Fiz.* **46**, pp. 1823–1827 (in Russian); *Sov. Phys. JETP* **19**, pp. 1228–1231.

Andrei, N. (1980). Diagonalization of the Kondo Hamiltonian, *Phys. Rev. Lett.* **45**, pp. 379–382.

Aryanpour, K., Hettler, M. H. and Jarrell M. (2002). Analysis of the dynamical cluster approximation for the Hubbard model, *Phys. Rev. B* **65**, pp. 153102–1-4.

Bardeen, J. (1962). Critical fields and currents in superconductors, *Rev. Mod. Phys.* **34**, pp. 667–681.

Bardeen, J., Cooper, L. and Schreiffer, J. R. (1957). Theory of superconductivity, *Phys. Rev.* **108**, pp. 1175–1204.

Baym, G. and Mermin, N. D. (1961). Determination of thermodynamic Green's functions, *J. Math. Phys.* **2**, pp. 232–234.

Bethe, H. (1935). Statistical theory of superlattices, *Proc. Roy. Soc. London A*, **150** pp. 552–558.

Binnig, G., Rohrer, H., Gerber, Ch. and Weibel, E. (1983). 7×7 Reconstruction on Si(111) resolved in real space, *Phys. Rev. Lett.* **50**, pp. 120–123.

Bogoliubov, N. N., Tolmachev, V. V. and Shirkov, D. V. (1958). *A new method in the theory of superconductivity* (Acad. Sci., USSR) (in Russian); (1959). (Consultants Bureau, Inc., New York).

Brandt, U. and Mielsch, C. (1989). Thermodynamics and correlation functions of the Falicov-Kimball model in large dimensions *Z. Phys. B: Condens. Matter* **75**, pp. 365–370.

Brent, R. P. (1973). *Algorithms for minimization without derivatives* (Prentice-Hall, Englewood Cliffs, New Jersey).

Browning, N. D., Chisholm, M. F., Pennycook, S. J., Norton D. P. and Lowndes, D. H. (1993). Correlation between hole depletion and atomic structure at high-angle $YBa_2Cu_3O_{7-\delta}$ grain boundaries, *Physica C* **212**, pp. 185-190.

Bulla, R. (1999). Zero temperature metal-insulator transition in the infinite-dimensional Hubbard model, *Phys. Rev. Lett.* **83**, pp. 136–139.

Bulla, R., Costi, T. A. and Vollhardt, D. (2001). Finite-temperature numerical renormalization group study of the Mott transition, *Phys. Rev. B* **64**, pp. 045103–1-9.

Bulla, R., Hewson, A. C. and Pruschke, Th. (1998). Numerical renormalization group calculations for the self-energy of the impurity Anderson model, *J. Phys.: Condens. Matter* **10**, pp. 8365–8380.

Bulla, R., Pruschke, Th. and Hewson, A. C. (1997). Anderson impurity in pseudo-gap Fermi systems, *J. Phys.: Condens. Matter* **9**, pp. 10463–10474.

Carnot, S. N. L. (1824). *Reflections on the motive power of fire and on proper machines to develop that motive power* (Bachelier, Paris).

Chen, K. and Jayaprakash, C. (1995). X-ray edge singularities with nonconstant density of states: A renormalization-group approach, *Phys. Rev. B* **52**, pp. 14436–14440.

Chen, W., Rylyakov, A. V., Patel, V., Lukens, J. E. and Likharev, K. K. (1999). Rapid single flux quantum T-flip flop operating up to 770 GHz, *IEEE Trans. Appl. Supercond.* **9**, pp. 3212-3215.

Chester, G. V. and Thellung, A. (1961). The law of Wiedemann and Franz, *Proc. Phys. Soc. (London)* **77**, pp. 1005–1013.

Chung, W. and Freericks, J. K. (1998). Charge-transfer metal-insulator transitions in the spin-one-half Falicov-Kimball model, *Phys. Rev. B* **57**, pp. 11955–11961.

Chung, W. and Freericks, J. K. (2000). Competition between phase separation and "classical" intermediate valence in an exactly solved model, *Phys. Rev. Lett.* **84**, pp. 2461–2464.

Dai, X., Savrasov, S. Y., Kotliar, G., Migliori, A., Ledbetter, H. and Abrahams, E. (2003). Calculated phonon spectra of Plutonium at high temperatures, *Science* **300**, pp. 953–955.

Datta, S. and Das, B. (1990). Electronic analog of the electro-optic modulator, *Appl. Phys. Lett.* **56**, pp. 665–667.

Demchenko, D. O., Joura, A. V. and Freericks, J. K. (2004). Effect of particle-hole asymmetry on the Mott-Hubbard metal-insulator transition, *Phys. Rev. Lett.* **92**, pp. 216401–1-4.

Dirac, P. A. M. (1958). *The Principles of Quantum Mechanics*, Fourth Edition (Clarendon Press, Oxford), pp. 136–139.

Domenicali, C. A. (1953). Irreversible thermodynamics of thermoelectric effects in inhomogeneous, anisotropic media, *Phys. Rev.* **92**, pp. 877–881.

Domenicali, C. A. (1954). Irreversible thermodynamics of thermoelectricity, *Rev. Mod. Phys.* **26**, pp. 237–275.

Drude, P. (1900). On the theory of electrons in metals I., *Ann. Phys. (Leipzig)* **1**, pp. 566–613.

Drude, P. (1900). On the theory of electrons in metals II., *Ann. Phys. (Leipzig)* **3**, pp. 369–402.

Dubos, P., Courtois, H., Pannetier, B., Wilhelm, F. K., Zaikin, A. D. and Schön, G. (2001). Josephson critical current in a long mesoscopic S-N-S junction, *Phys. Rev. B* **63**, pp. 064502–1-5.

Eckstein, J. N. and Bozović, I. (1995). High-temperature superconducting multilayers and heterostructures grown by atomic layer-by-layer molecular beam epitaxy, *Ann. Rev. Mater. Sci.* **25**, pp. 679–709.

Economou, E. N. (1983). *Green's functions in quantum physics* (Springer-Verlag, Berlin).

Edwards, J. T. and Thouless, D. J. (1972). Numerical studies of localization in disordered systems, *J. Phys. C* **5**, pp. 807–820.

Einstein, A. (1905). On the movement of small particles suspended in a stationary liquid demanded by the molecular-kinetic theory of heat, *Ann. Phys. (Leipzig)* **17**, pp. 549–560.

Esaki, L. (1958). New phenomenon in narrow Germanium p-n junctions, *Phys. Rev.* **109**, pp. 603–604.

Ewald, P. (1921). Calculation of optic and electrostatic lattice potential, *Ann. Phys. (Leipzig)* **64**, pp. 253–287.

Falicov, L. M. and Kimball, J. C. (1969). Simple model for semiconductor-metal transitions: SmB_6 and Transition-Metal Oxides, *Phys. Rev. Lett.* **22**, pp. 997-999.

Fermi, E. (1928). A statistical method for the determination of some atomic properties and the application of this method to the theory of the periodic system of elements, *Z. Physik* **48**, pp. 73–79.

Feynman, R. P. (1961). "There's plenty of room at the bottom" in *Minitiaturization,* ed. by Gilbert, H. D. (Reinhold, New York), pp. 282-296; the full text of his December 29, 1959 speech to the American Physical Society is also available at http://www.zyvex.com/nanotech/feynman.html .

Fick, A. (1855). On Diffusion, *Poggendorff's Ann. Phys. u. Chemie* **94**, pp. 59-86.

Fourier, J. B. J. (1822). *Theorie analytique de la chaleur* (Firman, Didot, Paris); English translation (1878). *The analytic theory of heat*, translated by Freeman, A. (Cambridge University Press, Cambridge).

Freericks, J. K. (2004a). Crossover from tunneling to incoherent (bulk) transport in a correlated nanostructure, *Appl. Phys. Lett.* **84**, pp. 1383–1385.

Freericks, J. K. (2004b). Dynamical mean-field theory for strongly correlated inhomogeneous multilayered nanostructures, *Phys. Rev. B* **70**, pp. 195342–1-14.

Freericks, J. K. (2005). Strongly correlated multilayered nanostructures near the Mott transition, *phys. stat. sol. b* **242**, pp. 189–195.

Freericks, J. K., Devereaux, T. P., Bulla, R. and Pruschke, Th. (2003a). Nonresonant inelastic light scattering in the Hubbard model, *Phys. Rev. B* **67**, pp. 155102–1-8.

Freericks, J. K., Demchenko, D. O., Joura, A. V. and Zlatić, V. (2003b). Optimizing thermal transport in the Falicov-Kimball model: binary-alloy picture, *Phys. Rev. B* **68**, pp. 195120–1-12.

Freericks, J. K., Devereaux, T. P. and Bulla, R. (2001). An exact theory for Raman scattering in correlated metals and insulators, *Phys. Rev. B* **64**, pp. 233114–1-4.

Freericks, J. K., Gruber, Ch. and Macris, N. (1999). Phase separation and the segregation principle in the infinite-U spinless Falicov-Kimball model, *Phys. Rev. B* **60**, pp. 1617–1626.

Freericks, J. K. and Falicov, L. M. (1990). Two-state one-dimensional spinless Fermi gas, *Phys. Rev. B* **41**, pp. 2163–2172.

Freericks, J. K., Jarrell, M. and Scalapino, D. J. (1993). Holstein model in infinite dimensions, *Phys. Rev. B* **48**, pp. 6302–6314.

Freericks, J. K. and Jarrell, M. (1995). Magnetic phase diagram of the Hubbard model, *Phys. Rev. Lett.* **74**, pp. 186–189.

Freericks, J. K., Lieb, E. H. and Ueltschi, D. (2002a). Phase separation due to quantum mechanical correlations, *Phys. Rev. Lett.* **88**, pp. 106401–1-4.

Freericks, J. K., Lieb, E. H. and Ueltschi, D. (2002b). Segregation in the Falicov-Kimball model, *Commun. Math. Phys.* **227**, pp. 243–279.

Freericks, J. K., Nikolić, B. K. and Miller, P. (2001). Tuning a Josephson junction through a quantum critical point, *Phys. Rev. B* **64**, pp. 054511-1–13; Erratum: (2003). *Phys. Rev. B* **68**, 099901–1-3.

Freericks, J. K., Nikolić, B. N. and Miller, P. (2002). Optimizing the speed of a Josephson junction with dynamical mean-field theory, *Int. J. Mod. Phys. B* **16**, pp. 531–561.

Freericks, J. K. and Nikolić, B. K. (2003). Temperature dependence of superconductor-correlated metal-superconductor Josephson junctions, *Appl. Phys. Lett.* **82**, pp. 970–972; Erratum: *Appl. Phys. Lett.* **83**, p. 1275.

Freericks, J. K., Tahvildar-Zadeh, A. N. and Nikolić, B. K. (2005). Use of a generalized Thouless energy in describing transport properties of Josephson junctions, *IEEE Trans. Appl. Supercond.* **15**, pp. 896–899.

Freericks, J. K. and Zlatić, V. (2001). Thermal transport in the Falicov-Kimball model, *Phys. Rev. B* **64**, pp. 245118–1-10; Erratum: *Phys. Rev. B* **66**, pp. 249901–1-2.

Freericks, J. K. and Zlatić, V. (2003). Exact dynamical mean-field theory of the Falicov-Kimball model, *Rev. Mod. Phys.* **75**, pp. 1333–1382.

Gajek, Z., Jędrzejewski, J. and Lemański, R. (1996). Canonical phase diagrams of the 1D Falicov-Kimball model at $T = 0$, *Physica A* **223**, pp. 175–192.

Gebhard, F. (1997). *The Mott Metal-Insulator Transition: Models and Methods* (Springer-Verlag, Berlin); Springer Tracts in Modern Physics, Vol. **137**.

Gell-Mann, M. and Brueckner, K. A. (1957). Correlation energy of an electron gas at high density, *Phys. Rev.* **106**, pp. 364–368.

de Gennes, P. G. (1966). *Superconductivity of metals and alloys* (Benjamin, New York).

Georges, A. and Kotliar, G. (1992). Hubbard model in infinite dimensions, *Phys. Rev. B* **45**, pp. 6479–6483.

Georges, A., Kotliar, G., Krauth, W. and Rozenberg, M. J. (1996). Dynamical mean-field theory of strongly correlated fermion systems and the limit of infinite dimensions, *Rev. Mod. Phys.* **68**, pp. 13–125.

Giaever, I. and Megerle, K. (1960). Study of superconductors by electron tunneling, *Phys. Rev.* **122**, pp. 1101–1111.

Ginzburg, V. L. and Landau, L. D. (1950). On the theory of superconductivity, *Zh. Eksp. Teor. Fiz.* **20**, pp. 1064-1082 (in Russian); Engl. translation Landau, L. D. (1965). *Men of Physics*, edited by ter Haar, D., Vol. **1**, (Pergamon Press, Oxford) pp. 138–167.

Golubov, A. A. and Kupriyanov, M. Yu. (1989). Josephson effect in SNINS and SNIS tunnel structures with finite transparency of the SN boundaries, *Zh. Eksp. Teor. Phys.* **96**, pp. 1420–1433 (in Russian); *Sov. Phys. JETP* **69**, pp. 805–812.

Golubov, A. A., Wilhelm, F. K. and Zaikin, A. D. (1997). Coherent charge transport in metallic proximity structures, *Phys. Rev. B* **55**, pp. 1123–1137.

Gor'kov, L. P. (1959). Microscopic derivation of the Ginzburg-Landau equations in the theory of superconductivity, *Zh. Eksp. Teoret. Fyz.* **36**, pp. 1918–1923 (in Russian); *Sov. Phys. JETP* **9**, pp. 1364–1367.

Gonzales-Buxton, C. and Ingersent, K. (1998). Renormalization-group study of Anderson and Kondo impurities in gapless Fermi systems, *Phys. Rev. B* **57**, pp. 14254–14293.

Gradshteyn, I. S. and Ryzhik, I. M. (1980). *Table of integrals, series, and products*, Corrected and Enlarged Edition (Academic Press, Orlando, Florida).

Greenwood, D. A. (1958). The Boltzmann equation in the theory of electrical conduction in metals, *Proc. Phys. Soc. (London)* **71**, pp. 585–596.

Grenzebach, C., Czycholl, G, Anders, F. B. and Pruschke, Th. (2006). Transport properties of heavy-fermion systems, *cond-mat/0603544* (preprint), pp. 1–20.

Guerrero, M. and Noack, R. M. (1996). Phase diagram of the 1-d Anderson lattice, *Phys. Rev. B* **53**, pp. 3707–3712.

Gurvitch, M., Washington, M. A. and Huggins, H. A. (1983). High quality refractory Josephson tunnel junctions utilizing thin aluminum layers, *Appl. Phys. Lett.* **42**, pp. 472–474.

Haller, K. and Kennedy, T. (2001). Periodic ground states in the neutral Falicov-Kimball model in two dimensions, *J. Stat. Phys.* **102**, pp. 15–34.

Hammerl, G., Schmehl, A., Schulz, R. R., Goetz, B., Bielefeldt, H., Schneider,

C. W., Hilgenkamp, H. and Mannhart, J. (2000). Enhanced supercurrent density in polycrystalline $YBa_2Cu_3O_{7-\delta}$ at 77 K from Calcium doping of grain boundaries, *Nature* **407**, pp. 162–164.

Held, K., Keller, G., Eyert, V., Vollhardt, D. and Anisimov, V. I. (2001). Mott-Hubbard metal-insulator transition in paramagnetic V_2O_3: An LDA+DMFT(QMC) study, *Phys. Rev. Lett.* **86**, pp. 5345-5348.

Hettler, M. H., Mukherjee, M., Jarrell, M. and Krishnamurthy, H. R. (1998). The dynamical cluster approximation: Nonlocal dynamics of correlated electron systems, *Phys. Rev. B* **61**, pp. 12739-12756.

Hettler, M. H., Tahvildar-Zadeh, A. N., Jarrell, M., Pruschke, Th. and Krishnamurthy, H. R. (1998). Nonlocal dynamical correlations of strongly interacting electron systems, *Phys. Rev. B* **58**, pp. R7475–R7479.

Hewson, A. C. (1993). *The Kondo problem to heavy fermions* (Cambridge University Press, Cambridge).

Hicks, L. D. and Dresselhaus, M. S. (1993). Effect of quantum-well structures on the thermoelectric figure of merit, *Phys. Rev. B* **47**, pp. 12727–12731.

Hirsch, J. E. (1983). Discrete Hubbard-Stratonovich transformation for fermion lattice models, *Phys. Rev. B* **28**, pp. 4059–4061; Erratum: *Phys. Rev. B* **29**, p. 4159.

Hirsch, J. E. (1993). Polaronic superconductivity in the absence of electron-hole symmetry, *Phys. Rev. B* **47**, pp. 5351-5358.

Hirsch, J. E. and Fye, R. M. (1986). Monte Carlo method for magnetic impurities in metals, *Phys. Rev. Lett.* **56**, pp. 2521–2524.

Hohenberg, P. and Kohn, W. (1964). Inhomogeneous electron gas, *Phys. Rev.* **136**, pp. B864–B871.

Hubbard, J. (1959). Calculation of partition functions, *Phys. Rev. Lett.* **3**, pp. 77-78.

Hubbard, J. (1963). Electron correlations in narrow energy bands, *Proc. R. Soc. (London)* Ser. A **276**, pp. 238–257.

Hubbard, J. (1965). Electron correlations in narrow energy bands: III. An improved solution, *Proc. R. Soc. (London)* Ser. A **281**, pp. 401–419.

Hurd, M. and Wendin, G. (1994). Andreev level spectrum and Josephson current in a superconducting ballistic point contact, *Phys. Rev. B* **49**, pp. 15258–15262.

Inoue, I. H., Hase, I., Aiura, Y., Fujimora, A., Morikawa, T., Haruyama, Y., Maruyama, T. and Nishihara, Y. (1994). Systematic change of spectral function observed by controlling electron correlation in $Ca_{1x}Sr_xVO_3$ with fixed $3d^1$ configuration, *Physica C* **235–240**, pp. 1007-1008.

Ioffe, A. F. and Regel, A. P. (1960). Non-crystalline, amorphous and liquid electronic semiconductors, *Prog. Semicond.* **4**, pp. 237–291.

Jarrell, M. (1992). Hubbard model in infinite dimensions: A quantum Monte Carlo study, *Phys. Rev. Lett.* **69**, pp. 168–171.

Jarrell, M. (1995). Symmetric periodic Anderson model in infinite dimensions, *Phys. Rev. B* **51**, pp. 7429-7440.

Jarrell, M. and Gubernatis, J. E. (1996). Bayesian inference and the analytic continuation of imaginary-time quantum Monte Carlo data, *Phys. Rep.* **269**, pp. 133–195.

Jarrell M. and Krishnamurthy, H. R. (2001). Systematic and causal corrections to the coherent potential approximation, *Phys. Rev. B* **63**, pp. 125102–1-10.

Jonson, M. and Mahan, G. D. (1980). Mott's formula for the thermopower and the Wiedemann-Franz law, *Phys. Rev. B* **21**, pp. 4223–4229.

Jonson, M. and Mahan, G. D. (1990). Electron-phonon contribution to the thermopower of metals, *Phys. Rev. B* **42**, pp. 9350–9356.

Josephson, B. D. (1962). Possible new effects in superconductive tunneling, *Phys. Lett.* **1**, pp. 251–253.

Joule, J. P. (1841). On the heat evolved by metallic conductors of electricity, *Phil. Mag.* **19**, pp. 260–265.

Joura, A. V., Demchenko, D. O. and Freericks, J. K. (2004). Thermal transport in the Falicov-Kimball model on a Bethe lattice, *Phys. Rev. B* **69**, pp. 165105–1-5.

Kadanoff, L. P. and Baym, G. (1962). *Quantum statistical mechanics* (W. A. Benjamin, New York)

Karski, M., Raas, C. and Uhrig, G. S. (2005). Electron spectra close to a metal-insulator transition, *Phys. Rev. B* **72**, pp. 113110–1-4.

Kaul, A. B., Whitely, S. R., Van Duzer, T., Yu, L., Newman, N. and Rowell, J. M. (2001). Internally shunted sputtered NbN Josephson junctions with a Ta_xN barrier for nonlatching logic applications, *Appl. Phys. Lett.* **78**, pp. 99–101.

Kehrein, S. (1998). Density of states near the Mott-Hubbard transition in the limit of large dimensions, *Phys. Rev. Lett.* **81**, pp. 3912–3915.

Keldysh, L. V. (1964). Diagram technique for nonequilibrium processes, *J. Exptl. Theoret. Phys. (USSR)* **47**, pp. 1515–1527 (in Russian); (1965). *Sov. Phys. JETP* **20**, pp. 1018–1026.

Keller, G., Held, K., Eyert, V., Vollhardt, D. and Anisimov, V. I. (2004). Electronic structure of paramagnetic V_2O_3 : Strongly correlated metallic and Mott insulating phase, *Phys. Rev. B* **70**, pp. 205116–1-14.

Kennedy, T. (1994). Some rigorous results on the ground states of the Falicov-Kimball model, *Rev. Math. Phys.* **6**, pp. 901-925.

Kennedy, T. (1998). Phase separation in the neutral Falicov-Kimball model, *J. Stat. Phys.* **91**, pp. 829-843.

Kennedy, T. and Lieb, E. H. (1986). An itinerant electron model with crystalline or magnetic long range order, *Physica A* **138**, pp. 320–358.

Kent, P. R. C., Jarrell, M., Maier, T. A. and Pruschke, Th. (2005). Efficient calculation of the antiferromagnetic phase diagram of the 3D Hubbard model, *Phys. Rev. B* **72**, pp. 060411–1-4 (2005).

Khurana, A. (1990). Electrical conductivity in the infinite-dimensional Hubbard model, *Phys. Rev. Lett.* **64**, p. 1990.

Kohn, W. and Sham, L. J. (1965). Self-consistent equations including exchange and correlation effects, *Phys. Rev.* **140**, pp. A1133–A1138.

Kondo, J. (1964). Resistance minimum in dilute magnetic alloys, *Prog. Theor. Phys.* **32**, pp. 37–48.

Kotliar, G. and Vollhardt, D. (March, 2004). Strongly correlated materials: insights from dynamical mean-field theory, *Phys. Today* **57**, pp. 53–59.

Kramers, H. A. (1930). General theory of parametric rotation in crystals, *Proc. Acad. Amsterdam* **33**, pp. 959–972.

Krishna-Murthy, H. R., Wilkins, J. R., and Wilson, K. G. (1980a). Renormalization-group approach to the Anderson model of dilute magnetic alloys. I. Static properties for the symmetric case, *Phys. Rev. B* **21**, pp. 1003–1043.

Krishna-Murthy, H. R., Wilkins, J. R., and Wilson, K. G. (1980b). Renormalization-group approach to the Anderson model of dilute magnetic alloys. II. Static properties for the asymmetric case, *Phys. Rev. B* **21**, pp. 1044–1083.

Kubo, R. (1957). Statistical-mechanical theory of irreversible processes. I. General theory and simple applications to magnetic and conduction problems, *J. Phys. Soc. Japan* **12**, pp. 570–586.

Kulik, I. O. and Omelyanchuk, A. N. (1977). Properties of superconducting microbridges in the pure limit, *Fiz. Nizk. Temp.* **3**, pp. 945–947 (in Russian); *Sov. J. Low Temp. Phys.* **3**, pp. 459–461.

Landau, L. D. (1956). The theory of a Fermi liquid, *Zh. Eksp. Teor. Fiz.* **30**, pp. 1058-1064 (in Russian); (1957). *Sov. Phys. JETP* **3**, pp. 920-925.

Lemański, R., Freericks, J. K. and Banach, G. (2002). Stripe phases in the two-dimensional Falicov-Kimball model, *Phys. Rev. Lett.* **89**, pp. 196403–1-4.

Lemański, R., Freericks, J. K. and Banach, G. (2004). Charge stripes due to electron correlations in the two-dimensional spinless Falicov-Kimball model, *J. Stat. Phys.* **116**, pp. 699–718.

Lemberger, P. (1992). Segregation in the Falicov-Kimball model, *J. Phys. A: Math. Gen.* **25**, pp. 715-733.

Lieb, E. H. and Wu, F. C. (1968). Absence of Mott transition in an exact solution of the short-range one-band model in one dimension, *Phys. Rev. Lett.* **20**, pp. 1445–1448.

Likharev, K. K. (2000). Superconducting devices for ultrafast computing, in *Applications of Superconductivity: Proceedings of NATO Advanced Study Institute on Superconductive Electronics*, edited by Weinstock, H. (Kluwer, Dordrecht), pp. 247–293.

Limelette, P., Wzietek, P., Florens, S., Georges, A., Costi, T. A., Pasquier, C., Jérome, D., Mézière, C. and Batail, P. (2003). Mott transition and transport crossovers in the organic compound κ-(BEDT-TTF)$_2$Cu[N(CN)$_2$]Cl, *Phys. Rev. Lett.* **91**, pp. 016401–1-4.

Lorenz, L. (1872). Determination of heat temperature in absolute units, *Ann. Phys. u. Chem (Leipzig)* **147**, pp. 429–451.

Luttinger, J. M. (1962). Fermi surface and some simple equilibrium properties of a system of interacting Fermions, *Phys. Rev.* **119**, pp. 1153–1163.

Luttinger, J. M. (1964). Theory of thermal transport coefficients, *Phys. Rev.* **135**, pp. A1505–A1514.

Macridin, A., Jarrell, M. and Maier, Th. (2004) Absence of the d-density-wave state from the two-dimensional Hubbard model, *Phys. Rev. B* **70**, 113105-1-4.

Mahan, G. D. (1990). *Many particle physics (Second edition)* (Plenum, New York).

Mahan, G. D. (1998). Good thermoelectrics, *Solid. St. Phys.* **51**, pp. 81–157.

Mahan, G. D. and Sofo, J. O. (1996). The best thermoelectric, *Proc. Nat. Acad. Sci. (USA)* **93**, pp. 7436-7439.

Maldague, P. F. (1977). Optical spectrum of a Hubbard chain, *Phys. Rev. B* **16**, pp. 2437–2446.

Mannhart, J. (2005). Interfaces in materials with correlated electron systems, in *Thin Films and Heterostructures for Oxide Electronics*, edited by Ogale, S. (Springer-Verlag, Berlin) pp. 251–278.

Mannhart, J. and Hilgenkamp, H. (1998). Possible influence of band bending on the normal state properties of grain boundaries in high-T_c superconductors, *J. Mat. Sci. Eng. B* **56**, pp. 77–85.

Mather, P. G., Perrella, A. C., Tan, E., Read, J. C. and Buhrman, R. A. (2005). Tunneling spectroscopy studies of treated aluminum oxide tunnel barrier layers, *Appl. Phys. Lett.*, **86**, pp. 242504–1-3.

McMillan, W. L. (1968). Tunneling model of the superconducting proximity effect, *Phys. Rev.* **175**, pp. 537–542.

Metzner, W. (1991). Linked-cluster expansion around the atomic limit of the Hubbard model, *Phys. Rev. B* **43**, pp. 8549–8563.

Metzner, W. and Vollhardt, D. (1989). Correlated lattice Fermions in $d = \infty$ dimensions, *Phys. Rev. Lett.* **62**, pp. 324–327.

Miller, P. and Freericks, J. K. (2001). Microscopic self-consistent theory of Josephson junctions including dynamical electron correlations, *J. Phys.: Conden. Mat.* **13**, pp. 3187–3213.

Mo, S.-K., Denlinger, J. D., Kim, H.-D., Park, J.-H., Allen, J. W., Sekiyama, A., Yamasaki, A., Kadono, K., Suga, S., Saitoh, Y., Muro, T., Metcalf, P., Keller, G., Held, K., Eyert, V., Anisimov, V. I. and Vollhardt, D. (2003). Prominent quasiparticle peak in the photoemission spectrum of the metallic phase of V_2O_3, *Phys. Rev. Lett.* **90**, pp. 186403–1-4.

Mott, N. F. (1949). The basis of the electron theory of metals, with special reference to the transition metals, *Proc. Phys. Soc. (London)* **A62**, pp. 416–422.

Mukhanov, O. A., Semenov, V. K. and Likharev, K. K. (1987). Ultimate performance of the RSFQ logic circuits, *IEEE Trans. Magn. Mater.* **MAG-23**, pp. 759–762.

Müller, D. (1956). A method for solving algebraic equations using an automatic computer, *Mathematical Tables and Computations* **10**, pp. 208–215.

Müller-Hartmann (1989a). Correlated Fermions on a lattice in high dimensions, *Z. Phys. B: Condens. Matter* **74**, pp. 507–512.

Müller-Hartmann (1989b). The Hubbard model at high dimensions: some exact results and weak coupling theory, *Z. Phys. B: Condens. Matter* **76**, pp. 211–217.

Müller-Hartmann (1989c). Fermions on a lattice in high dimensions, *Int. J. Mod. Phys. B* **3**, pp. 2169–87.

Nagamatsu, J., Nakagawa, N., Muranaka, T., Zenitani, Y. and Akimitsu, J. (2001). Superconductivity at 39 K in Magnesium Diboride, *Nature* **410**, pp. 63–64.

Nagaoka, Y. (1966). Ferromagnetism in a narrow, almost half-filled *s* band, *Phys. Rev.* **147**, pp. 392–405.

Nambu, Y. (1960). Quasi-particles and gauge invariance in the theory of superconductivity, *Phys. Rev.* **117**, pp. 648–663.

Nekrasov, I. A., Keller, G., Kondakov, D. E., Kozhevnikov, A. V., Pruschke, Th., Held, K., Vollhardt, D. and Anisimov, V. I. (2005). Comparative study of correlation effects in $CaVO_3$ and $SrVO_3$, *Phys Rev. B* **72**, 155106–1-6.

Nernst, W. H. (1889). The electromotive effectiveness of ions: I. The theory of diffusion, *Z. Phys. Chem.* **4**, pp. 129–181.

Nikolić, B. K., Freericks, J. K. and Miller, P. (2002a). Equilibrium properties of double-screened-dipole-barrier SINIS Josephson junctions, *Phys. Rev. B* **65**, pp. 064529–1-11.

Nikolić, B. K., Freericks, J. K. and Miller, P. (2002b). Suppression of the "quasiclassical" proximity gap in correlated-metal-superconductor structures, *Phys. Rev. Lett.* **88**, pp. 077002–1-4.

Nolas, G. S., Sharp, J. and Goldsmid J. (2001). *Thermoelectrics: Basic Principles and New Materials Developments* (Springer-Verlag, Berlin).

Nozieres, P. (1985). Magnetic impurities and the Kondo effect, *Ann. Phys. (Paris)* **10**, pp. 19–35.

Nozieres, P. (1998). Some comments on Kondo lattices and the Mott transition *Eur. Phys. J. B* **6**, pp. 447–457.

Obermeier, T., Pruschke, T. and Keller, J. (1997). Ferromagnetism in the large-U Hubbard model, *Phys. Rev. B* **56**, pp. R8479-R8482.

Ohm, G. S. (1827). *The Galvonic current investigated mathematically* (J. G. F. Kniestädt, Berlin).

Ohtomo, A., Muller, D. A., Grazul, J. L. and Hwang, H. Y. (2002). Artificial charge-modulation atomic-scale perovskite titanate superlattices, *Nature* **419**, pp. 378–380.

Okamoto, S. and Millis, A. J. (2004a). Electronic reconstruction at an interface between a Mott insulator and a band insulator, *Nature* **428**, pp. 630–633.

Okamoto, S. and Millis, A. J. (2004b). Spatial inhomogeneity and strong correlation physics: A dynamical mean-field theory study of a model Mott-insulator–band-insulator heterostructure, *Phys. Rev. B* **70**, pp. 241104(R)–1-4.

Onnes, H. K. (1911). Further experiments with liquid helium. G. On the electrical resistance of pure metals, etc. VI. On the sudden change in the rate at which the resistance of mercury disappears, *Communications from the Physical Laboratory of the University of Leiden* No. 124c.

Onsager, L. (1931a). Reciprocal relations in irreversible processes, I., *Phys. Rev.* **37**, pp. 405–426.

Onsager, L. (1931b). Reciprocal relations in irreversible processes, II., *Phys. Rev.* **38**, pp. 2265–2279.

Peltier, J. C. A. (1834). Investigation of the heat developed by electric currents in homogeneous materials and at the junction of two different conductors, *Ann. Chim. Phys.* **56**, pp. 371–386.

Penn, D. R. (1966). Stability theory of the magnetic phases for a simple model of the transition metals, *Phys. Rev.* **142**, pp. 350–365.

Pennycook, S. J. (2002). Structural determination through Z-contrast microscopy, in *Advances in imaging and electron physics*, Vol. **123**, edited by Merli, P. G., Calestani, G. and Vittori-Antisari, M. (Academic Press, San Diego) pp. 173–205.

Perrella, A. C., Rippard, W. H., Mather, P. G., Plisch, M. J. and Buhrman, R. A. (2002). Scanning tunneling spectroscopy and ballistic electron emission studies of aluminum-oxide surfaces, *Phys. Rev. B* **65**, pp. 201403(R)–1-4.

Pilgram, S., Belzig, W. and Bruder, C. (2000). Excitation spectrum of mesoscopic proximity structures, *Phys. Rev. B* **62**, pp. 12462–12467.

Pines, D. (1953). A collective description of electron interactions: IV. Electron interaction in metals, *Phys. Rev.* **92**, pp. 626–636.

Potthoff, M. (2002). Metal-insulator transitions at surfaces, *Adv. Solid State Phys.* **42**, 121–131.

Potthoff, M. and Nolting, W. (1999a). Surface metal-insulator transition in the Hubbard model, *Phys. Rev. B* **59**, pp. 2549–2555.

Potthoff, M. and Nolting, W. (1999b). Effective mass at the surface of a Fermi liquid, *Physica B* **259-261**, pp. 760–761.

Potthoff, M. and Nolting, W. (1999c). Metallic surface of a Mott insulator—Mott insulating surface of a metal, *Phys. Rev. B* **60**, pp. 7834–7849.

Potthoff, M. and Nolting, W. (1999d). Dynamical mean-field study of the Mott transition in thin films, *Eur. Phys. J. B* **8**, pp. 555–568.

Progrebnyakov, A. J., Redwing, J. M., Raghaven, S., Vaithyanathan, V., Schlom, D. G., Xu, S. Y., Li, Q., Tenne, D. A., Soukiassian, A., Xi, X. X., Johannes, M. D., Kasinathan, D., Pickett, W. E., Wu, J. S. and Spence, J. C. H. (2004). Enhancement of the superconducting transition temperature of MgB_2 by a strain-induced bond-stretching mode softening, *Phys. Rev. Lett.* **93**, pp. 147006–1-4.

Pruschke, Th., Bulla, R. and Jarrell, M. (2000). Low-energy scale of the periodic Anderson model, *Phys. Rev. B* **61**, pp. 12799–12809.

Rashba, E. I. (1960). Properties of semiconductors with an extremum loop. 1. Cyclotron and combinational resonance in a magnetic field perpendicular to the plane of the loop, *Fiz. Tverd. Tela (Leningrad)* **2**, pp. 1224–1238 (in Russian); *Sov. Phys. Solid State* **2**, pp. 1109–1122.

Regis, E. (1995). *Nano: The emerging science of nanotechnology: Remaking the world-molecule by molecule,* (Little Brown and Company, New York); Ch. 4 has a discussion of the history behind Feynman's talk.

Rippard, W. H., Perrella, A. C., Albert, F. J. and Buhrman, R. A. (2002). Ultrathin aluminum oxide tunnel barriers, *Phys. Rev. B* **88**, pp. 046805–1-4.

Rontani, M. and Sham, L. J. (2001). Thermoelectric properties of junctions between metal and strongly correlated semiconductor, *Appl. Phys. Lett.* **77**, pp. 3033–3035.

Rowell, J. M., Gurvitch, M. and Geerk, J. (1981). Modification of tunneling barriers on Nb by a few monolayers of Al, *Phys. Rev. B* **24**, pp. 2278–2281.

Saint-James, D. (1964). Elementary excitations in the neighborhood of an interface between a normal and a superconducting metal, *J. Phys. (Paris)* **25**, pp. 899–905.

Sakai, O. and Kuramoto, Y. (1994). Application of the numerical renormalization-group method to the Hubbard model in infinitie dimensions, *Solid St. Commun.* **89**, pp. 307–311.

Savrasov, S., Kotliar, G. and Abrahams, E. (2001). Electronic correlations in metallic Plutonium within dynamical mean-field picture, *Nature* **410**, pp. 793–795.

Schottky, W. (1940). Deviations from Ohm's law in semiconductors, *Phys. Z.* **41**, pp. 570–573.

Schwartz, L. and Siggia, E. (1972). Pair effects in substitutional alloys. I. Systematic analysis of the coherent-potential approximation, *Phys. Rev. B* **5**, pp. 383-396.

Seebeck, T. J. (1823). Evidence of the thermal current of the combination Bi-Cu by its action on magnetic needle, *Abhandlung der Deutschen Akademie der Wissenschaft zu Berlin*, pp. 265–373.

Sekiyama, A., Fujiwara, H., Imada, S., Suga, S., Eisaki, H., Uchida, S. I., Takegahara, K., Harima, H., Saitoh, Y., Nekrasov, I. A., Keller, G., Kondakov, D. E., Kozhevnikov, A. V., Pruschke, Th., Held, K., Vollhardt, D. and Anisimov, V. I. (2004). Mutual experimental and theoretical validation of bulk photoemission spectra of $Sr_{1-x}Ca_xVO_3$, *Phys. Rev. Lett.* **93**, 156402–1-4.

Sharvin, Yu. V. (1965). A possible method for studying Fermi surfaces, *Zh. Eksp. Teor. Phys.* **48**, 984–985 (in Russian); *Sov. Phys. JETP* **21**, p. 655.

von Smoluchowski, M. (1906). The kinetic theory of Brownian molecular motion and suspensions *Ann. Phys. (Leipzig)* **21**, pp. 756–780.

Sommerfeld, A. (1927). On the theory of electrons in metals, *Naturwiss.* **15**, 825–832.

Stratonovitch, R. L. (1957). A method for calculating quantum distribution functions, *Doklady Akad. Nauk S.S.S.R.* **115**, pp. 1097–1100 (in Russian); (1958). *Soviet Phys. Doklady* **2**, pp. 416–419.

Tahvildar-Zadeh, A. N., Freericks, J. K. and Nikolić, B. K. (2006). Thouless energy as a unifying concept for Josephson junctions tuned through the Mott metal-insulator transition, *Phys. Rev. B* **73**, pp. 075108–1-15.

Tahvildar-Zadeh, A. N., Jarrell, M. and Freericks, J. K. (1997). Protracted screening in the periodic Anderson model, *Phys. Rev. B* **55**, pp. R3332–R3335.

Tahvildar-Zadeh, A. N., Jarrell, M. and Freericks, J. K. (1998). Low-temperature coherence in the periodic Anderson model: Predictions for photoemission of heavy Fermions, *Phys. Rev. Lett.* **80**, pp. 5168–5171.

Tahvildar-Zadeh, A. N., Jarrell, M., Pruschke, Th. and Freericks, J. K. (1999). Evidence for exhaustion in the conductivity of the infinite-dimensional periodic Anderson model, *Phys. Rev. B* **60**, pp. 10782–10787.

Thomas, L. H. (1927). The calculation of atomic fields, *Proc. Cambridge Philos. Soc.* **23**, pp. 542–548.

Thomson, W. [Lord Kelvin] (1851). On a mechanical theory of thermoelectric currents, *Proc. Royal Soc. (Edinburgh)*, pp. 91–98; reprinted in *Mathematical and Physical Papers*, (C. J. Clay and Son, London, 1882), pp. 316–323.

Thomson, W. [Lord Kelvin] (1854). Thermoelectric currents, *Trans. Royal Soc. (Edinburgh)* **21**, I, pp. 123–182; reprinted in *Mathematical and Physical Papers*, (C. J. Clay and Son, London, 1882), pp. 232–291.

Thorpe, M. F. (1981). Bethe lattices, in *Excitations in disordered systems*, ed. by M. F. Thorpe (Plenum, New York), pp. 85–107.

Thouless, D. J. (1974). Electrons in disordered systems and the theory of localization, *Phys. Rep.* **13**, pp. 93–142.

Tinkham, M. (1975). *Introduction to superconductivity*, (McGraw-Hill, New York).

Turkowski, V. M. and Freericks, J. K. (2006). Spectral moment sum rules for strongly correlated electrons in time-dependent electric fields, *Phys. Rev. B* **73**, pp. 075108–1-15.

Uhrig, G. S. (1996). Conductivity in a symmetry-broken phase: Spinless Fermions with $1/d$ corrections, *Phys. Rev. B* **54**, pp. 10436–10451.

Varela, M., Lupini, A. R., Pena, V., Sefrioui, Z., Arslan, I., Browning, N. D., Santamaria, J. and Pennycook, S. J. (2005). Direct measurement of charge transfer phenomena at ferromagnetic/superconducting oxide interfaces, *cond-mat/0508564* (preprint).

Varela, M., Lupini, A. R., Pennycook, S. J., Sefrioui, Z. and Santamaria, J. (2003). Nanoscale analysis of $YBa_2Cu_3O_{7-x}/La_{0.67}Ca_{0.33}MnO_3$ interfaces, *Solid. St. Electronics* **47**, pp. 2245–2248.

Wang, H., McCartney, M. R., Smith, D. J., Jiang, X., Wang, R., van Dijken, S. and Parkin, S. S. P. (2005). Structural characterization of base/collector interfaces for magnetic tunnel transistors grown on Si(001), *J. Appl. Phys.* **97**, pp. 104514–1-6.

Warusawithana, M., Colla, E., Eckstein, J. N. and Weissman, M. B. (2003). Artificial dielectric superlattices with broken inversion symmetry, *Phys. Rev. Lett.* **90**, pp. 036802–1-4.

Wiedemann, G. and Franz, R. (1853). On the heat conductivity of metals, *Ann. Phys. u. Chem. (Leipzig)* **89**, pp. 497–531.

Wiegmann, P. B. and Tsvelick, A. M (1983). Exact solution of the Anderson model, *J. Phys. C: Sol. State Phys.* **16**, pp. 2281–2319.

Wilson, K. G. (1975). The renormalization group: Critical phenomena and the Kondo problem, *Rev. Mod. Phys.* **47**, pp. 775–840.

Wong, J., Krisch, M., Farber, D. L., Occelli, F., Schwartz, A. J., Chiang, T. C., Wall, M., Boro, C. and Xu, R. Q. (2003). Phonon dispersions of fcc δ-Plutonium-Gallium by inelastic X-ray scattering, *Science* **301**, pp. 1078–1080.

Wong, J., Krisch, M., Farber, D. L., Occelli, F., Xu, R., Chiang, T. C., Clatterbuck, D., Schwartz, A. J., Wall, M. and Boro, C. (2005). Crystal dynamics of δ fcc Pu-Ga alloy by high-resolution inelastic X-ray scattering, *Phys. Rev. B* **72**, pp. 064115–1-12.

Yu, L., Stampfl, C., Marshall, D., Eshrich, T., Narayanan, V., Rowell, J. M., Newman, N. and Freeman, A. J. (2002). Mechanism and control of the metal-to-insulator transition in rocksalt Tantalum Nitride, *Phys. Rev. B* **65**, pp. 245110–1-5.

Yu, L. Gandikota, R., Singh, R., Gu, L., Smith, D., Meng, X., Van Duzer, T., Rowell, J. and Newman, N. (2006). Internally shunted Josephson junctions with barriers tuned near the metal-insulator transition for RSFQ logic applications, (unpublished).

Yuval, G. and Anderson, P. W. (1970). Exact results for the Kondo problem: One-body theory and extension to finite temperature, *Phys. Rev. B* **1**, pp. 1522–1528.

Zeng, X., Progrebnyakov, A. V., Kotcharov, A., Jones, J. E., Xi, X. X., Lysczek, E. M., Redwing, J. M., Xu, S., Li, Q., Lettieri, J., Schlom, D. G., Tian, W., Pan, X. and Liu, Z.-K. (2002). *In situ* epitaxial MgB_2 thin films for superconducting electronics, *Nature Materials* **1**, pp. 35–38.

Zhang, X. Y., Rozenberg, M. J. and Kotliar, G. (1993). Mott transition in the $d = \infty$ Hubbard model at zero temperature, *Phys. Rev. Lett.* **70**, pp. 1666–1669.

Zhou, F., Charlat, P., Spivak, B. and Pannetier, B. (1998). Density of states in superconductor-normal metal-superconductor junctions, *J. Low Temp. Phys.* **110**, pp. 841–850.

Index